INDUCTION, ALGORITHMIC LEARNING THEORY, AND PHILOSOPHY

AF788176

LOGIC, EPISTEMOLOGY, AND THE UNITY OF SCIENCE

VOLUME 9

Editors
Shahid Rahman, *University of Lille III, France*
John Symons, *University of Texas at El Paso, U.S.A.*

Editorial Board
Jean Paul van Bendegem, *Free University of Brussels, Belgium*
Johan van Benthem, *University of Amsterdam, the Netherlands*
Jacques Dubucs, *University of Paris I-Sorbonne, France*
Anne Fagot-Largeault, *Collège de France, France*
Bas van Fraassen, *Princeton University, U.S.A.*
Dov Gabbay, *King's College London, U.K.*
Jaakko Hintikka, *Boston University, U.S.A.*
Karel Lambert, *University of California, Irvine, U.S.A.*
Graham Priest, *University of Melbourne, Australia*
Gabriel Sandu, *University of Helsinki, Finland*
Heinrich Wansing, *Technical University Dresden, Germany*
Timothy Williamson, *Oxford University, U.K.*

Logic, Epistemology, and the Unity of Science aims to reconsider the question of the unity of science in light of recent developments in logic. At present, no single logical, semantical or methodological framework dominates the philosophy of science. However, the editors of this series believe that formal techniques like, for example, independence friendly logic, dialogical logics, multimodal logics, game theoretic semantics and linear logics, have the potential to cast new light on basic issues in the discussion of the unity of science.

This series provides a venue where philosophers and logicians can apply specific technical insights to fundamental philosophical problems. While the series is open to a wide variety of perspectives, including the study and analysis of argumentation and the critical discussion of the relationship between logic and the philosophy of science, the aim is to provide an integrated picture of the scientific enterprise in all its diversity.

Induction, Algorithmic Learning Theory, and Philosophy

Edited by

Michèle Friend
George Washington University, Washington, D.C., U.S.A.

Norma B. Goethe
National University of Cordoba, Cordoba, Argentina

Valentina S. Harizanov
George Washington University, Washington, D.C., U.S.A.

Springer

A C.I.P. Catalogue record for this book is available from the Library of Congress.

ISBN 978-90-481-7544-4
ISBN 978-1-4020-6127-1 (e-book)

Published by Springer,
P.O. Box 17, 3300 AA Dordrecht, The Netherlands.

www.springer.com

Printed on acid-free paper

Cover image: Adaptation of a Persian astrolabe (Brass 1712-13), from the collection of the Museum of the History of Science, Oxford. Reproduced by permission

All Rights Reserved
© 2007 Springer
Softcover reprint of the hardcover 1st edition 2007
No part of this work may be reproduced, stored in a retrieval system, or transmitted
in any form or by any means, electronic, mechanical, photocopying, microfilming,
recording or otherwise, without written permission from the Publisher, with the exception
of any material supplied specifically for the purpose of being entered
and executed on a computer system, for exclusive use by the purchaser of the work.

To Hilary Putnam

Contents

Editors' Preface

The idea of the present volume emerged in 2002 from a series of talks by Frank Stephan in 2002, and John Case in 2003, on developments of algorithmic learning theory. These talks took place in the Mathematics Department at the George Washington University. Following the talks, Valentina Harizanov and Michèle Friend raised the possibility of an exchange of ideas concerning algorithmic learning theory. In particular, this was to be a mutually beneficial exchange between philosophers, mathematicians and computer scientists.

Harizanov and Friend sent out invitations for contributions and invited Norma Goethe to join the editing team. The Dilthey Fellowship of the George Washington University provided resources over the summer of 2003 to enable the editors and some of the contributors to meet in Oviedo (Spain) at the 12th International Congress of Logic, Methodology and Philosophy of Science. The editing work proceeded from there.

The idea behind the volume is to rekindle interdisciplinary discussion. Algorithmic learning theory has been around for nearly half a century. The immediate beginnings can be traced back to E.M. Gold's papers: "Limiting recursion" (1965) and "Language identification in the limit" (1967).

However, from a logical point of view, the deeper roots of the learning-theoretic analysis go back to Carnap's work on inductive logic (1950, 1952). In the context of his goal to construct a system of inductive logic that may take its rightful place beside the formal systems of deductive logic, Carnap asked whether inductive procedures are less regulated by exact rules than deductive procedures. In the Schilpp volume devoted to Carnap's philosophy (1963), Reichenbach's student H. Putnam published his seminal article " 'Degree of confirmation' and inductive logic". According to Carnap's reply to Putnam (included by Schilpp), the young critic had already advanced his objections in conversation in 1953; the paper was written around 1955–6.

Relying upon recursion theory, Putnam argued that Carnap's project of defining a *quantitative* concept of 'degree of confirmation' is completely misguided. Thus, a radically new, alternative perspective on induction saw the light for the first time. In the same year, Putnam published a second article "Probability and confirmation" (1963). Relying on his previous results, Putnam now called Carnap's goal into question, on the ground that his 'universal learning machine' is "a learning device of very low power". Nonetheless, Putnam concluded optimistically that even if present-day "inductive logics are learning devices of very low power", in the future, "the development of a powerful mathematical theory of inductive inference, of 'machines that learn', and of better mathematical models for human learning may all grow out of this enterprise".

Much of the subsequent work done in the area of algorithmic learning theory grows out of this spirit. Only two years later, Putnam published a third important paper closely related to the new perspective on induction, entitled "Trial and error predicates and a solution to a problem of Mostowski" (1965), which coincidentally appears in the same issue of the *Journal of Symbolic Logic* as Gold's first paper.

The reaction to the paper from philosophers partly consisted in tempering the imagination of popularisers of the technical work. The popularisers had started making claims about computers learning, and that it followed that computers have a brain, or something approaching free will. This led to speculation about moral obligations towards computers (as autonomous beings with rights). One can see, in retrospect at least, why these speculations needed reining in. The mathematicians and computer scientists have moved on. They have been mapping out the territory of limitations to learning, developing definitions, combining ideas and changing the parameters of the discussion to give an intricate and rigorous understanding of the phenomenon we call "learning".

There has been some serious philosophical work in the area since Putnam and Gold published their papers, for instance, by K.T. Kelly, C. Glymour, O. Schulte, and others. Otherwise, the thread of philosophical reactions to the developments in the mathematical sphere has been much thinner than the potential significance the work warrants. It is time to thicken the thread.

Harizanov, Goethe and Friend diagnosed that one of the contributing factors to the lack of communication between the mathematical and philosophical communities has been the plethora of technical terms which have accompanied the development of the mathematics. We address this problem here by giving some definitions explicitly in many of the papers. We hope that this will help bridge the gap between the mathematicians and the philosophers.

Acknowledgments

We should like to acknowledge the funding received from the Dilthey Faculty Fellowship Award of the George Washington University. This is an award whose primary purpose is to stimulate interdisciplinary work that requires crossing disciplines to achieve an integrated perspective and analysis. The funding was enough to spare Michèle Friend and Valentina Harizanov their teaching duties in the summer of 2003. Norma B. Goethe should like to acknowledge the support of her research by a grant of FONCyT (Buenos Aires, Argentina).

We should also like to thank the contributors for their valuable and prompt contributions, and all the reviewers. We should especially like to thank John Case for his useful advice and support, especially at the beginning of the project. We should like to thank Sarah Pingrey for help with the final proofreading of several papers, and for her valuable assistance with TeX and assembling the volume. Sarah Pingrey was partially supported by Harizanov's George Washington University UFF grant. We should also like to thank Ed Canavan for his continuing support. Finally, Harizanov is most grateful to her daughters, Sofija and Kalina, for their constant encouragement and endless inspiration.

Washington, DC — V. S. H., M. F.
Cordoba, R. A. — N. B. G.
June 2006

Contributors

Michèle Friend, Department of Philosophy, George Washington University, Washingon DC 20052, USA, michele@gwu.edu

Clark Glymour, Department of Philosophy, Carnegie Mellon University, Pittsburgh, PA 15213, USA, cg09@andrew.cmu.edu

Norma B. Goethe, School of Philosophy, National University of Cordoba, 5000 Cordoba, Argentina, ngoethe@ffyh.unc.edu.ar

Valentina Harizanov, Department of Mathematics, George Washington University, Washington DC 20052, USA, harizanv@gwu.edu

Iraj Kalantari, Department of Mathematics, Western Illinois University, Macomb, IL 61455, USA, i-kalantari@wiu.edu

Kevin T. Kelly, Department of Philosophy, Carnegie Mellon University, Pittsburg, PA 15213, USA, kk3n@andrew.cmu.edu

Brendan Larvor, School of Humanities, University of Hertfordshire, Hatfield AL10 9AB, Hertfordshire, United Kingdom, b.p.larvor@herts.ac.uk

Eric Martin, School of Computer Science and Engineering, National ICT Australia, UNSW Sydney, NSW 2052, Australia, emartin@cse.unsw.edu.au

Oliver Schulte, Department of Philosophy and School of Computing Science, Simon Fraser University, Burnaby, B.C. V5A 1S6, Canada, oschulte@sfu.ca

Arun Sharma, Division of Research and Commercialisation, Queensland University of Technology, 2 George Street, GPO Box 2434, Brisbane QLD 4001, Australia, Arun.Sharma@qut.edu.au

Frank Stephan, Department of Mathematics and School of Computing, National University of Singapore, Singapore 117543, fstephan@comp.nus.edu.sg

1

INTRODUCTION TO THE PHILOSOPHY AND MATHEMATICS OF ALGORITHMIC LEARNING THEORY

VALENTINA S. HARIZANOV[1], NORMA B. GOETHE[2] AND MICHÈLE FRIEND[3]

[1]*Department of Mathematics, George Washington University, Washington, D.C. 20052, U.S.A., harizanv@gwu.edu*

[2]*School of Philosophy, National University of Cordoba, 5000 Cordoba, Argentina, ngoethe@ffyh.unc.edu.ar*

[3]*Department of Philosophy, George Washington University, Washington, D.C. 20052, U.S.A., michele@gwu.edu*

1 DESCRIPTION AND PURPOSE OF THE VOLUME

Algorithmic learning theory is a mathematically precise, general framework for studying the existence of computational strategies for converging to the truth in empirical questions. As such, algorithmic learning theory has immediate implications for the philosophy of science and induction and for empirical methodology in general. Indeed, the subject was independently co-invented by E. Mark Gold [16], a mathematician/computer scientist and by Hilary Putnam [44], a philosopher/logician. But it is fair to say that the subject has seen the bulk of its development within mathematics and computer science. The purpose of this volume is to bolster discussion among mathematicians and philosophers, both to encourage the development of new and more philosophically relevant theoretical results and to bring existing results to a wider philosophical audience.

The volume was occasioned by two series of lectures on algorithmic learning theory, one by Frank Stephan and one by John Case. It took

M. Friend, N.B. Goethe and V.S. Harizanov (eds.), Induction, Algorithmic Learning Theory, and Philosophy, 1–24.
© 2007 *Springer.*

shape when Harizanov, Goethe and Friend invited philosophers, mathematicians, and computer scientists to engage with the recent literature concerning induction in algorithmic learning theory. The contributing authors were asked primarily to focus on the interface between the mathematical theory and traditional issues that arise in the philosophy of induction and scientific inference. In fact, several of the authors have gone farther than that to actually extend the standard framework of algorithmic learning theory to deeper philosophical applications. Accordingly, the volume will appeal to a broad audience of computer scientists, mathematicians, and philosophers—or anyone interested in a precise, mathematical investigation of the systematic connections between scientific method, truth, and computability.

2 A BRIEF INTRODUCTION TO ALGORITHMIC LEARNING THEORY

The basic situation studied by algorithmic learning theorists is simple. The *learner* is situated in an environment that presents a potentially unending *stream of inputs*. Computer scientists often think of a learner as a *Turing machine*. We do not have to make this assumption. A learner which is not a Turing machine is called a *general learner*. Some *question* is posed (the theory does not ask how). In interesting empirical questions, the right answer is not known in advance because the learner does not know in advance which input stream he, or she, is receiving inputs from, although he, or she, might have some *background knowledge* about its nature. A *learning function* maps finite sequences of inputs to possible answers to the question (or, more generally, to degrees of belief in possible answers). The learning functions entertained may be restricted either by *feasibility* (e.g., Turing computability) or by *method* (e.g., production only of answers consistent with the current data). Any such restriction is called a *strategy*. A learning function *converges to the truth in* an input stream just in case it stabilizes, in some specified sense, to the state of producing the right answer to the question in that input stream. An *empirical problem* consists in a choice of question and knowledge. A learning function *solves* a given problem according to a given sense of convergence just in case it converges (in the specified sense) to a correct answer in each input stream that satisfies background knowledge. Solutions to a problem can also be compared by efficiency, measured in terms of number of errors, number of mind-changes (changes of output), or total elapsed time prior to convergence to the truth, an idea already proposed by

Putnam [45]. That's all there is to it! Nonspecialists are urged to seek the outlines of this simple picture through the mathematical publications in the area.

For a simple illustration of how it all works, consider the *computable function identification paradigm*, which was proposed as a model of inductive inference by both Gold and Putnam. Intuitively, the aim is to stabilize to a computer program that predicts the successive data one will encounter in the future exactly. This paradigm models a kind of highly idealized, instrumentalistic science, in which the inferred program is a predictive procedure which specifies, once and for all, every observation in the input stream. It also models, again quite ideally, the practical problem of "automated computer programming", in which a computer tries to "synthesize" (produce) a correct program by observing increasing amounts of the input-output behavior the inferred program is supposed to have. In mathematical terms, the input stream f is assumed to be a computable sequence of natural numbers in which the ith entry is given by $f(i)$. This is more general than it sounds, since the numbers could effectively encode more "interesting" sorts of inputs such as attribute values, state descriptions, rational-valued meter readings, etc. Let φ_i denote the partial function computed by the Turing machine with Gödel number i.[1] Say that i is a correct answer for f just in case i is a Gödel number (code) for a correct program for f (i.e., just in case $f = \varphi_i$). Background knowledge B is some collection of total, computable functions. A learning function responds to each finite input sequence (of natural numbers) with a Gödel code specifying some Turing machine. The relevant sense of convergence is to eventually stabilize to some i such that the actual input stream is the total computable function φ_i. Now each set B of total computable functions determines a learning problem in the computable function identification paradigm. If the corresponding problem is solvable in the limit, we say that B is *learnable*.

Another heavily studied paradigm, also proposed by Gold in [16], is the *language learning paradigm*, which was originally introduced to model a child learning his, or her, native language.[2] In the language learning paradigm, a possible language is modeled as a computably verifiable set of

[1] For a review of the relevant material concerning computable functions, cf. [40] or [50].

[2] Gold's project was motivated by Chomsky's [10] conception of empirical linguistics. Chomsky's idea was that children cannot possibly learn the grammar of an arbitrarily concocted language, since any finite amount of data would be consistent with infinitely many languages. Therefore, there must be some constraint on the grammars children can learn in principle and that constraint—universal grammar—is the nature of natural language. In early implementations of the program, Chomsky proposed various computational formalisms as specifications of universal grammar and that raised a question of

numbers (usually interpreted as effective code numbers for finite strings of words) and the learner is required to infer a grammar (i.e., a computable positive test). Note that the grammar inferred may be interpreted as a latent disposition rather than as an occurrent belief state that the learner could articulate. Gold distinguished two distinct formats for the data presented to the learner. Presentation by *text* means that the learner receives an input stream that lists all and only the code numbers of strings in the language to be learned (as though a child were passively listening to competent speakers converse). Presentation by an *informant* provides a list of all possible strings, each of which is labelled with 1 or 0 indicating membership or non-membership in the language. Sometimes text is called *positive* information, whereas an informant is said to provide both positive and *negative* information. It turns out that learning certain sorts of languages cannot be done from text; they can only be learned from an informant.

The function and language learning paradigms have attracted a great deal of attention in the algorithmic learning literature, partly because they both appeared in Gold's original presentation of the subject and partly because they are extremely simple settings in which to investigate potentially deep and complex interactions between learnability and computability. Other, more flexible paradigms have been studied that may be more plausible as models of scientific inference. For example, one can express empirical questions as a relation between input streams and code numbers of potential answers [24]. Then language learning and function identification are just two examples of correct relations. In the *logical* paradigm [24], [31], the problem is to infer the complete theory in some specified fragment of a first-order language from observations drawn from an underlying model of the language. In a similar spirit, Stephan and Ventsov [51] initiated the study of the classes of natural algebraic structures. They showed that, in the case of learning all algorithmically generated ideals of a computable ring, learnability from text has strong connections to the algebraic properties of the ring. Harizanov and Stephan [18] studied learning algorithmically generated subspaces of a computable vector space. Here, it also turns out that the notions of learnability from positive data, as well as from switching have corresponding algebraic characterizations. On the other hand, learning from an informant no longer has a nice algebraic characterization. The paper by

consistency: if arbitrary grammars expressible in such a formalism are not learnable even in the limit, then children cannot do so in the short run either. However, Gold's language learning paradigm is too strict in some respects to make it plausible as a necessary check on hypotheses regarding universal grammar.

Harizanov in this volume addresses this interaction between learning theory and algebra.

Once a paradigm is fixed, one can use the framework to address a number of interesting methodological questions:

(i) Most concretely, one can ask what a learning function says at a given time with respect to the inputs received by that time.
(ii) Generalizing with respect to time, one can ask whether a given learning function converges to the truth on a given input stream.
(iii) Generalizing, further, with respect to input streams, one can ask whether a given learning function solves a given problem.
(iv) Generalizing with respect to learning functions, one can ask whether a given problem is solvable.
(v) Generalizing with respect to problems, one can ask whether two given success criteria are equivalent (in terms of the problems solvable according to each).
(vi) Generalizing with respect to criteria of success, one can ask what the strongest feasible sense of success is that a given problem is solvable in and, hence, whether a given solution solves it in the best feasible sense.

The normative and explanatory core of the theory centers on non-learnability—as does the need for general mathematics. For it is a nice thing to say that a method is guaranteed to converge to the truth in the limit, but it is hardly a compelling recommendation, for one might hope to do better. The justification for a truth-finding method is clinched, however, by showing that no better performance is *possible*, which amounts to showing that the problem is not solvable in stronger, more desirable, senses. For example, suppose that the hypothesis is that a black box under observation will emit exactly one zero at some unspecified time in the future. One can decide this hypothesis in the limit with just two mind-changes: say "no" until the first occurrence of a zero, say "yes" when exactly one zero has been observed, and halt with "no" when a second zero is observed. One might hope to do better, but no learning function can. For let an arbitrary learning function that converges to the truth be given. By feeding it units forever, you can force it to say "no" (on pain of not converging to the truth), after which it can be shown a single zero followed by all units until it says "yes" (on pain of not converging to the truth), after which one can feed it another zero followed by all units, forcing it back to "no". So the two retraction learning function is optimal and, hence, justified, so far as finding the truth is concerned.

Non-learnability results are also crucial for comparing the relative stringency of paradigms, which amounts to an objective comparison of the relative informational value of disparate concepts of convergence or of different assumptions about how data are presented. For example, Gold showed in [16] that it is harder to learn from text than from an informant. Every countable collection of languages is at least ideally learnable by informant. An ideal agent can use her ideal powers to enumerate a collection of acceptance algorithms covering exactly the languages in the countable collection of possible languages. She can then output, at each stage, the first program in the enumeration that agrees with the labelled inputs so far. This is called the *enumeration* method.[3] Evidently, if the actual input stream is an informant for one of the languages in the collection to be learned, then there is some first position at which a correct program occurs. By some finite stage, each preceding (incorrect) acceptance program is rejected, after which the enumeration method converges to a correct acceptance program.

But now suppose that only positive information (text) is received and the collection of possible languages includes all finite languages plus one infinite language. Suppose, for the purposes of a *reductio* argument, that some (ideal) learning function learns an arbitrary language in the collection. You can enumerate the infinite language and present the first element x_0 until the learning function produces (on pain of failing to converge to the truth on a text for a finite language) a positive test for finite language $\{x_0\}$. Then you can present x_1 until the learning function produces a positive test for language $\{x_0, x_1\}$, and so forth. In the limit, the learning function fails to converge on a text for the infinite language in the collection. Contradiction.

A third sort of insight gained from non-learnability results concerns the impact of computability on solvability, for it is often the case that problems solvable ideally in the limit are not solvable by computable agents. Such results are sober medicine for the venerable philosophical tradition of "ideal agent" epistemology, where an ideal agent is assumed not to see the future but to be capable of settling any formal question instantaneously. By way of illustration, consider the very first unsolvability result in the subject, proved independently by Gold [16] and essentially by Putnam [44]. The theorem is that the set of all computable input streams is learnable by an ideal (non-computable) learning function but is not learnable by any computable learning function.

That the collection of total computable functions is ideally learnable is quite intuitive: just ideally enumerate the set of all programs for total computable functions and test them in sequence, as discussed above. But in the case of computable learning functions, the situation becomes much more

[3] Caution: philosophers often use the term "induction by enumeration" to refer to the straight rule of probability estimation. In this case, the learning theoretic usage is more natural.

interesting, so it is worth reviewing the argument.[4] Suppose, for the purposes of a *reductio* argument, that a computable learning function M identifies all total computable functions in the limit. It suffices for contradiction to produce an algorithm for an input stream ε on which the learning function M fails to converge to any answer whatever. An effective procedure defines $\varepsilon(n)$ as follows. At stage 0, let e_0 be the empty sequence. At stage n, finite input sequence e_n of length k has been given. Any input stream that converges to all 0s is computable (use a look-up table for the non-zero segment). Hence, there is some number i of zeros we can add to e_n to get M to converge to a program for $e_n * 0^\infty$, which denotes e_n concatenated with an infinite string of zeros. Hence, there is some number i of zeros we can add to e_n to form $e_n * 0^i$ on which M produces a program that predicts zero at position $k + i + 1$ (the first position after the presented 0s). Hence, this program halts with 0 on input $k + i + 1$ in some number j of computational steps. Therefore, we can effectively enumerate all pairs of natural numbers and rest assured that we will find the first pair (i, j) such that upon seeing $e_n * 0^i$, the function M produces a program that halts in j steps with output 0 on input $k + i + 1$. At that point, define $e_{n+1} = e_n * 0^i * 1$, so that the prediction of M's output program at $k + i + 1$ is wrong. Let ε be the input stream defined by the preceding procedure in the limit. To compute $\varepsilon(x)$, run the preceding procedure until the xth position of ε is defined and then announce it. Observe that, by construction, M produces incorrect programs infinitely often along computable input stream ε. Contradiction. Hence, the set of all total, computable input streams is ideally, but not computably, learnable.

A proof is a proof, but part of understanding a negative learning theoretic argument is to *try* to do the impossible and to observe what breaks. Suppose, for example, that you were to try to implement the enumeration method on a computer. Testing total programs poses no difficulty—each total program either produces outputs agreeing with the current data or disagreeing with them, resulting in a crisp decision procedure for consistency with the data. So it must be that simply *enumerating* a set of total programs covering all the total computable input streams is computationally impossible. And that is, indeed, a standard result in the theory of computability. So any attempt to implement the enumeration method would either miss some possible input streams (in which case it would fail to converge) or it would include some partial programs (that would run forever when tested). Either way, the method would fail to converge to the right answer. Hence, some empirical problems

[4] The following argument is a refinement in [9] of Gold's original version, which only constructs an input stream on which the given learning function fails to converge to a fixed hypothesis.

are unsolvable simply because it is impossible to effectively *consider* the right possibilities.[5]

Notice that the preceding proof shows more than the theorem claims, namely, that an arbitrary, computable method can be forced to produce mistaken programs infinitely often along some computable input stream. Hence, there is no computable learning function that is guaranteed to eventually produce correct programs (even if we do not require that it stabilize to a unique, correct program). This weaker sense of convergence, in which for all but finitely often the learner produces some correct program or other is called BC- (for "behaviourally correct") convergence. This type of convergence was introduced by Case and Lynes [7] and independently by Osherson and Weinstein [42]. By way of contrast, the requirement that there exist a correct program that is produced all but finitely often is called EX- (for "explanatory") convergence. The difference is that EX-convergence requires convergence to a unique program, whereas BC-convergence requires convergence only to a correct input-output behaviour. The irrelevant allusion to explanation in the EX moniker is unfortunate but unavoidably entrenched in the literature—just treat EX as a technical abbreviation.

It is, therefore, natural to ask whether EX-identification is equivalent to or harder than BC-identification for computable learners in the computable function identification paradigm. This question is also answered affirmatively (and ingeniously) in [9]. For consider the function identification problem in which the relevant possibilities are the "almost self-describing data streams". A unit variant of a data stream is a partial computable function that is just like the data stream except that it may disagree or be undefined in at most one position. A data stream is *almost self-describing* just in case it is a unit variant of the function computed by the program whose code number occurs in the data stream's first position. In other words, an "almost self-describing" data stream "gives away" a nearly correct program, but it doesn't say where the possible mismatch might be. A BC-learner can succeed by continually patching the "given away" program with ever larger lookup tables specifying what has been seen so far, since eventually the lookup table corrects the mistake in the "given away" program. But a EX-learner would have to know *when* to stop patching, and this information was not given away. The only remaining technical issue concerns the existence of almost self-describing input streams, which is guaranteed by the fixed point

[5] Putnam's moral was to recommend enumeration procedures capable of receiving extra hypotheses from some oracular source (e.g., "creative intuition"). But that raises further questions about how one could determine whether "creative intuition" presents only total programs—a worse problem than the one we started with [27].

theorems of computability theory, which are often deployed in learning theoretic arguments of this sort.[6]

In the problem just described, it is trivial to EX-identify an almost correct program (just output the first datum) whereas no computable learner can EX-identify an exactly correct program. Indeed, for each finite number of allowed errors there is a learning problem that is computably solvable under that error allowance but not with one fewer error. This result, known as the *anomaly hierarchy theorem* [9], can be established by means of functions that are self-describing up to n possible errors.

To continue this review of standard results would require a book in itself (e.g., [22]). For the purposes of this introduction, however, it is time to turn our attention to potential philosophical applications of the approach.

3 ALGORITHMIC LEARNING THEORY AND THE PHILOSOPHY OF INDUCTION

It isn't fair to speak of a "philosophical" approach to inductive inference, since philosophy characteristically fosters a broad range of perspectives on the subject including algorithmic learning theory. But there is, nonetheless, a dominant viewpoint, which N.B. Goethe calls the *classical* perspective in her paper in this volume, to which algorithmic learning theory constitutes a radical alternative—an alternative whose full import will become all the more apparent if the sorts of interdisciplinary avenues witnessed by the papers in this volume are pursued.

According to the classical perspective, inductive rationality is a short-run constraint on relations between the outputs of ones' learning function and the inputs from which they are generated. The name and precise nature of this constraint depend upon which classical school one adheres to (e.g., confirmation, Bayesian updating, abduction, belief revision, etc.). It is taken as obvious that such rational conclusions are justified. The nature of the connection between rationality and truth-finding is a more troublesome question whose resolution is not at the top of the classical agenda. In fact, it is often presupposed by the classical agenda. The hope or vague promise is that rationality is a mark or sign of the truth, in the sense that rational conclusions are somehow objectively correlated with truth. But since the data themselves are compatible with infinitely many alternative opinions in typical inductive problems (the problem of *theoretical*

[6] One might plausibly object, however, that self-referential problems of the sort invoked to prove the theorem are a far cry from any real problem science will ever encounter.

underdetermination), this desired correlation must be explained, if at all, by some hidden causal mechanism that we may never be in a position to know (Providence according to Descartes and Leibniz or, more recently, the hidden details of our evolutionary past [14]).[7] Advocates of the classical viewpoint prefer to downplay the question of truth in favor of the purely semantic task of *explicating* inductive rationality, which is done by mining intuitive reactions to concrete examples of scientific practice (i.e., case studies) and by massaging away examples in which intuition runs generally counter to the proposed theory. Of course, it is a delicate question when to trust one's explication of rationality and when to trust the intuitions that conflict with it. Striking the right balance is called "reflective equilibrium" which means, more or less, that you are satisfied with the resulting analysis in spite of the counterexamples. For some, "reflective equilibrium" is a term of art. It is neither an objective equilibrium, nor a settling on truth, except in the sense of a revisable social construct.

Algorithmic learning theory, on the other hand, understands the relevant objects of justification to be learning functions rather than particular conclusions, and learning functions are justified entirely in terms of their truth-finding efficacy. No further principles of short-run "rationality" play any irreducible explanatory role. As long as a method solves a problem efficiently and in the best feasible sense, it doesn't matter how implausible it sounds—novelty is all for the better as long as it is accompanied by stronger performance. This fundamental difference cascades into what amounts to an alternative conceptual scheme or paradigm for the philosophy of induction. Let us consider the significance of the shift issue by issue.

3.1 Underdetermination as the Justification of Induction

As we have mentioned, the classical approach to inductive inference flirts with the notion that rationality somehow indicates or is objectively correlated with truth. The fact that many potential answers to an empirical question are compatible with the available evidence (i.e., the underdetermination of the correct answer by evidence) implies either that there is no such correlation (if it is grounded on the data themselves) or that the correlation is grounded in a mysterious, "pre-established harmony" between the true conclusion and its rationality (by our lights). Hence, underdetermination is a *problem* for the classical approach. One can *redefine* underdetermination in terms of

[7] Philosophers are hardly unique in this respect, for some prominent texts in machine learning (e.g., [34]) appeal to evolution to explain the efficacy of otherwise mysterious methodological principles like Ockham's razor.

the existence of alternative conclusions *rationally* compatible with the data, but that hardly explains how rationality sniffs out the truth, since, *prima facie*, there is an infinite number of alternative possibilities compatible with experience.

In algorithmic learning theory, underdetermination *justifies* induction, rather than undermining it. Underdetermination can be explicated both intuitively and with mathematical precision in terms of an empirical problem's intrinsic *complexity*, defined as the strongest sense in which the problem can be solved [24]. The most strongly determined questions are those in which one can halt with the correct answer (e.g., "is the current input zero?"). At the next level of underdetermination, it is possible to halt, eventually, with the right answer in light of further experience (e.g, "will the next input be zero?"). In the preceding cases it is possible to succeed with no mind-changes because the inputs eventually guarantee the truth of some answer. Moving up a level, the problem may require at least one mind-change prior to convergence to the truth (e.g., "will every input be zero?"). In such cases, the data may never uniquely determine the correct answer (if the correct answer is "yes"), but at least the answer "no" is crisply verifiable and the answer "yes" is crisply refutable. Moving up another level, the question may require two mind-changes (e.g., "will exactly one input be zero"). In this case, neither "yes" nor "no" is either verifiable or refutable, so the connection between experience and the correct answer is weaker than in the preceding cases. The same is true of each successive number of mind-changes. Less determined still are the answers to problems that cannot be solved under any finite bound on mind-changes (e.g., "given that the input stream will converge to zero or one, which will it converge to?"). Nor does underdetermination end there, for some problems allow for convergence to the truth only for some answers and not others (e.g., "will the input stream converge to some value in the limit?").

Given the preceding, learning theoretic explication of empirical underdetermination, it is underdetermination that justifies induction, for, if a problem is not underdetermined at all, no mind-changes are necessary to solve it (wait for the right answer to be determined uniquely), whereas methods that risk inductive conclusions open themselves to extra mind-changes (i.e., are sub-optimal solutions). But if the problem is underdetermined, then inductive leaps beyond the current data are necessary to solve it and, as long as the performance of the inductive learning function is optimal for the problem, it is justified, along with its leaps, in the usual way that any procedure is justified: as an optimal solution to the problem it is intended to solve. Thus, algorithmic learning theory turns the usual argument for inductive skepticism on its head: underdetermination *justifies* induction because underdetermined problems require inductive inferences of their optimal solutions.

3.2 The Logic of Discovery as a Logic of Justification

The classical notion that inductive conclusions are justified by a relation of "rationality" between theory and evidence leads to the conclusion, familiar in the philosophy of science, that there is no "logic of discovery". The argument for this vague and surprising doctrine is as follows [30]. Since (in the classical view) justification is conferred by a relation of rational support or confirmation holding between theory and evidence, any procedure for producing justified conclusions does so in virtue of producing conclusions that satisfy the relation with respect to the available data. Hence, the philosophy of induction is over once that relation is satisfied. All that remains is the engineering problem of how to search for a justified hypothesis in light of the data.

The learning theoretic response is immediate [25]. Since justification is a matter of a learning function's truth-finding efficacy, justification pertains as immediately to learning functions that solve decision problems (problems with binary answers) as it does to discovery problems (problems with multiple or even infinitely many possible answers). Indeed, the real aim of science is to find a true theory, not to test a given theory, so in that sense, discovery methods are more directly justified than testing methods, which are usually embedded in a larger discovery enterprise.

Indeed, algorithmic learning theory has something precise to say about when the existence of a test method suffices for a discovery method. For example, the original Gold/Putnam result shows that the existence of a procedure for testing total computer programs does not suffice for the existence of an automatic programmer. More generally, there exists a discovery procedure for a correctness relation just in case there is a single, effective, limiting verification procedure for an effectively enumerable collection of answers covering background knowledge. Hence, effective refutability of individual answers is not necessary [24].

3.3 Relations of Ideas as Matters of Fact

Since the time of Hume [20], philosophers have distinguished sharply between formal questions, which were thought to be decidable in principle, and empirical questions, which were thought to preclude infallible decision procedures due to the unavoidable dependence of truth on as-yet unseen facts. Hume's challenge was to find some other justification for induction, to which the classical paradigm responds with a rational method that confers justification ("confirmation" or "empirical support") upon its outputs. So there is a sharp, methodological bifurcation according to which the aim of *formal* inquiry is truth-finding, but the aim of *empirical* inquiry is to produce rational (e.g., confirmed) outputs. Now we know that infinitely many formal problems are noncomputable, so that halting with the truth in such problems is no more realistic than in empirical problems. Even in form, the two sorts of undecidable problems appear similar: (e.g., "the computation will never halt" sounds for all the world like "a white raven will never appear"). But the classical viewpoint responds to the problem of inductive inference by proposing accounts of rationality (e.g., relations of confirmation, Bayesian coherence and updating, and more recently, theories of rational belief revision) that presuppose noncomputable idealizations like logical omniscience. In the schizophrenic philosophy that results, empirical undecidability is deemed an unavoidable and hence forgivable aspect of the human condition whereas formal undecidability is a sin against rationality.

In algorithmic learning theory, computable agents who fall short of ideal rationality are justified in drawing the conclusions that they do as long as they find the truth as well as a computable agent possibly could. The same is true of formal problems: if the problem is not solvable by a method that infallibly halts with the right answer, seek the strongest sense in which the problem is solvable in the limit, as in the empirical case. This approach to formal and empirical undecidability is entirely evenhanded, in the sense that both are analyzed in terms of the existence of procedures that converge to the true answer (formal or empirical) in the best feasible sense [26]. There is no noncomputable notion of inductive rationality in the background to cause trouble for computable agents either. Hence, there is no problem of "logical omniscience".

3.4 The Truth-Conduciveness of Irrationality

The preceding discussion of uncomputability leads to one of the most interesting and fruitful philosophical applications of algorithmic learning theory: restrictiveness results. The classical viewpoint would have one believe, on vague grounds, that rationality is "truth-conducive". But for algorithmic learning theorists, truth-conduciveness is a matter of solving a given empirical problem in the strongest feasible sense. But a restriction on learning functions couldn't possibly make a given empirical problem *easier* to solve, any more than restricting the possible movements of your queen helps you win at chess. It might even *impede* the search for truth by preventing you from selecting one of the best learning functions—a point urged on purely historical grounds by the philosopher P. Feyerabend [12]. In a similar, and yet purely mathematical spirit, learning theorists say that a proposed constraint on learning strategies is *restrictive* just in case there exists a learning problem solvable in a given sense that is no longer solvable in that sense given that learning function satisfies the constraint. Insofar as justification is a matter of truth-finding efficacy, any alleged notion of "inductive rationality" that is determined to be restrictive not only fails to justify its conclusions but is, itself, unjustified so far as truth-finding is concerned. We illustrate with some examples from the literature.

Consistent learning. No methodological principle seems more obvious than that one's hypothesis should be consistent with the data currently available. Indeed, for ideal learners, the principle makes perfectly good sense, since an answer inconsistent with current information couldn't be true and any solution to a problem can (ideally) be converted into a consistent solution by (ideally) weeding out and replacing the inconsistent outputs with consistent ones. But consistent learning is very restrictive when algorithmic. If a computably enumerable language is consistently EX-learnable by an algorithmic learner, then it must be computable. Moreover, there exists a hypothesis that can be computably decided with just one mind-change by an inconsistent, computable learner, but that cannot be decided even in the limit by a consistent, computable learner [24].

Reliable learning. A learner is *reliable* if it is not allowed to converge incorrectly. That is, if we assume that a learner M learns a class $\mathcal{L}$ of computably enumerable languages, we say that M is *reliable* if whenever M converges on the text for some language L, it will give in the limit a correct algorithm for L, even if L does not belong to $\mathcal{L}$.[8] The learner M might never converge on the text for a language not in $\mathcal{L}$. It is not hard to show that the class of all finite languages can be EX-learned reliably. However, it turns out that this notion of reliability is a very strong one, since if a class $\mathcal{L}$ is reliably EX-learnable from text, then every language in $\mathcal{L}$ must be finite. Thus, Minicozzi [33] further introduced the following notion of reliability for classes $\mathcal{C}$ of (total) computable functions. A learner M for such $\mathcal{C}$ is *total-function reliable for* $\mathcal{C}$ if, in addition to learning $\mathcal{C}$, if M converges on data for any computable function, M must converge accurately. It can be shown that not every EX-learnable class of computable functions can be learned by a total-function reliable learner. Note that this definition of "reliable learning" is not the same as what Kelly refers to as "reliable learning" in his work.

[8] Philosophers baffled by this use of the term "reliable" might find some solace in this: if "convergence" were replaced with "halting", then "reliability" would amount to infallibility. So "reliability" is a kind of convergent analogue of infallibility. Or, just memorize the definition without taking the terminology too seriously.

Refuting learning. This type of learning, introduced by Mikouchi and Arikawa [37], is a sharpened version of reliable learning. A reliable learner either converges to a correct program or diverges. In the refuting learning paradigm, instead of diverging, which is an infinite process, the learner has to give a special refutation symbol after finitely many steps. This symbol signifies that the learner recognizes that the language underlying presented data will be outside its learning capability. Since this notion turned out to be rather restrictive, further variants of refuting learning were studied by various researchers. Recently, Merkle and Stephan [32] introduced limit-refuting learning as an intermediate notion, strictly between reliable learning and refuting learning. While a refuting learner continues to output the refutation symbol after it first outputs it, a limit-refuting learner is allowed to output either guesses for languages or the refutation symbol before starting to converge. Merkle and Stephan showed that for a class of languages closed under subsets, limit-refuting learnability and the so-called *confident* learnability coincide.

Popperian learning.This notion was introduced by Case and Ngo Manguelle [8] for classes of total functions. Let f be a computable (hence total) function. A learner is *Popperian* just in case it produces only total programs in response to any data. The intended connection to Popper is that Popper insisted that scientists produce crisply falsifiable hypotheses. In the function learning paradigm, computable scientists can only crisply test programs that halt with a prediction for each input. So a counterfactual Popper who thought about the analogy between falsifiability of empirical hypotheses and the testability of computer programs might have advocated the production of total programs. Unlike partial computable functions, (total) computable functions can be tested against finite sequences of data given to the learner, since a total function is guaranteed to be defined on any input. It can be shown that not every algorithmically EX-learnable class of computable functions can be learned by a Popperian learner, so Popper's recommendation, translated into a computational context, is bad for computable learners.

Decisiveness prohibits a learner from re-introducing an answer rejected in the past. It turns out that decisive EX-learning from text is not restrictive in the context of general learners. That is, every class of languages EX-learnable from text by a general learner is EX-learnable by a decisive general learner. Moreover, decisive EX-learning from text is not restrictive for algorithmic EX-learners from text of computable functions. Namely, Schäfer-Richter [48] showed that every class of computable functions learnable by an algorithmic EX-learner from text is learnable by a decisive algorithmic EX-learner. For computably enumerable languages decisiveness reduces the power of algorithmic learning from text, both for EX-learning, as shown by Baliga, Case, Merkle and Stephan [3], and for BC-learning, as shown by Fulk, Jain and Osherson [13].

U-shaped learning. This notion, motivated by developmental and cognitive psychology, was introduced by Baliga, Case, Merkle and Stephan [3]. For example, there is a phenomenon observed by psycho-linguists studying how children learn language. Apparently, children will usually conjugate even irregular verbs correctly when they are first learning to speak a language. They then go through a phase of treating irregular verbs as regular (so they 'forget' how to conjugate), and then they start to conjugate verbs correctly again. This is called U-shaped learning by psycho-linguists and learning theorists, because the children learn, unlearn and learn again. While non-U-shaped learning from text does not restrict EX-learning, it does restrict BC-learning (see [3]). Recently, Carlucci, Case, Jain and Stephan [5] obtained some surprising results concerning non-U-shaped learning.

3.5 Ockham's Razor, Realism, and Truth

One of the most puzzling questions in the philosophy of science concerns the truth-conduciveness of *Ockham's razor*, the principle that one should choose the simplest answer compatible with experience. *Scientific realists* argue that the simplest answer is best confirmed, so belief in it is justified [15]. *Anti-realists* [52] argue that more complex theories are also compatible with experience and, hence, might also be correct, so belief in the simplest answer is not justified. The debate is intractable because the realist and the anti-realist both lack a clear conception of "truth-conduciveness" (cf. the above discussion of underdetermination).

However, algorithmic learning theory has a natural such notion: convergence as efficient as possible in the best possible sense. In his first paper in this volume, Kelly shows that, under a plausible interpretation of simplicity, the policy of never producing an answer other than the simplest one compatible with the data is *necessary* for minimizing mind-changes prior to convergence to the truth. In other words, no strategy that violates Ockham's razor is optimally truth-conducive. The idea is related to U-shaped learning (discussed above), for Ockham methods never perform U-turns. Moreover, violators can be forced to do so. Results of this kind not only prescribe uniquely rational inductive methods but also *explain* them in terms of the aim of finding the truth as efficiently as possible. By way of contrast, the classical, Bayesian account of Ockham's razor simply begs the question at hand by presupposing that the prior probability of a complex world is lower than that of a simple world, which amounts to saying that you should use Ockham's razor *if* you happen to be so inclined.

3.6 Goodman's New Riddle of Induction

Another outstanding puzzle in the philosophy of science concerns Goodman's [17] "grue" example. An emerald is observed to be *grue*$[k]$ at stage n just in case it is observed to be green at n and $n \leq k$ or it is observed to be blue at n and $n > k$. The rub for the classical perspective on induction is this: if anything seems rational, it is that after seeing sufficiently many (say a billion) objects with property Φ, it is rational to conclude that all objects have Φ. Thus, for example, after seeing that a billion emeralds are green, it is rational to conclude that all emeralds are green. But a billion green emeralds are also a billion grue[billion] emeralds. So it is also rational to conclude that all emeralds are grue[billion], which implies that the next observed emerald will be blue rather than green. And there is no point objecting that the grue concept is more complex than the concept green, for a speaker of the grue language would say the same of "green", for "green at n can be defined as 'grue up to k and bleen thereafter' where 'bleen' means blue up to k and green thereafter". The example has led to widespread skepticism that there is no principled sense to be made of empirical simplicity; much less that simplicity could be truth-conducive.

According to Kelly's learning theoretic account of simplicity, the simplicity of an input stream is essentially relative to the question asked. In some problems each world has a determinate simplicity degree (agreeing, intuitively, with the number of "free parameters" it requires to be set by the data), whereas in other problems the degrees of simplicity may be quite different or even undefined. Indeed, if the problem is to identify the input stream exactly (i.e., function identification), simplicity is undefined and any enumeration method is as good as any other. But if the problem resembles many concrete problems in science, such as finding the true form of a polynomial law, then simplicity exists, agrees with the usual intuitions, and must be favored on pain of inefficient convergence to the truth. So the philosophical skeptics are right—but so are the scientists, since their questions, unlike those constructed by the philosophers, determine a well-defined notion of simplicity.

3.7 Collapsing the Naturalistic Inductive Regress

Naturalistic philosophers of science hold that the success of scientific methods is itself a scientific question that science is in the best position to answer (using scientific methods) [14]. This raises the specter of a circle or regress of applications of scientific method and of explaining what the point of such a circle or regress amounts to, so far as finding the truth is concerned. In his second paper in this volume, Kelly provides a learning theoretic analysis of material regresses of this sort. Suppose that a learning problem is unsolvable. Then an arbitrary learning function succeeds on some input streams and fails on others. Call the set of all input streams on which a learning function succeeds the *empirical presupposition* of the learning function. Now the question arises whether the presupposition is true. In response to that challenge, one can apply another, meta-learning function to the problem of deciding the presupposition. But if that problem is unsolvable, the meta-learning function succeeds only under some meta-presupposition, and so forth. In that way, one builds a finite or infinite *regress* of methods checking the presuppositions of methods, all the way down to the original method that addresses the original problem. Kelly shows how to collapse such infinite regresses of methods by showing that the existence of the regress is equivalent to the existence of a single method that succeeds in a given sense with respect to the original problem. For example, an infinite sequence of refutation methods (learning functions that never retract more than once and that never take back a rejection) in which each successive method has a presupposition entailed by that of its predecessor is equivalent to a single refutation method that succeeds over the disjunction of all the assumptions of the learning functions in the regress. An infinite regress of verification methods, on the other hand, is equivalent only to a single method that refutes the original hypothesis in the limit, meaning that it converges to "no" if and only if the original hypothesis is false. Philosophers are wont to say that some regresses are *vicious* whereas others are not. Kelly suggests that *vicious* regresses are regresses that cannot be collapsed into a single method with the best feasible truth-finding performance. Thus, Kelly's paper formulates a crisp distinction between vicious and non-vicious regresses.

The preceding points provide a mere, introductory sample of how a flexible approach to algorithmic learning theory can yield crisp results directly relevant to central, contemporary issues in the philosophy of science and induction. Much more remains to be done, however, and it is the hope of Harizanov, Goethe and Friend, that this volume will serve to encourage further mathematical and philosophical interaction.

4 SUGGESTED READING ORDER

The order of the papers in this volume suggests an order of reading. However, each paper is self-contained, so none is a necessary pre-condition for reading a second. The general introduction to the volume is followed by four technical papers (Part I). The first three papers of Part I can be read as introductory papers of three different, complementary, and mutually inspiring approaches. These papers can be read as introductions, in the sense that the reader can still gain a good understanding of the overall subject of learning theory; even if he/she ignores some of the technical details. The fourth paper in this first section is devoted to extending the proof technique of mathematical induction from the (countable) set of natural numbers to the (uncountable) set of real numbers. The remaining papers in the second section (Part II) serve to bring various philosophical problems into focus. The following are summaries of the papers, in the order in which they appear in the volume.

Harizanov's paper presents many of the key learning theoretic concepts and gives depth to knowledge of various parameters of a learning paradigm. The paper first introduces basic concepts of computability theory and inductive inference in the limit, such as computable and computably enumerable languages, and learning from text (positive information). It then discusses different convergence criteria used to determine learning success: explanatory learning versus behaviourally correct learning. The paper also presents learning with negative information, in particular, learning from an informant, and learning from switching, which has been only recently studied. The second part of the paper is concerned with generalizing inductive inference from sets, which are collections of independent elements, to mathematical structures in which various elements are interrelated. This is a new direction in formal learning theory [51, 18], which connects it with the theory of algebraic structures.

The paper by Martin, Sharma and Stephan presents a unified logic (parametric logic) for discussing the parameters of inference in algorithmic learning theory. Parametric logic shows that deduction and induction are not incompatible forms of reasoning, but can be embedded in a common framework where they are particular instances of a generalized notion of logical consequence, which is a function of a few parameters. For some particular values of the parameters, this generalized notion of logical consequence is the classical one, totally captured by deduction, and induction is an "empty notion". For other values of the parameters, the generalized notion of logical consequence leaves scope not only to deduction and induction, but also to more complex forms of reasoning, whose complexity can be captured by a degree of uncertainty related to the notion of mind change and classification in the limit from inductive inference, and to the notion of difference and Borel hierarchies from topology. This paper should be of interest to philosophers because it provides alternative definitions of induction and deduction than those current in the philosophical literature.

The papers by Kelly provide novel learning theoretic solutions to two of the most troublesome open questions in the philosophy of science. The first paper by Kelly (in Part I) discusses the relationship between simplicity and truth in scientific inference. Kelly argues that a systematic preference for the simplest theory compatible with experience does not point at or indicate the truth, but nonetheless keeps science on the straightest path hereto.

Kalantari's paper pursues the idea that induction, in essence, is a method for proving a universal sentence over some well-founded, partially ordered universe. When this partial order is "less than or equal to" over the natural numbers, we have "ordinary induction"; when the partial order is over a tree, we have "structural induction"; when the partial order includes infinite ordinals, we have "transfinite induction"; when the partial order is over some abstract universe we have (through the Axiom of Choice) "Zorn's Lemma". The paper focuses on "induction over the continuum", that is, "induction" over a subinterval of the real line of the form "from a to b" (including a but not b). The analogy between "induction over natural numbers" and its "well ordering principle" carries over to "induction over the interval" and "the greatest lower bound principle" for reals to reveal interesting similarities. The new formulation of induction over the continuum is put to use to untangle arguments and give transparent proofs (avoiding indirect reasoning) of results in real analysis.

Schulte's paper follows Kelly's [24] work in presenting formal learning theory as a general inductive epistemology and a foundation for the philosophy of science. His paper offers a friendly introduction to the general concepts and definitions of reliable convergence to the truth, which explains Putnam's conception of "empirical success" for inductive methods and addresses some of the standard objections. Against this background, Schulte discusses two leading examples, one of them Goodman's Riddle of Induction, to illustrate learning theoretic ideas and contrast them with other notions familiar from philosophical epistemology.

Friend's paper presents an explicitly philosophical investigation into the limitations of method. Friend offers a philosophical assessment concerning one of the types of learning identified by formal learning theorists as "confident learning". Such philosophical enquiry increases our sensitivity to the parameters of learning, and counsels caution when making claims about learning the truth.

The second paper by Kelly (in Part II) provides insight into the epistemic significance of infinite regresses of methods. This paper answers the question "what is an infinite regress worth" by collapsing regresses into single methods and checking whether the collapsed method finds the truth in the best feasible sense. Both papers by Kelly invoke the notion of retractions prior to convergence, which was introduced by Putnam in [45], and thus may be seen as a further development of these notions.

Glymour's paper considers trade-offs between the reliability and the informativeness of inductive inference methods. Methods that give unambiguous conclusions often do so by imposing strong tacit assumptions, implicitly sacrificing reliability, while more broadly reliable methods may more often return no conclusion. Glymour proposes that these trade-offs, while often tacit, are ubiquitous in science, for example, in computerized data analysis in applied statistics. He illustrates his claims, and the thesis that these trade-offs are clarified by placing them in the framework of formal learning theory, through a series of hypothetical but nevertheless plausible cases. Glymour does not claim that formal learning theory allows us to decide among these trade-offs. Rather, his point is that having seen these trade-offs articulated in formal learning theory we are better able to recognize them in concrete cases.

Goethe's paper discusses two approaches to inductive reasoning in philosophy, hitherto largely distinct: the classical way of conceiving induction and the formal learning theoretic perspective. In the first part of her paper she presents modern discussions of knowledge acquisition and proof from Hume through Kant as attempts to determine the cognitive bounds of human cognition. In the second part of the paper, Goethe pays attention to the origins of the formal learning theoretic approach, tracing it back to seminal works by Carnap [6] and by Putnam [44, 45]. This approach, as exemplified in Putnam's work and further developed by Kelly [24, 25] exploits three main ideas. First, it is based on an analogy between human observation and algorithmic procedures and sees inductive reasoning as a way of gaining knowledge by analyzing a stream of data. Second, it conceives of inquiry as governed by means-ends imperatives rather than categorical imperatives. Third, truth is identified with success in the limit of inquiry, in a clear echo of pragmatist notions of truth. In so doing, formal learning theory blurs together the descriptive and normative questions about induction inherited from the classical way of thinking about induction.

In his paper Larvor argues that there cannot be a comprehensive account of scientific enquiry, for two reasons. The first is that our understanding of scientific enquiry depends on disciplines with incompatible standards of rigour and modes of explanation: Formal Logic and Historiography. The second reason is that the rigour and success of empirical science depend on the particular features of: (i) domains of enquiry, and (ii) research programs. Consequently, Larvor argues, our understanding of enquiry is more like a wisdom-tradition than a science, to which formal learning theory may be seen as a valuable addition.

ACKNOWLEDGMENTS

The authors wish to thank Kevin Kelly, Brendan Larvor, and Oliver Schulte for proofreading and valuable suggestions.

REFERENCES

[1] Angluin, D. (1980). "Inductive Inference of Formal Languages from Positive Data", *Information and Control* 45, 117–135.
[2] Angluin, D. and Smith, C.H. (1983). "Inductive Inference: Theory and Methods", *Computing Surveys* 15, 237–269.
[3] Baliga, G., Case, J., Merkle, W. and Stephan, F. (2000). "Unlearning Helps", in Montanari, U., Rolim, J.D.P. and Welzl, E. [35], 844–855.
[4] Blum, L. and Blum, M. (1975). "Toward a Mathematical Theory of Inductive Inference", *Information and Control* 28, 125–155.
[5] Carlucci, L., Case, J., Jain, S. and Stephan, F. (forth.). "U-Shaped Learning May Be Necessary", *Journal of Computer and System Sciences.*
[6] Carnap, R. (1950). *Logical Foundations of Probability*, Chicago: University of Chicago Press.
[7] Case, J. and Lynes, C. (1982). "Machine Inductive Inference and Language Identification", in Nielsen, M. and Schmidt, E.M. [39], 107–115.
[8] Case, J. and Ngo Manguelle, S. (1979). "Refinements of Inductive Inference by Popperian Machines", Technical Report 152, Buffalo: State University of New York at Buffalo.
[9] Case, J. and Smith, C. (1983). "Comparison of Identification Criteria for Machine Inductive Inference", *Theoretical Computer Science* 25, 193–220.
[10] Chomsky, N. (1965). *Aspects of the Theory of Syntax*, Cambridge (Mass.): MIT Press.
[11] Curd, M. and Cover, J.A. (eds.) (1998). *Philosophy of Science: The Central Issues*, New York: W.W. Norton.
[12] Feyerabend, P. (1975). *Against Method*, London: Verso.
[13] Fulk, M., Jain, S. and Osherson, D.N. (1994). "Open Problems in 'Systems That Learn'", *Journal of Computer and System Sciences* 49, 589–604.
[14] Giere, R. (1985) "Philosophy of Science Naturalized", *Philosophy of Science* 52, 331–356.
[15] Glymour, C. (1980). *Theory and Evidence*, Princeton: Princeton University Press.
[16] Gold, E.M. (1967). "Language Identification in the Limit", *Information and Control* 10, 447–474.
[17] Goodman, N. (1983). *Fact, Fiction, and Forecast*, 4th ed., Cambridge (Mass.): Harvard University Press.
[18] Harizanov, V.S. and Stephan, F. (2007). "On the Learnability of Vector Spaces", *Journal of Computer and System Sciences* 73, 109–122.
[19] Hull, D., Forbes, M. and Burian, R. (eds.) (1994). *Proceedings of the 1994 Biennial Meeting of the Philosophy of Science Association*, East Lansing: Philosophy of Science Association.
[20] Hume, D. (1984). *An Inquiry Concerning Human Understanding*, Hendell, C. (ed.), New York: Bobbs-Merrill.

[21] Jain, S. and Stephan, F. (2003). "Learning by Switching Type of Information", *Information and Computation* 185, 89–104.
[22] Jain, S., Osherson, D., Royer, J.S. and Sharma, A. (1999). *Systems That Learn: An Introduction to Learning Theory*, 2nd ed., Cambridge (Mass.): MIT Press.
[23] Jantke, K.P., Kobayashi, S., Tomita, E. and Yokomori, T. (eds.) (1993). *Algorithmic Learning Theory: Proceedings of the 4th International Workshop*, Lecture Notes in Computer Science 744, Berlin: Springer-Verlag.
[24] Kelly, K.T. (1996). *The Logic of Reliable Inquiry*, Oxford: Oxford University Press.
[25] Kelly, K.T. (2000). "The Logic of Success", *The British Journal for the Philosophy of Science*, Special Millennium Issue 51, 639–666.
[26] Kelly, K.T. (2004). "Uncomputability: The Problem of Induction Internalized", *Theoretical Computer Science* 317, 227–249.
[27] Kelly, K. and Juhl, C. (1994). "Realism, Convergence, and Additivity", in Hull, D., Forbes, M. and Burian, R. [19], 181–190.
[28] Lakatos, I. (1976). *Proofs and Refutations*, Cambridge: Cambridge University Press.
[29] Lakatos, I. (1998). "Science or Pseudo-Science", in Curd, M. and Cover, J.A [11], 20–26.
[30] Laudan, L. (1980). "Why Was the Logic of Discovery Abandoned?", in Nickles, T. [38], 173–183.
[31] Martin, E. and Osherson, D. (1998). *Elements of Scientific Inquiry*, Cambridge (Mass.): MIT Press.
[32] Merkle, W. and Stephan, F. (2003). "Refuting Learning Revisited", *Theoretical Computer Science* 298, 145–177.
[33] Minicozzi, E. (1976). "Some Natural Properties of Strong-Identification in Inductive Inference", *Theoretical Computer Science* 2, 345–360.
[34] Mitchell, T.M. (1997). *Machine Learning*, New York: McGraw-Hill.
[35] Montanari, U., Rolim, J.D.P. and Welzl, E. (eds.) (2000). *Automata, Languages and Programming. Proceedings of the 27th International Colloquium* (ICALP 2000), Lecture Notes in Computer Science 1853, Berlin: Springer-Verlag.
[36] Mostowski, M. (2001). "On Representing Concepts in Finite Models", *Mathematical Logic Quarterly* 47, 513–523.
[37] Mukouchi, Y. and Arikawa, S. (1993). "Inductive Inference Machines That Can Refute Hypothesis Spaces", in Jantke, K.P., Kobayashi, S., Tomita, E. and Yokomori, T. [23], 123–136.
[38] Nickles, T. (ed.) (1980). *Scientific Discovery, Logic, and Rationality*, Dordrecht: Reidel.
[39] Nielsen, M. and Schmidt, E.M. (eds.) (1982). *Automata, Languages and Programming: Proceedings of the 9th International Colloquium*, Lecture Notes in Computer Science 140, Berlin: Springer-Verlag.
[40] Odifreddi, P. (1989). *Classical Recursion Theory*, Amsterdam: North-Holland.
[41] Osherson, D.N., Stob, M. and Weinstein, S. (1986). *Systems That Learn: An Introduction to Learning Theory for Cognitive and Computer Scientists*, Cambridge (Mass.): MIT Press.
[42] Osherson, D.N. and Weinstein, S. (1982). "Criteria of Language Learning", *Information and Control* 52, 123–138.
[43] Popper, K. (1963). *Conjectures and Refutations: The Growth of Scientific Knowledge*, London: Routledge.
[44] Putnam, H. (1963). "'Degree of Confirmation' and Inductive Logic", in Putnam, H. [47], 270–292.

[45] Putnam, H. (1965). "Trial and Error Predicates and the Solution to the Problem of Mostowski", *Journal of Symbolic Logic* 30, 49–57.

[46] Putnam, H. (1975). "Probability and Confirmation", in Putnam, H. [47], 293–304.

[47] Putnam, H. (1975). *Mathematics, Matter, and Method*, Cambridge: Cambridge University Press.

[48] Schäfer-Richter, G. (1984). *Über Eingabeabhängigkeit und Komplexität von Inferenzstrategien*, Aachen, Germany: PhD Dissertation, Rheinisch-Westfälische Techniche Hochschule.

[49] Sharma, A. (1998). "A Note on Batch and Incremental Learnability", *Journal of Computer and System Sciences* 56, 272–276.

[50] Soare, R.I. (1987). *Recursively Enumerable Sets and Degrees. A Study of Computable Functions and Computably Generated Sets*, Berlin: Springer-Verlag.

[51] Stephan, F. and Ventsov, Yu. (2001). "Learning Algebraic Structures from Text", *Theoretical Computer Science* 268, 221–273.

[52] Van Fraassen, B. (1981). *The Scientific Image*, Oxford: Clarendon Press.

I

TECHNICAL PAPERS

2

INDUCTIVE INFERENCE SYSTEMS FOR LEARNING CLASSES OF ALGORITHMICALLY GENERATED SETS AND STRUCTURES

VALENTINA S. HARIZANOV

Department of Mathematics, George Washington University, Washington, D.C. 20052, U.S.A., harizanv@gwu.edu

Abstract: Computability theorists have extensively studied sets A the elements of which can be enumerated by Turing machines. These sets, also called *computably enumerable* sets, can be identified with their Gödel codes. Although each Turing machine has a unique Gödel code, different Turing machines can enumerate the same set. Thus, knowing a computably enumerable set means knowing one of its infinitely many Gödel codes. In the approach to learning theory stemming from E.M. Gold's seminal paper [9], an inductive inference learner for a computably enumerable set A is a system or a device, usually algorithmic, which when successively (one by one) fed data for A outputs a sequence of Gödel codes (one by one) that at a certain point stabilize at codes correct for A. The convergence is called semantic or *behaviorally correct*, unless the same code for A is eventually output, in which case it is also called syntactic or *explanatory*. There are classes of sets that are semantically inferable, but not syntactically inferable.

Here, we are also concerned with generalizing inductive inference from sets, which are collections of distinct elements that are mutually independent, to mathematical structures in which various elements may be interrelated. This study was recently initiated by F. Stephan and Yu. Ventsov. For example, they systematically investigated inductive inference of the ideals of computable rings. With F. Stephan we continued this line of research by studying inductive inference of computably enumerable vector subspaces and other closure systems.

In particular, we showed how different convergence criteria interact with different ways of supplying data to the learner. Positive data for a set A are its elements, while negative data for A are the elements of its complement. Inference from *text* means that only positive data are supplied to the learner. Moreover, in the limit, all positive data are given. Inference from *switching* means that changes from positive to negative data or *vice*

M. Friend, N.B. Goethe and V.S. Harizanov (eds.), Induction, Algorithmic Learning Theory, and Philosophy, 27–54.
© 2007 *Springer*.

versa are allowed, but if there are only finitely many such changes, then in the limit all data of the eventually requested type (either positive or negative) are supplied. Inference from an *informant* means that positive and negative data are supplied alternately, but in the limit all data are supplied. For sets, inference from switching is more restrictive than inference from an informant, but more powerful than inference from text. On the other hand, for example, the class of computably enumerable vector spaces over an infinite field, which is syntactically inferable from text does not change if we allow semantic convergence, or inference from switching, but not both at the same time. While many classes of inferable algebraic structures have nice algebraic characterizations when learning from text or from switching is considered, we do not know of such characterizations for learning from an informant.

1 COMPUTABLY ENUMERABLE LANGUAGES

In theoretical computer science, inductive inference introduced by Gold [9] in 1967 is a framework for learning theory. Learning is viewed as a dialogue between a learner and a teacher. The learner is trying to learn a family of computably enumerable sets of natural numbers. A set of natural numbers A to be learned can be viewed as coding a language L by identifying, in an algorithmic manner, the grammatically correct sentences of L with the elements of A.

A set is *computably enumerable* if there is an algorithm, a computable function, which generates it by enumerating (listing) its elements. Computably enumerable sets are often abbreviated by c.e. They are often called recursively enumerable and abbreviated by r.e. In other words, a language, which is the set of all correct sentences in a certain alphabet, is c.e. if there is an algorithmic grammar that generates all correct sentences. A c.e. language is also called a Chomsky language. Chomsky languages are further classified according to the restrictiveness of their grammars as regular, context-free, context-sensitive, and unrestricted.

Regular languages have the most restrictive grammars and are context-free. However, there are context-free languages that are not regular. Similarly, context-free languages are context-sensitive, but not all context-sensitive languages are context-free. There are unrestricted languages that are not context-sensitive. These various classes of Chomsky languages also have machine equivalents. While regular languages coincide with languages recognized by finite automata, context-free languages coincide with those recognized by push-down automata. Context-sensitive languages are the same as languages recognized by linear-bounded Turing machines. An unrestricted language can be characterized as the domain of a partial computable function. That is, for such a language there is a Turing machine that on any input x halts (converges) if x is a correct sentence, and computes forever (diverges) if x is an incorrect sentence.

All c.e. sets can be simultaneously algorithmically listed by systematically listing (coding) all Turing machines. Let

$$P_0, P_1, P_2, \ldots, P_e, \ldots$$

be a fixed algorithmic (computable) enumeration of all Turing machines. For a Turing machine P_e, let φ_e be the unary partial function it computes. Then

$$\varphi_0, \varphi_1, \varphi_2, \ldots, \varphi_e, \ldots$$

is an algorithmic enumeration of all unary partial computable functions. As we already mentioned, it can be shown that a set A is c.e. if and only if there is a partial computable function φ such that the domain of φ is A. In other words, the set of all outputs of an algorithm is the set of all inputs on which some algorithm halts.

The domain of a partial computable function φ_e is the c.e. set denoted by W_e. We call the index e its Gödel code. Thus,

$$W_0, W_1, W_2, \ldots, W_e, \ldots$$

is an algorithmic enumeration of all c.e. sets. We can also think of this enumeration as an enumeration of all Chomsky languages. Every partial computable function and every c.e. set have infinitely many Gödel codes, because for every Turing machine there are infinitely many distinct Turing machines that compute the same function. Hence knowing a c.e. language is knowing one of its infinitely many Gödel codes. Similarly, every Chomsky language is generated by infinitely many grammars. We say that such grammars are equivalent.

In the next subsection, we introduce an important subclass of c.e. languages: decidable, or computable languages. In Section 2, we discuss learning classes of c.e. sets using positive and negative information, and also using different convergence criteria. In Section 3, we generalize learning to classes of c.e. algebraic structures. (We can think of sets as being special algebraic structures with the identity relations only.) Many subsections contain detailed proofs with explanations of the relevant computability theoretic and algebraic facts.

1.1 Computable and Noncomputable Sets

Some c.e. sets are *computable*, also called *recursive* or *decidable*. More precisely, a set is computable if and only if both the set and its complement are c.e. From these two computable (algorithmic) enumerations, we can devise an algorithm (decision procedure) for recognizing the elements of the set and, equivalently, the nonelements of the set. That is, a set is computable exactly when it has a computable characteristic function. The *characteristic function* of a set A is the function that outputs 1 on the elements of the set A, and outputs 0 on the elements of the complement of A. Every context-sensitive language is computable.

Clearly, the complement of a computable set is computable. On the other hand, there are c.e. sets with complements that are not c.e., namely, noncomputable c.e. sets. For example, the *halting set* K is c.e., but not computable. The set K consists of all inputs e on which the Turing machine with Gödel code e halts. Equivalently, K consists of those e that appear in W_e:

$$K = \{e : P_e(e) \text{ halts}\} = \{e : \varphi_e(e) \downarrow\} = \{e : e \in W_e\}.$$

The set K is c.e. because it is enumerated by the procedure that simultaneously runs

$$P_0(0), P_1(1), \dots, P_e(e), \dots,$$

and enumerates those e for which $P_e(e)$ converges, as soon as the convergence occurs. Here, simultaneously means that, say, at each step we add (activate) a new Turing machine and also run all activated machines for an additional computational step.

The complement of K, the *divergence set* $\overline{K}$, is not c.e. If $\overline{K}$ were c.e., then for some e_0,

$$\overline{K} = W_{e_0} = \{e : e \notin W_e\}.$$

Hence, for this particular e_0, we have

$$e_0 \in \overline{K} \Leftrightarrow e_0 \in W_{e_0} \Leftrightarrow e_0 \notin W_{e_0},$$

which is a contradiction.

Computable sets can also be viewed as c.e. sets with elements that can be algorithmically enumerated in an increasing order. Finite initial segments of such an enumeration give additional negative information that certain "small" elements that have not been enumerated so far will never be enumerated into the set.

A sequence of computable sets $A_0, A_1, A_2, \ldots$ is *uniformly computable* if there is a binary computable function f such that for every i, the unary function g defined by $g(x) = f(x, i)$ is the characteristic function of A_i. We also write $A_i(x) = f(x, i)$. For more on computability theory, see Odifreddi [19] and Soare [23].

2 INDUCTIVE INFERENCE OF COMPUTABLY ENUMERABLE LANGUAGES: IDENTIFICATION IN THE LIMIT

Although we are mainly interested in algorithmic learners, which can be thought of as Turing machines or computable functions, a learner can also be general. A general learner can be thought of as a function (not necessarily algorithmic) that maps finite sequences of numbers (coding sentences in a language) to numbers (Gödel codes for c.e. languages). A learner receives an infinite stream of data $x_0, x_1, x_2, \ldots$ of a c.e. set A to be learned, and outputs a sequence of hypothesized codes

$$e_0, e_1, e_2, \ldots, e_n, \ldots,$$

which, in case the learner successfully learns A, converges to a "description" of the set A. In addition to Gödel codes, the learner is also allowed to output a special non-numerical symbol, say "?".

2.1 Learning from Text

One way to feed data for a c.e. set (language) to the learner is via a text. A *text* t for a set A is any infinite sequence of its elements

$$t = a_0, a_1, a_2, \ldots,$$

possibly with repetitions, such that $A = \{a_0, a_1, a_2, \ldots\}$. The set A is the *content* of t. Hence every member (positive datum) of A appears in t at least once, but no nonmember of A (negative datum) is in t. If A is finite, then some element must appear infinitely often in t. One-element sets have only one text, while any set with at least two elements has uncountably many texts. A learner M *converges on* t if after every finite sequence of data $a_0, \ldots, a_n$ the learner outputs a code e_n, in symbols $M(a_0, \ldots, a_n) = e_n$, such that eventually (i.e., after finitely many steps) every output e_n

codes the correct language. In the *strong* sense, which is the assumption we make in this subsection, eventually coding A means that for some n, we have

$$e_n = e_{n+1} = e_{n+2} = \ldots, \text{ where } A = W_{e_n}.$$

A learner learns A *from text* if the learner converges on any text for A. A learner learns a partial computable function φ if it learns its graph $\{(x, \varphi(x)) : \varphi(x) \text{ halts}\}$. Partial computable functions have c.e. graphs. A learner learns a class $\mathcal{L}$ of languages if it learns every language A in $\mathcal{L}$. Clearly, if the learner learns a class $\mathcal{L}$, then it also learns every subclass $\mathcal{L}_0 \subseteq \mathcal{L}$. In general, we do not care what happens to the languages that are not in the class to be learned. On the other hand, we call a learner for $\mathcal{L}$ *confident* if it always converges to some hypothesis, even when given a text for a language that does not belong to the class $\mathcal{L}$. However, the learner does not have to converge accurately on the languages that are not in $\mathcal{L}$. Not every learnable class of c.e. languages can be learned by a confident learner. Moreover, there is a class that is learnable by an algorithmic learner, and is learnable by another confident learner, but cannot be learned by an algorithmic and confident learner (see [21]). A learner for a class $\mathcal{L}$ is called *class-comprising* if the learner always guesses a language in $\mathcal{L}$.

A class consisting of a single language is always learnable since there is a learning algorithm that always outputs the same index for the single set in the class. An example of an infinite learnable class is the class of all finite sets of natural numbers. It is learnable since the learner can guess that the language consists exactly of the elements that are seen up to that point. Since every language in the class is finite, the learner will be correct in the limit. However, it can be shown that this learner cannot be confident (see [21]). The class of all sets of natural numbers missing exactly one number is also learnable. The learner guesses that the missing number is the least number not seen so far. Again, clearly, such a learning strategy will be correct in the limit.

For a sequence σ, the range of σ, $rng(\sigma)$, is the set of all elements of σ. We use the symbol ^ for concatenation of sequences. A learner M from text is *consistent* (Angluin [1]) if on every sequence σ of data, M guesses a c.e. set with code $M(\sigma)$, which contains $rng(\sigma)$, that is, $W_{M(\sigma)} \supseteq rng(\sigma)$. A learner M from text is *conservative* (Angluin [1]) if whenever $M(\sigma\char`^\tau) \neq M(\sigma)$, then it must be that $rng(\sigma\char`^\tau) \nsubseteq W_{M(\sigma)}$. Thus, a conservative learner makes only justified changes to its hypotheses. While conservatism is not restrictive for general learning, it does restrict algorithmic learning.

We can also allow blanks (pause symbols) in the texts—these are pauses in the presentation of data (see [3]). More precisely, we fix a new symbol, $\sharp$, and allow any number of occurrences of this pause symbol in a text. This does not change learnability properties, but allows the teacher not to give any data at certain steps. Furthermore, we can also assign the following text to the empty language:

$\sharp, \sharp, \sharp, \sharp, \ldots$

However, we agree that for a sequence σ that might contain the symbol $\sharp$, the set $rng(\sigma)$ does not contain $\sharp$.

If we have a class of computable sets, then we may also want to consider learning characteristic functions of these sets. That is, instead of guessing codes for enumerating algorithms of the set to be learned, the learner tries to guess a code (index) for its characteristic function.

2.2 Different Convergence Criteria

Gold [9] introduced a strong convergence criterion for identification in the limit. Assume that the learner's sequence of hypotheses for a c.e. set A is

$$e_0, e_1, e_2, \ldots, e_n, e_{n+1}, e_{n+2}, \ldots.$$

The strong convergence criterion requires that after finitely many steps, the hypotheses are the *same* and correct:

$$e_0, e_1, e_2, \ldots, e, e, e, \ldots, \text{ and}$$
$$A = W_e.$$

This convergence is also called *syntactic*. The corresponding learning is often called *intensional* or *explanatory*, and is abbreviated by EX.

A weaker notion of convergence, introduced by Case and Lynes [4] and independently by Osherson and Weinstein [20], allows the hypotheses to be *distinct*, although they must be correct after finitely many steps. That is, there is some n such that

$$A = W_{e_n} = W_{e_{n+1}} = W_{e_{n+2}} = \ldots.$$

This convergence is also called *semantic*. The corresponding learning is often called *extensional* or *behaviorally correct*, and is abbreviated by BC.

It can be shown that the class of all computable functions is EX-learnable from text by a general learner. On a given finite subset of the graph of a computable function to be learned, the learner guesses that it is the c.e. set with the least index, which contains this finite set. On the other hand, Gold [9] proved that the class of all computable functions is not EX-learnable from text by an algorithmic learner (see the introductory paper in this volume).

The identification in the limit does not require that the learner signals or confirms convergence. However, it yields the following result for a general EX-learner, due to L. Blum and M. Blum [3].

Proposition 2.1 (*L. Blum and M. Blum*) *If an EX-learner can learn a c.e. set A from text, then there is a finite sequence σ of elements of A, called a* locking sequence *for A, onto which the learner "locks" its conjecture for A. That is, if the learner outputs e after seeing σ, then $A = W_e$, and after seeing any sequence of elements from A extending σ, the learner outputs e again.*

For example, Proposition 2.1 implies that the class $\mathcal{L}$ of all finite sets enlarged by adding the set $\mathbb{N}$ of all natural numbers is not EX-learnable. The reason is that no learner will be able to distinguish between the set $\mathbb{N}$ and the finite set given by the locking sequence for $\mathbb{N}$, which also belongs to the class $\mathcal{L}$. A similar example of a class that is not EX-learnable is the collection consisting of an infinite c.e. set together with all of its finite subsets. Hence, as established by Gold [9], for example, the class of all regular languages and the class of all computable languages are not EX-learnable.

The following result in [1] gives an important characterization of EX-learnability from text for a general learner.

Theorem 2.2 (*Angluin*) *Let $\mathcal{L}$ be a class of c.e. sets. Then $\mathcal{L}$ is EX-learnable from text if and only if for every A in $\mathcal{L}$, there is a finite set $D \subseteq A$ such that for no U in $\mathcal{L}$ we can have*

$$D \subseteq U \subset A.$$

Proof. Assume that the class $\mathcal{L}$ is EX-learnable and that A is in $\mathcal{L}$. Then the corresponding set D will be the set of all elements of the locking sequence σ for A. Now if U is the language in $\mathcal{L}$ containing D and contained in A, then for any text for U extending σ, the learner will guess A, so $U = A$ (thus, U cannot be properly contained in A).

Conversely, assume that for every A in $\mathcal{L}$, there is a finite set $D \subseteq A$ such that for no U in $\mathcal{L}$ we can have $D \subseteq U \subset A$. Choose such D_A for $A \in \mathcal{L}$. Then, after receiving a finite subsequence σ of the text, the learner outputs the least index of a language A in $\mathcal{L}$ such that D_A is contained in $rng(\sigma)$, and $rng(\sigma)$ is contained in A: $D_A \subseteq rng(\sigma) \subseteq A$. If such A does not exist, the learner outputs an arbitrary fixed index. Note that the learner is not necessarily algorithmic.

We will now show that this learner identifies $\mathcal{L}$. Let i be the least index of a language A in $\mathcal{L}$, such that its text t is fed to the learner. There is a step n by which enough of t is given to the learner—a sequence σ so that $rng(\sigma)$ contains D_A but $rng(\sigma)$ is not contained in any language W_j in $\mathcal{L}$ among $W_0, \ldots, W_{i-1}$ such that $A \not\subseteq W_j$. We claim that the learner correctly conjectures W_i at every step after n. The required condition for W_i is satisfied for a corresponding τ:

$$D_A \subseteq rng(\sigma) \subseteq rng(\sigma^\wedge\tau) \subseteq A.$$

Furthermore, the learner will not conjecture any W_j in $\mathcal{L}$ among $W_0, \ldots, W_{i-1}$, since otherwise, for some σ' extending σ, we have $D_{W_j} \subseteq rng(\sigma') \subset W_j$. Hence $rng(\sigma)$ is contained in W_j, so $A \subset W_j$. Therefore,

$$D_{W_j} \subseteq rng(\sigma') \subseteq A \subset W_j.$$

However, $A \in \mathcal{L}$ and this contradicts the choice of D_{W_j}. ∎

2.3 *EX*-Learnability is More Restrictive than *BC*-Learnability

Clearly, EX-learnability implies BC-learnability. It can be shown, by finding examples, that algorithmic EX-learnability is more restrictive than algorithmic BC-learnability. We will now present one of the first such examples for learning from text (see [21]).

Example 2.3 *Recall that K is the halting set. Let $\mathcal{L}$ be the following class of sets:*

$$\mathcal{L} = \{K \cup D : D \textit{ is a finite set of natural numbers}\}.$$

The class $\mathcal{L}$ is algorithmically BC-learnable from text, but not algorithmically EX-learnable from text.

Proof. Let us first show that $\mathcal{L}$ is not algorithmically EX-learnable. To obtain a contradiction, assume that $\mathcal{L}$ is EX-learnable via some algorithmic learner M. Fix a locking sequence σ for K. We will use this locking sequence to show that the complement $\overline{K}$ is c.e. Since K is c.e., we can fix its algorithmic enumeration:

$$K = \{k_0, k_1, k_2, \ldots\}.$$

Given an input n, if $n \in K$, then $K = K \cup \{n\}$; otherwise, $K \neq K \cup \{n\}$, but the learner still correctly learns $K \cup \{n\}$ because it belongs to $\mathcal{L}$. More precisely, n is enumerated in $\overline{K}$ if and only if there is a sufficiently long sequence $k_0, \ldots, k_m$ such that the learner M converges on

$$\sigma\hat{\ }n\hat{\ }k_0\hat{\ }\ldots\hat{\ }k_m$$

to an index different from M's hypothesis on σ. This is a contradiction, since $\overline{K}$ is not c.e.

On the other hand, the class $\mathcal{L}$ is algorithmically BC-learnable from text because, by the s-m-n Theorem of computability theory (Chapter 1, Theorem 3.5 in [23]), there is an algorithm which for every finite sequence $a_0, \ldots, a_n$ outputs a Gödel code for the c.e. set $K \cup \{a_0, \ldots, a_n\}$. The outcome of this algorithm depends on the code for K and on the sequence $a_0, \ldots, a_n$. For some n_0, the sequence $a_0, \ldots, a_{n_0}$ will include a complete text for the finite set D. However, we cannot algorithmically find such n_0. ■

BC-learnability is much more powerful than EX-learnability, even when learning with anomalies (mistakes) is allowed, as showed by Bardzin and independently by Case, Smith and Harrington (see [5]). Case, Smith and Harrington established that a specific class of algorithmically BC-learnable computable functions is not algorithmically EX^*-learnable, where EX^*-learnability allows the learner to eventually guess a function finitely different from the one for which the data (i.e., the elements of the graph) are being fed to the learner.

2.4 Positive versus Negative Information. Learning from Text versus Learning from an Informant

So far, we have considered only learning from text, that is, when the learner requests only positive data (elements of the set to be learned), and the teacher eventually provides all of them. Learning from text will be abbreviated by Txt.

Learning from an *informant* is when the learner alternately requests positive and negative data (negative data are elements of the complement of a set, or incorrect sentences of a language), and the teacher eventually provides every element (with type label 1) of the set to be learned, and every element of its complement (with type label 0). Again, blanks (pauses) are also allowed in the presentation of data. Learning from an informant will be abbreviated by Inf.

It can be shown, by finding examples, that learning from text is more restrictive than learning from an informant. For example, the collection $\mathcal{L}$ consisting of $\mathbb{N}$ together with all of its finite subsets can be learned from an informant, but not from text. We already saw that $\mathcal{L}$ is not EX-learnable from text. This class $\mathcal{L}$ is EX-learnable from an informant because the learner, in addition to positive information, also obtains complete negative information in the limit. Thus, the learner guesses that the set to be learned is $\mathbb{N}$, until the learner sees a nonelement. After that, the learner guesses that the set to be learned is the finite set consisting of the elements given so far. Moreover, Gold [9] showed that the class of all context-sensitive languages is algorithmically learnable from an informant.

The class of all c.e. sets is learnable by a general learner from an informant. On any finite sequence of positive and negative data, the learner's hypothesis is the least Gödel code of a c.e. set that is consistent with the given sequence. On the other hand, Gold [9] proved that the class of all c.e. sets is not algorithmically learnable from an informant. Hence general learning from an informant is more powerful than algorithmic learning from an informant.

In [22], Sharma showed that combining learning from an informant with a restrictive convergence requirement that the first numerical (different from ?) hypothesis is already the correct one implies learnability from text. The convergence criterion where the learner is allowed to make only one conjecture, which has to be correct, is known as *finite identification* (see [9]). Since finite identification can be viewed as a model for batch learning, Sharma's result shows that batch learnability from both positive and negative data coincides with incremental learnability (i.e., identification in the limit) from positive data only. For more on learning theory of c.e. languages from text or from an informant see Case and Smith [5], Osherson, Stob and Weinstein [21], and Jain, Osherson, Royer and Sharma [13].

2.5 Learning from Switching Type of Data

Motoki [17], and later Baliga, Case and Jain [2] considered different ways of supplying to the learners finite sets of negative data for a c.e. set A. For example, there might be a finite set of negative data $F \subseteq \overline{A}$ such that the learner succeeds in learning the set A from F, in addition to a text for A. However, there is an algorithmic learner that learns all c.e. sets in this sense. Thus, the following framework that Baliga, Case and Jain introduced in [2] seems more interesting. There is a finite set $F \subseteq \overline{A}$ such that the learner always succeeds in learning the set A from a text for A, plus a text for a set C containing F and disjoint from A, that is, satisfying $F \subseteq C \subseteq \overline{A}$. In any case, Baliga, Case and Jain established that these finite amounts of negative information result in a strong increase of learning power, and also lead to increased speed of learning.

In [12], Jain and Stephan treated positive and negative data symmetrically and introduced several ways of learning from all positive and some negative data, or from all negative and some positive data. We will focus on the following framework in [12] and call the corresponding learning criterion, abbreviated by Sw, learning from *switching* (type of data). The learner is allowed to request positive or negative data for A, but when the learner after finitely many switches always requests data of the same type, the teacher eventually gives all elements in A (if the type is positive), or all elements in its complement $\overline{A}$ (if the type is negative). If the learner switches its requests infinitely often, then it still has to learn the language accurately.

One motivation for learning from switching comes from computability theory, precisely, from the *n-c.e. sets* for $n \geq 1$ (see p. 58 in [23]). Recall that for a set A, we write $A(x) = 1$ if $x \in A$, and $A(x) = 0$ if $x \notin A$. For an n-c.e. set A, the function $A(x)$ is *limit-computable*, $A(x) = \lim_{s \to \infty} f(x, s)$ for a binary computable function $f(x, s)$, and, assuming that $f(x, 0) = 0$, the limit must be achieved after at most n changes. Hence 1-c.e. sets are exactly the c.e. sets, and 2-c.e. sets are the set-theoretic differences of two c.e. sets. n-c.e. sets further generalize to *α-c.e. sets* for an arbitrary computable ordinal α.

The following example, due to Jain and Stephan [12], shows that switching type of information provides more learning power than giving positive information only. A set is co-finite if its complement is finite.

Example 2.4 *Let $\mathcal{L}$ be the class of all finite and all co-finite sets of natural numbers.*

(i) *The class $\mathcal{L}$ is not EX-learnable from text.*
(ii) *The class $\mathcal{L}$ is EX-learnable from switching.*

Proof. (i) The class $\mathcal{L}$ is not EX-learnable from text since the subclass of $\mathcal{L}$ consisting of all finite sets plus the set of all natural numbers (which is co-finite) is not EX-learnable from text.

(ii) The following learner can learn $\mathcal{L}$ from switching. If the number of positive data (elements) exceeds the number of negative data (nonelements) given to the learner, then the learner requests a negative datum. Moreover, it conjectures the co-finite set excluding exactly the nonelements given so far. In the other case, the learner requests a positive datum, and conjectures the finite set consisting of all elements given so far. The learner is correct in the limit. ■

Jain and Stephan [12] also showed that learning from switching is weaker than learning from an informant. The following result from [11] gives a general sufficient condition for non-$SwBC$-learnability.

Theorem 2.5 (*Harizanov and Stephan*) *Let $\mathcal{L}$ be a class of c.e. sets. Assume that there is some set A in $\mathcal{L}$ such that for every finite set D, there are U, U' in $\mathcal{L}$ with:*

$$U \subset A \subset U'$$

(U approximates A from below, and U' approximates A from above), and

$$D \cap U = D \cap U'$$

(U and U', and hence A, coincide on D). Then the class $\mathcal{L}$ cannot be BC-learned from switching, even by a general learner.

Proof. Let M be a given general $SwBC$-learner. We will show that M cannot learn $\mathcal{L}$. More precisely, we will show that there is a way of presenting data, according to the Sw-protocol, for a language in $\mathcal{L}$ on which M will be confused and will not be able to BC-infer the correct language. We will now describe such a sequence of data given to M.

(i) If the current hypothesis of M is a correct index for A, and there is a finite sequence $\overrightarrow{x} = x_1, x_2, \ldots, x_k$ of data of some length k, corresponding to the sequence of requests $y_1, y_2, \ldots, y_k$ of M ($y_1, y_2, \ldots, y_k \in \{0, 1\}$), such that after concatenating the sequence $\overrightarrow{x}$ to the current sequence of data presented to M, the hypothesis of M will become an incorrect index for A, then the first datum x_1 from one of the shortest such sequences will be given to M. (After k steps for the least k, a whole such sequence will be given to M.)

(ii) If the current hypothesis of M is an incorrect index for A, and y is M's current data type request, then M is given the least x that has not yet appeared in the data sequence and such that $A(x) = y$, where $A(\cdot)$ is the characteristic function of A.

(iii) In the remaining case, we have that all future hypotheses of M, when given appropriate data (positive or negative) consistent with A, result in hypotheses for A. We consider the following two subcases.

(a) If the pair U, U' has not been chosen yet, it will be chosen at this step as follows. Let D be the set of positive data given so far to the learner M. Choose U, U' so that:

$$U \subset A \subset U' \text{ and } D \cap U = D \cap A = D \cap U'.$$

The learner will receive the pause symbol $\sharp$.

(b) If the pair U, U' has been chosen, take the least x that has not yet appeared in the data sequence given to M, and which satisfies $x \in U$ in the case when the learner M requests a positive datum, and $x \notin U'$ in the case when M requests a negative datum. If such x does not exist (when $U = \emptyset$ or $U' = \omega$), then M is given the pause symbol $\sharp$.

For the proof, we first assume that M infinitely often conjectures a hypothesis incorrect for A. Then case (ii) applies infinitely often and M is given either all elements of A, or all nonelements of A. Hence M fails to BC-learn A from switching.

Otherwise, we assume that the learner M eventually ends up in case (iii), and at a certain step, U and U' are chosen so that $U \subset A \subset U'$. If infinitely often M requests positive data, then M is given all elements of U and some nonelements of U' (and hence nonelements of U). If infinitely often M requests negative data, then M is given all nonelements of U' and some elements of U (and hence elements of U'). In the first case, M is expected to learn U, and in the second case, M is expected to learn U'. However, in both cases, M almost always conjectures an index for the set A. Hence M does not learn $\mathcal{L}$ from switching. ■

3 LEARNING CLASSES OF ALGEBRAIC STRUCTURES

More recently, Stephan and Ventsov [24] extended learning theory from sets to algebraic structures. They investigated learnability from text for classes of c.e. substructures of certain computable algebraic structures. In particular, they considered c.e. submonoids and c.e. subgroups of a computable group, and c.e. ideals of a computable commutative ring. Several of these learnable classes have nice algebraic characterizations. Stephan and Ventsov also studied ordinal bounds on the learner's mind changes when learning classes of algebraic substructures from text. In addition, Stephan and Ventsov showed that learnability of algebraic structures can greatly depend on the semantic knowledge given at the synthesis of the learner. Since we deal with structures, semantic knowledge may consist of programs that compute structural operations. Harizanov and Stephan [11] further investigated learnability of classes of c.e. vector spaces. Again, some learnable classes coincide with natural algebraic classes. For these classes, in addition to learning from text, Harizanov and Stephan also considered learning with negative data.

3.1 Algebraic Structures

An *algebraic structure* is a nonempty set, called the domain, together with some operations and relations satisfying certain axioms. We often use the same symbol for a structure and its domain. A *semigroup* $(G, *)$ is a structure with the domain G and an associative binary operation $*$:

$$(x * y) * z = x * (y * z) \text{ for all } x, y, z \in G.$$

A semigroup is *abelian* or *commutative* if its binary operation is commutative. A *monoid* is a semigroup $(G, *)$ that contains an *identity* element e:

$$x * e = e * x = x \quad \text{for all } x \in G.$$

It can be shown that the identity element is unique. An example of a monoid is $(\{0, 1, 2, \ldots\}, +)$. A *group* is a monoid $(G, *)$ such that for every $x \in G$, there exists an *inverse* element $x^{-1} \in G$:

$$x * x^{-1} = x^{-1} * x = e.$$

An example of a group is the additive structure of integers $(Z, +)$. Here, $e = 0$ and $x^{-1} = -x$. The structure $(\{\ldots, -2, 0, 2, 4, \ldots\}, +)$ is a subgroup of $(Z, +)$. Another important example of a group is the multiplicative structure of nonzero rationals $(Q\backslash\{0\}, \cdot)$. For this group, $e = 1$ and $x^{-1} = \frac{1}{x}$.

A *ring* is a structure $(R, +, \cdot)$ with two binary operations, usually called addition and multiplication, such that $(R, +)$ is a commutative group, the multiplication is associative, and

$$x \cdot (y + z) = x \cdot y + x \cdot z \text{ and } (y + z) \cdot x = y \cdot x + z \cdot x.$$

It can be shown that

$$x \cdot 0 = 0 \cdot x = 0.$$

If for a ring $(R, +, \cdot)$ the multiplication $\cdot$ is commutative, then the ring is said to be commutative. If there is an element $1 \in R$ such that for all $x \in R$,

$$x \cdot 1 = 1 \cdot x = 1,$$

then $(R, +, \cdot)$ is a ring with identity 1. If $(R, +, \cdot)$ is a nontrivial (containing at least two elements) commutative ring with identity and, in addition, every nonzero element has an inverse with respect to multiplication, then $(R, +, \cdot)$ is a *field*. For example, the ring of integers $(Z, +, \cdot)$ is not a field, while the ring of rationals $(Q, +, \cdot)$ is a field. There are finite fields (also called Galois fields). For a prime number p, the ring $(Z_p, +_p, \cdot_p)$ of residue classes modulo p is a field.

Let

$$RJ =_{def} \{r \cdot j : r \in R \wedge j \in J\}.$$

A subring $(J, +, \cdot)$ of $(R, +, \cdot)$ is an *ideal* if and only if

$$RJ \subseteq J \wedge JR \subseteq J.$$

For example, $\{\ldots, -2, 0, 2, 4, \ldots\}$ is an ideal of $(Z, +, \cdot)$. Two obvious ideals of $(R, +, \cdot)$ are the unit ideal—the ring itself, and the trivial ideal with the domain $\{0\}$. A set $D \subseteq J$ *generates* the ideal J, in symbols $J = I(D)$, if J is the smallest ideal (with respect to the set-theoretic inclusion) that contains D. For more on rings and ideals, see [15].

A *vector space* over a field F is an additive commutative group $(V, +)$, together with scalar multiplications $\cdot_\gamma$ for every element $\gamma \in F$, which satisfies additional axioms for vector addition $+$ and scalar multiplication. The scalar product of $\gamma \in F$ and a vector $x \in V$ is usually denoted by γx. The axioms are:

$$\gamma(x + y) = \gamma x + \gamma y,$$
$$(\gamma + \delta)x = \gamma x + \delta x,$$
$$\gamma(\delta x) = (\gamma \delta)x,$$
$$1x = x,$$

where $x, y \in V$ and $\gamma, \delta, 1 \in F$.

A *basis* for a vector space consists of a maximal set of linearly independent vectors. All bases have the same size, called the *dimension* of the space. An example of an important infinite dimensional vector space is the vector space Q^∞ over the field Q of rationals. We can think of the elements of Q^∞, the vectors, as infinite sequences of rational numbers with only finitely many nonzero components. We have pointwise vector addition and pointwise scalar multiplication, e.g.,

$$(6, 3, -4, 0, 0, \ldots) + (-1, 4, 0, 0, 0, \ldots) = (5, 7, -4, 0, 0, \ldots),$$

and

$$6(1, \frac{1}{2}, -\frac{2}{3}, 0, 0, \ldots) = (6, 3, -4, 0, 0, \ldots).$$

For example, the three vectors

$$(1, \frac{1}{2}, -\frac{2}{3}, 0, 0, \ldots), (-\frac{1}{2}, 2, 0, 0, 0, \ldots), (5, 7, -4, 0, 0, \ldots)$$

are linearly dependent since

$$6(1, \frac{1}{2}, -\frac{2}{3}, 0, 0, \ldots) + 2(-\frac{1}{2}, 2, 0, 0, 0, \ldots) = (5, 7, -4, 0, 0, \ldots).$$

The following vectors

$$(1, 0, 0, 0, \ldots), (0, 1, 0, 0, \ldots), (0, 0, 1, 0, \ldots), \ldots$$

are linearly independent. They form a *standard basis* for Q^∞.

3.2 Computable Structures and Their Computably Enumerable Substructures

A countable structure is *computable* if its domain is a computable subset of natural numbers, and its relations and operations are uniformly computable. For example, a group $(G, *)$ is computable if its domain G is computable and the operation $*$ is computable. It then follows that the unary function $^{-1}$ assigning to each element g its inverse g^{-1} is also computable. Clearly, every finite group is computable. Examples of computable infinite groups include the additive group of integers $(Z, +)$, the additive group of rationals $(Q, +)$, and the multiplicative group of rationals $(Q\backslash\{0\}, \cdot)$, under the standard representations, that is, coding of elements and operations by the natural numbers. A ring $(R, +, \cdot)$ is computable if the domain R is a computable set and the binary operations $+, \cdot$ are computable. An example of a computable ring is the ring of integers $(Z, +, \cdot)$ under the standard representation.

Metakides and Nerode [16] showed that the study of algorithmic vector spaces can be reduced to the study of V_∞, an infinite dimensional vector space over a computable field F, consisting of all finitely nonzero infinite sequences of elements of F, under pointwise operations. Clearly, these operations can be performed algorithmically. Every element in F can be identified with its Gödel code, which is a natural number. Unless we explicitly state otherwise, we will assume that F is infinite. In that case, we can even assume, without loss of generality, that F is the field of rationals $(Q, +, \cdot)$ and identify V_∞ with Q^∞. A *dependence algorithm* for a vector space decides whether any finite set of its vectors is linearly dependent. Since the standard basis for V_∞ is computable, V_∞ has a dependence algorithm.

Roughly speaking, a c.e. vector space over a computable field is one where the set of vectors is c.e., operations of vector addition and scalar multiplication are partial computable, and the vector equality is a c.e. relation. Metakides and Nerode [16] showed that any c.e. vector space is isomorphic to the *quotient space* $\frac{V_\infty}{V}$, where V is a c.e. subspace of V_∞. In $\frac{V_\infty}{V}$, the equality of vectors is *modulo* V. That is, two vectors u and w are equal if their difference $u - w$ is 0 *modulo* V, i.e., belongs to V. Thus, the class of all c.e. subspaces (substructures) of $\frac{V_\infty}{V}$ can be viewed as the class $\mathcal{L}(V)$ of all c.e. subspaces of V_∞ that contain V, the superspaces of V. Let $V_0, V_1, V_2, \ldots$ be an *algorithmic list* of all c.e. subspaces of V_∞ indexed by their Gödel codes. For example, we can assume that V_e is generated by the c.e. set W_e of independent vectors in V_∞, where $W_0, W_1, W_2, \ldots$ is a fixed *algorithmic list* of all c.e. subsets of independent vectors. This is possible since V_∞ has a computable basis, which can be identified with the set of natural numbers.

3.3 Learning Classes of Ring Ideals from Text

A commutative ring with identity $(R, +, \cdot)$ is *Noetherian* if it does not have an infinite strictly ascending chain of ideals. Equivalently, $(R, +, \cdot)$ is Noetherian if and only if every ideal of $(R, +, \cdot)$ is *finitely generated* (see [15]). Hence every ideal of a computable commutative Noetherian ring is c.e. It can be shown that it is even computable (see [25]). An example of a computable commutative Noetherian ring with identity is the ring $Q[x_1, \ldots, x_n]$ of polynomials with rational coefficients and variables $x_1, \ldots, x_n$. On the other hand, the ring $Q[x_1, x_2, x_3, \ldots]$ of rational polynomials in infinitely many variables is not Noetherian.

The following result from [24] algebraically characterizes BC-learnability from text of ideals of a computable commutative ring.

Theorem 3.1 (*Stephan and Ventsov*) *Let $(R, +, \cdot)$ be a computable commutative ring with identity. Let $\mathcal{L}$ be the set of all ideals of R. Then the following two statements are equivalent.*

(i) *The ring $(R, +, \cdot)$ is Noetherian.*
(ii) *The family $\mathcal{L}$ (of c.e. ideals) is algorithmically BC-learnable from text.*

Proof. (i)$\Rightarrow$ (ii) For a given finite set D of positive data, the learner hypothesizes a code for the ideal generated by D. There is a stage by which the learner has seen all elements of a finite set that generates the ideal to be learned. Hence, from that stage on the learner correctly guesses a code for the ideal to be learned.

(ii)⇒ (i) Assume that M is a BC-learner for $\mathcal{L}$. We will show that every ideal of $(R, +, \cdot)$ is finitely generated. Let I be an ideal in $\mathcal{L}$. Fix a locking sequence σ for I. Consider the ideal J generated by $rng(\sigma)$. For any sequence τ of elements in J, after seeing the sequence $\sigma^\wedge\tau$ of positive data, the learner M guesses an index for I. Since $J \in \mathcal{L}$ and M learns $\mathcal{L}$, it follows that $I = J$. Hence I is generated by the finite set $rng(\sigma)$. ■

For certain computable commutative Noetherian rings with identity, the classes of their ideals are EX-learnable from text, even by well-behaved learners. An example of such a ring is the polynomial ring $Q[x_1, \ldots, x_n]$ under the standard representation. This example generalizes to the following result (see [24]).

Theorem 3.2 (*Stephan and Ventsov*) *Let $(R, +, \cdot)$ be a computable commutative Noetherian ring with identity. Let $\mathcal{L}$ be the set of all ideals of R. Then the following two statements are equivalent.*

(i) *There is a uniformly computable sequence of all ideals in $\mathcal{L}$.*
(ii) *The set $\mathcal{L}$ is EX-learnable from text by a class-comprising, consistent, and conservative algorithmic learner.*

Stephan and Ventsov [24] constructed a ring that allowed them to establish the following negative result about EX-learnability.

Theorem 3.3 (*Stephan and Ventsov*) *There is a computable commutative Noetherian ring with identity $(R, +, \cdot)$ for which the class of all ideals is not EX-learnable from text. Moreover, this is also true for the class of all ideals in every computable isomorphic copy of $(R, +, \cdot)$.*

Stephan and Ventsov [24] also constructed a computable ring $(R, +, \cdot)$ with the class of all ideals EX-learnable from text, while in some computable isomorphic copy of $(R, +, \cdot)$ the class of ideals is BC-learnable, but not EX-learnable.

The following result from [24] gives a computability theoretic characterization of EX-learnability of Noetherian ring ideals. This characterization is in terms of dominating time for enumeration of elements in the (finitely generated) ideals.

Theorem 3.4 (*Stephan and Ventsov*) *Let $(R, +, \cdot)$ be a computable commutative Noetherian ring with identity, and let $\mathcal{L}$ be the set of all ideals of $(R, +, \cdot)$. Then $\mathcal{L}$ is EX-learnable from text if and only if there is a computable function f with the property that for every $J \in \mathcal{L}$, there is a finite set D such that $J = I(D)$ and every $x \in I(D)$ is enumerated in $I(D)$ within $f(x)$ computation steps.*

A commutative ring is *Artinian* if it does not have an infinite strictly descending chain of ideals. Every commutative Artinian ring with identity is Noetherian. Hence every commutative Artinian ring with identity has a finite *length* n, where n is the maximum number such that there is a chain of $(n + 1)$ ideals.

Theorem 3.5 (*Stephan and Ventsov*) *Let* $(R, +, \cdot)$ *be a computable commutative ring with identity, and let* $\mathcal{L}$ *be the set of all ideals of* $(R, +, \cdot)$. *Then* $\mathcal{L}$ *is learnable from text with a constant bound on the number of mind changes if and only if the ring* $(R, +, \cdot)$ *is Artinian. The constant bound is exactly the length of the ring.*

Since every ideal of a computable commutative Noetherian ring is computable, Stephan and Ventsov also investigated learning Gödel codes for characteristic functions of these ideals (see [24]).

Theorem 3.6 (*Stephan and Ventsov*) *Let* $(R, +, \cdot)$ *be a computable commutative Noetherian ring with identity. Let* $\mathcal{L}$ *be the set of all ideals of* R.

(*i*) *If the class* $\mathcal{L}$ *is* EX*-learnable from text, then the characteristic function indices of the ideals in* $\mathcal{L}$ *can be* EX*-learned by a confident learner.*

(*ii*) *If the characteristic function indices of the ideals in* $\mathcal{L}$ *are* BC*-learnable from text, then these characteristic function indices can be also* EX*-learned by a confident learner.*

3.4 Characterizing Text-Learnable Classes of Vector Spaces

We will show that the classes $\mathcal{L}(V)$ of c.e. vector spaces that are syntactically inferable from text have a nice algebraic characterization. Moreover, this characterization does not change if we allow either semantic convergence or inference by switching.

Theorem 3.7 (*Harizanov and Stephan*) *Assume that the dimension of* $\frac{V_\infty}{V}$ *is finite. Then the class* $\mathcal{L}(V)$ *is* EX*-learnable from text by an algorithmic learner.*

Proof. The learner guesses that the space to be learned is generated by $V \cup \{x_0, \ldots, x_{n-1}\}$, where $x_0, \ldots, x_{n-1}$ is the sequence of positive data given to the learner so far. Hence this guess will change whenever the learner receives a new element. However, since the dimension of $\frac{V_\infty}{V}$ is finite, every space A in $\mathcal{L}(V)$ is generated by $V \cup D$ for some finite set D of elements independent over V. If a text for such a space A is given to the learner, then for some m, the set $\{x_0, \ldots, x_{m-1}\}$ will include all of D. Thus, the learner will syntactically identify A in the limit.

Now, we have to show that this can be done in an algorithmic manner. Notice that, since the dimension of $\frac{V_\infty}{V}$ is finite, the space V is computable. In addition, there is an algorithm which checks for every finite set D of vectors in V_∞ and every vector v, whether v is in the linear closure of $V \cup D$ (i.e., in the space generated by $V \cup D$). Such an algorithm exists, since V_∞ has a computable basis consisting of vectors from a (computable) basis for V plus finitely many additional vectors.

Let i be a fixed index for a c.e. set of independent vectors that generates V. Assume that at some stage the learner is given $x_0, \ldots, x_{n-1}$. The learner first algorithmically checks whether x_0 belongs to V. If it does, the learner omits it from the sequence, and next checks whether x_1 belongs to V. If x_0 does not belong to V, then the learner keeps x_0 in the sequence, and algorithmically checks whether x_1 belongs to the linear closure of $V \cup \{x_0\}$. If it does, the learner omits it from the sequence, and next checks whether x_2 belongs to the linear closure of $V \cup \{x_0\}$. If x_1 does not belong to the linear closure of $V \cup \{x_0\}$, then the learner keeps x_0, x_1 in the sequence, and algorithmically checks whether x_2 belongs to the linear closure of $V \cup \{x_0, x_1\}$. After n rounds, the learner ends up with a subsequence $x'_0, \ldots, x'_{k-1}$ $(k \leq n)$, each element of which is linearly independent from the previous ones together with V. Now the learner executes an algorithm that on input $(i, x'_0, \ldots, x'_{k-1})$ outputs a code e for the space V_e generated by $V \cup \{x_0, \ldots, x_{n-1}\}$. Moreover, the learner changes its hypothesis only when the space generated by $V \cup \{x_0, \ldots, x_n\}$ properly contains the previous space $V \cup \{x_0, \ldots, x_{n-1}\}$, where for $n = 0$ the previous space is V. Since the sequence $x'_0, \ldots, x'_{k-1}$ eventually stabilizes, the learner will keep outputting the same code e and will EX-learn A. Thus, the number of mind changes that the learner makes is bounded by the dimension of $\frac{V_\infty}{V}$. ■

Theorem 3.8 (*Harizanov and Stephan*) *Assume that the class $\mathcal{L}(V)$ is BC-learnable from text by an algorithmic learner. Then the dimension of $\frac{V_\infty}{V}$ is finite.*

Proof. Let $v_0, v_1, \ldots$ be a computable enumeration of V_∞. We define U_n for $n \geq 0$ to be the vector space generated by $V \cup \{v_0, v_1, \ldots, v_n\}$. Since $U_{n+1} = U_n \cup \{v_{n+1}\}$, the dimension of U_{n+1} is either the same as for U_n or increases by 1. Clearly, V_∞ is the ascending union of all spaces U_n:

$$V_\infty = U_0 \cup U_1 \cup U_2 \cup \ldots, \text{ and}$$

$$U_0 \subseteq U_1 \subseteq U_2 \subseteq \ldots.$$

If follows from a learning theoretic result that such a class $\mathcal{L}(V)$ can be BC-learned from text only if there are finitely many distinct sets in this ascending chain. Hence, there is m such that

$$U_m = U_{m+1} = U_{m+2} = \ldots.$$

It follows that

$$V_\infty = U_0 \cup \ldots \cup U_m = U_m.$$

Hence the dimension of $\frac{V_\infty}{V}$ is at most $m + 1$, and thus finite. ■

Theorem 3.9 (*Harizanov and Stephan*) *Assume that the class $\mathcal{L}(V)$ is EX-learnable from switching by an algorithmic learner. Then the dimension of $\frac{V_\infty}{V}$ is finite.*

Proof. Assume that $V \neq V_\infty$. We will first prove the following claim.

Claim. *If $\mathcal{L}(V)$ is EX-learnable from switching by an algorithmic learner, then V is computable.*

Assuming the claim, we will now prove the theorem. To obtain a contradiction, we suppose that $\frac{V_\infty}{V}$ is infinite dimensional. Then, since V is computable, we can find a computable basis $\{u_0, u_1, \ldots\}$ of a vector space U such that $U \cap V = \{0\}$. Recall that K denotes the halting set. Let W be the linear closure of

$$V \cup \{u_n : n \in K\}.$$

The space W is not computable, so it follows by the claim that $\mathcal{L}(W)$ is not EX-learnable from switching. Since W contains V, the class $\mathcal{L}(W)$ is contained in the class $\mathcal{L}(V)$. Hence, $\mathcal{L}(V)$ is also not EX-learnable from switching, which is a contradiction.

To prove the claim, assume that M is an algorithmic $SwEX$-learner for $\mathcal{L}(V)$. Fix a computable ordering of the elements of V_∞, as well as of all finite sequences of these elements.

(i) If the current hypothesis of M is old (i.e., the index that M guesses is the same as the previous one) and there is a finite sequence $\overrightarrow{x} = x_1, x_2, \ldots, x_k$ of data consistent with V, corresponding to requests $y_1, y_2, \ldots, y_k$ of M, such that M will change its hypothesis after seeing $\overrightarrow{x}$, then M is given the first datum x_1 from the first shortest such sequence.

(ii) If the current hypothesis of M is new (i.e., the index that M guesses is different from the previous one), then after M requests a datum of type y ($y \in \{0, 1\}$), M is given the least x that has not yet been given to it such that $V(x) = y$.

Clearly, the learning process either goes finitely or infinitely many times both through case (i) and case (ii). First assume that the cases (i)–(ii) apply infinitely many times. Then, it follows that the learner M has made infinitely many different hypotheses. However, the learner has been given either all elements of V (if M requests all type 1 data after some step), or all elements of $\overline{V}$ (if M requests all type 0 data after some step). Thus, the learner is given information about V according to the Sw-protocol without converging syntactically. This contradicts the assumption of the claim. Thus, both (i) and (ii) are executed finitely many times.

Since (i) is executed only finitely many times, there is a stage in the learning process without M's further mind change, provided M is given data consistent with V (that is, elements of V upon requests of type 1, and elements of $\overline{V}$ upon requests of type 0). We can even assume that M's hypothesis from some point on is a fixed index for V, since otherwise, M would not EX-learn V when given information about V according to the Sw-protocol. Now, with this assumption, we further consider two cases.

(a) Assume that in every possible situation when the learner M requests a datum of type 1, it is given an element of V such that at some later stage M requests a datum of type 0. This allows us to take some proper c.e. superspace W of V (since $V \neq V_\infty$), and at every request of M for a datum of type 0, give M the least element x in $\overline{W}$, which M had not seen so far. Then M does not infer the space W, although M is given data for W.

(b) In the remaining possible case, M is fed finitely many data consistent with V such that M never later requests any datum of type 0. Consider the stage after which M requests only data of type 1. Let D be the set of all data of type 0 given to M up to that stage. We can now algorithmically enumerate $\overline{V}$ as follows: $x \in \overline{V}$ iff either M changes its mind while being fed (positive) data from the linear closure of $V \cup \{x\}$, or M requests a datum of type 0, or some element of D is enumerated into the linear closure of $V \cup \{x\}$. Hence $\overline{V}$ is c.e., and thus V is computable. ■

We can now summarize the previous results as follows.

Corollary 3.10 *The following statements are equivalent for a c.e. subspace V of V_∞.*

(i) *The dimension of $\frac{V_\infty}{V}$ is finite.*
(ii) *The class $\mathcal{L}(V)$ is EX-learnable from text by an algorithmic learner.*
(iii) *The class $\mathcal{L}(V)$ is BC-learnable from text by an algorithmic learner.*
(iv) *The class $\mathcal{L}(V)$ is EX-learnable from switching by an algorithmic learner.*

Proof. (i) $\Rightarrow$ (ii), (iii) $\Rightarrow$ (i) and (iv) $\Rightarrow$ (i): These follow from the above results.

(ii) $\Rightarrow$ ((iii) & (iv)): These follow directly from the definitions, since EX-learnability from text implies BC-learnability from text, as well as EX-learnability from switching. ■

Next, we will investigate the learnability of the class $\mathcal{L}(V)$ when the dimension of $\frac{V_\infty}{V}$ is infinite.

3.5 Characterizing *SwBC*-Learnable Classes of Vector Spaces

Assume that V is a c.e. subspace of V_∞ such that the dimension of $\frac{V_\infty}{V}$ is infinite. Let k be a natural number, possibly 0. In [14], Kalantari and Retzlaff introduced the following notion of a k-thin space.

Definition 3.11 *A space $V \in \mathcal{L}(V_\infty)$ with infinite dimension of $\frac{V_\infty}{V}$ is called k-thin, $k \geq 0$, if for every c.e. subspace W of V_∞ such that $W \supseteq V$:*

(i) *either the dimension of $\frac{V_\infty}{W}$ is at most k, or*
(ii) *the dimension of $\frac{W}{V}$ is finite,*

and there exists $U \in \mathcal{L}(V_\infty)$ such that the dimension of $\frac{V_\infty}{U}$ is exactly k.

Kalantari and Retzlaff proved that for every $k \geq 0$, there exists a k-thin subspace of V_∞.

Theorem 3.12 (*Harizanov and Stephan*) *The following statements are equivalent for a c.e. subspace V of V_∞.*

(i) *The class $\mathcal{L}(V)$ is BC-learnable from switching by an algorithmic learner.*
(ii) *The dimension of $\frac{V_\infty}{V}$ is finite, or V is 0-thin, or V is 1-thin.*

Proof. To prove $\neg$(ii) $\Rightarrow$ $\neg$(i), we will apply Theorem 2.5. Assume that $\frac{V_\infty}{V}$ has infinite dimension, and that V is neither 0-thin nor 1-thin. Then there is a c.e. space W such that $V \subset W \subset V_\infty$, the quotient space $\frac{W}{V}$ has infinite dimension, and $\frac{V_\infty}{W}$ has dimension at least 2. In particular, there are vectors x_1, x_2 such that $x_1 \notin W$, $x_2 \notin W$, and x_1, x_2 are linearly independent over W. Now, for every finite set D of vectors, we can choose a positive integer n such that none of the vectors in $D - W$ is in the linear closure of $W \cup \{x_1 + nx_2\}$. Furthermore, the linear closure of $V \cup (W \cap D)$ has finite dimension over V, and thus is different from W. So the condition of Theorem 2.5 is satisfied, and hence $\mathcal{L}(V)$ is not BC-learnable from switching.

To prove the converse, (ii) $\Rightarrow$ (i), we have to consider only the cases of 0-thin and 1-thin spaces, since Theorem 3.7 deals with the case when the dimension of $\frac{V_\infty}{V}$ is finite. In these two cases, there is a space W such that $V \subseteq W$ and $\frac{W}{V}$ is infinite dimensional and the following conditions are satisfied. If V is 0-thin, we have that $W = V_\infty$. If V is 1-thin, we have that $W \subset V_\infty$, and there is no other such c.e. vector space U with the quotient space $\frac{U}{V}$ of infinite dimension. This property allows us to give the following learning algorithm for $\mathcal{L}(V)$.

- A learner M requests data of type 0 until such a datum is enumerated into W. The learner's hypothesis is V_∞ as long as no datum of type 0 (except the pause symbol) is given, and W otherwise.
- If some datum of type 0 appears in W, then M requests data of type 1, and guesses the linear closure of $V \cup E$, where E is the set of all data of type 1 seen so far.

In the case when M guesses V_∞ or W, the learner M requests only data of type 0. If none are supplied, the hypothesis V_∞ is correct, and if some negative data are given, but they are in the complement of W, then the hypothesis W is correct. Otherwise, the vector space to be learned is the linear closure of $V \cup D$ for some finite set D. Since this space cannot contain all of W, a datum of type 0 and in W shows up and causes that from that time on M requests only data of type 1. So, the learner M is eventually given all elements of the linear closure of $V \cup D$. Since the space to be learned by M is finite dimensional over V, beginning at some stage, D is contained in the linear closure of $V \cup E$, where E is the set of positive data given to M by that stage. ■

If the space V_∞ is over a *finite* field, and V is a c.e. subspace of V_∞, then the quotient space $\frac{V_\infty}{V}$ has a dependence algorithm. Hence we obtain a different result on BC-learnablity from switching. On the other hand, Corollary 3.10 remains the same since its proof does not depend on the fact that the field is infinite.

Proposition 3.13 (*Harizanov and Stephan*) *Assume that V_∞ is over a finite field of scalars. Let V be a c.e. subspace of V_∞, and let k be any natural number. If V is k-thin, then $\mathcal{L}(V)$ is BC-learnable from switching by an algorithmic learner.*

Learning from an informant does not have a nice algebraic characterization, at least not one in terms of thin vector spaces, as the following result from [11] shows.

Theorem 3.14 (*Harizanov and Stephan*)

(i) *There is a 0-thin vector space V_1 such that the class $\mathcal{L}(V_1)$ is EX-learnable from an informant by an algorithmic learner.*
(ii) *There is a 0-thin vector space V_2 such that the class $\mathcal{L}(V_2)$ is not EX-learnable from an informant by an algorithmic learner.*

4 CONCLUSION

We presented inductive inference of classes of c.e. sets, c.e. ring ideals, and c.e. vector subspaces. With respect to the convergence criterion for the learner's hypotheses, we considered both syntactic inference (EX-identification) and semantic inference (BC-identification). We also considered different ways of giving positive and negative data to the learner. For classes of ring ideals, we considered only inference from text, while for classes of c.e. sets and c.e. vector spaces, we also considered inference from switching, and inference from an informant. It will be interesting to also study inference from switching and from an informant for classes of ring ideals.

Furthermore, it would be worthwhile to systematically investigate inductive inference of c.e. substructures for other important classes of computable algebraic structures. Some specific results in this direction have already been obtained. Stephan and Ventsov [24] studied inference of classes of *monoids*. For example, they showed that for the standard representation of the integers $(Z, +)$, the class of all submonoids of $(Z, +)$ can be EX-learned from text with the learner's mind change complexity ω^2. Moreover, this bound is optimal. Stephan and Ventsov also studied inference of classes of closed sets of c.e. partially ordered structures. A structure $(P, \sqsubseteq)$ is *a partial ordering* if the binary relation $\sqsubseteq$ is reflexive ($x \sqsubseteq x$) and transitive ($(x \sqsubseteq y \wedge y \sqsubseteq z) \Rightarrow x \sqsubseteq z$). A subset A of P is *closed* if for every $x \in A$ and $y \sqsubseteq x$, we have $y \in A$. For example, Stephan and Ventsov showed that there is a c.e. linear (i.e., without incomparable elements) ordering such that the family of closed sets is BC-inferable, but not EX-inferable from text.

Sets and vector spaces can be considered as examples of *closure systems*, which are abstract mathematical structures with dependence relations satisfying certain axioms (see [7]). Such dependence relations generalize linear dependence of vectors in vector spaces, and the identity of elements in sets. Harizanov and Stephan [11] studied inference of classes of c.e. substructures of computable closure systems. For example, Harizanov and Stephan showed that there is a computable closure system with the class of all c.e. closure subsystems that can be EX-learned from switching, but not with any bound on the number of switches.

Learning with access to an *oracle* (external set with given information about its elements and nonelements) is also important. Some results have already been obtained. For example, Harizanov and Stephan [11] showed that there is a computable closure system with the class of all c.e. subsystems that can be BC-learned from switching with oracle for the halting set K, but not with any oracle to which K is not Turing reducible. Stephan and Ventsov also studied learning of ring ideals with access to an oracle. In particular, they showed in [24] that the class of all ideals of any computable Noetherian ring is EX-learnable from text with oracle K.

For inference from text or from switching, many classes of inferable algebraic structures have natural algebraic characterizations. It would be interesting to also find algebraic properties of structures, which exactly correspond to the inferability from an informant, both for EX-learning and BC-learning.

ACKNOWLEDGMENT

The author thanks John Case for helpful comments and many valuable discussions on algorithmic learning theory.

REFERENCES

[1] Angluin, D. (1980). "Inductive Inference of Formal Languages from Positive Data", *Information and Control* 45, 117–135.

[2] Baliga, G., Case, J. and Jain, S. (1995). "Language Learning with Some Negative Information", *Journal of Computer and System Sciences* 51, 273–285.

[3] Blum, L. and Blum, M. (1975). "Toward a Mathematical Theory of Inductive Inference", *Information and Control* 28, 125–155.

[4] Case, J. and Lynes, C. (1982). "Machine Inductive Inference and Language Identification", in Nielsen, M. and Schmidt, E.M. [18], 107–115.

[5] Case, J. and Smith, C. (1983). "Comparison of Identification Criteria for Machine Inductive Inference", *Theoretical Computer Science* 25, 193–220.

[6] Cesa-Bianchi, N., Numao, M. and Reischuk, R. (eds.) (2002). *Algorithmic Learning Theory: 13th International Conference*, Lecture Notes in Artificial Intelligence 2533, Berlin: Springer-Verlag.
[7] Downey, R.G. and Remmel, J.B. (1998). "Computable Algebras and Closure Systems: Coding Properties", in Ershov, Yu.L., Goncharov, S.S., Nerode, A. and Remmel, J.B. [8], 977–1039.
[8] Ershov, Yu.L., Goncharov, S.S., Nerode, A. and Remmel, J.B. (eds.) (1998). *Handbook of Recursive Mathematics* 2, Amsterdam: Elsevier.
[9] Gold, E.M. (1967). "Language Identification in the Limit", *Information and Control* 10, 447–474.
[10] Griffor, E.R. (ed.) (1999). *Handbook of Computability Theory*, Amsterdam: Elsevier.
[11] Harizanov, V.S. and Stephan, F. (2002). "On the Learnability of Vector Spaces", in Cesa-Bianchi, N., Numao, M. and Reischuk, R. [6], 233–247.
[12] Jain, S. and Stephan, F. (2003). "Learning by Switching Type of Information", *Information and Computation* 185, 89–104.
[13] Jain, S., Osherson, D.N., Royer, J.S. and Sharma, A. (1999). *Systems That Learn: An Introduction to Learning Theory*, 2nd ed., Cambridge (Mass.): MIT Press.
[14] Kalantari, I. and Retzlaff, A. (1977). "Maximal Vector Spaces Under Automorphisms of the Lattice of Recursively Enumerable Vector Spaces", *Journal of Symbolic Logic* 42, 481–491.
[15] Kaplansky, I. (1974). *Commutative Rings*, Chicago: The University of Chicago Press.
[16] Metakides, G. and Nerode, A. (1977). "Recursively Enumerable Vector Spaces", *Annals of Mathematical Logic* 11, 147–171.
[17] Motoki, T. (1991). "Inductive Inference from All Positive and Some Negative Data", *Information Processing Letters* 39, 177–182.
[18] Nielsen, M. and Schmidt, E.M. (eds.) (1982). *Automata, Languages and Programming: Proceedings of the 9th International Colloquium*, Lecture Notes in Computer Science 140, Berlin: Springer-Verlag.
[19] Odifreddi, P. (1989). *Classical Recursion Theory*, Amsterdam: North-Holland.
[20] Osherson, D.N. and Weinstein, S. (1982). "Criteria of Language Learning", *Information and Control* 52, 123–138.
[21] Osherson, D.N., Stob, M. and Weinstein, S. (1986). *Systems That Learn: An Introduction to Learning Theory for Cognitive and Computer Scientists*, Cambridge (Mass.): MIT Press.
[22] Sharma, A. (1998). "A Note on Batch and Incremental Learnability", *Journal of Computer and System Sciences* 56, 272–276.
[23] Soare, R.I. (1987). *Recursively Enumerable Sets and Degrees. A Study of Computable Functions and Computably Generated Sets*, Berlin: Springer-Verlag.
[24] Stephan, F. and Ventsov, Yu. (2001). "Learning Algebraic Structures from Text", *Theoretical Computer Science* 268, 221–273.
[25] Stoltenberg-Hansen, V. and Tucker, J.V. (1999). "Computable Rings and Fields", in Griffor, E.R. [10], 363–447.

3

DEDUCTION, INDUCTION, AND BEYOND IN PARAMETRIC LOGIC

ERIC MARTIN[1], ARUN SHARMA[2], AND FRANK STEPHAN[3]

[1]*School of Computer Science and Engineering, National ICT Australia, UNSW Sydney, NSW 2052, Australia, emartin@cse.unsw.edu.au*

[2]*Division of Research and Commercialisation, Queensland University of Technology, 2 George Street, GPO Box 2434, Brisbane QLD 4001, Australia, arun.sharma@qut.edu.au*

[3]*School of Computing, National University of Singapore, Singapore 117543, fstephan@comp.nus.edu.sg*

Abstract: With parametric logic, we propose a unified approach to deduction and induction, both viewed as particular instances of a generalized notion of logical consequence. This generalized notion of logical consequence is Tarskian, in the sense that if a sentence φ is a generalized logical consequence of a set T of premises, then the truth of T implies the truth of φ. So in particular, if φ can be induced from T, then the truth of T implies the truth of φ. The difference between deductive and inductive consequences lies in the process of deriving the truth of φ from the truth of T.

If φ is a deductive consequence of T, then φ can be conclusively inferred from T with absolute certainty that φ is true. If φ is an inductive consequence of T, then φ can be (correctly, though only hypothetically) inferred from T, but will also be (incorrectly, still hypothetically and provisionally only) inferred from other theories T' that might not have φ as an inductive consequence (but that have enough in common with T to 'provisionally force' the inference of φ). The hallmark of induction is that given such a theory T', φ can actually be refuted from T': $\neg\varphi$ can be inferred from T' with absolute certainty that $\neg\varphi$ is true.

As a consequence, deduction and induction are both derivation processes that produce 'truths,' but the deductive process is characterized by the possibility of committing no mind change, whereas the inductive process is characterized by the possibility of committing one mind change at most. More generally, when a sentence φ is a generalized logical consequence of a set T of premises, it might be possible to infer the truth of φ from the truth of T with fewer than β

M. Friend, N.B. Goethe and V.S. Harizanov (eds.), Induction, Algorithmic Learning Theory, and Philosophy, 55–110.
© 2007 *Springer.*

mind changes, for the least nonnull ordinal β, or it might be possible to infer the truth of φ from the truth of T in the limit, though without any mind change bound, or it might be possible to discover the truth of φ from the truth of T thanks to still more complex notions of inference. 'Discovering with an ordinal mind change bound' and 'discovering in the limit' are concepts from formal learning theory , an illustration of the fact that parametric logic puts together fundamental notions from mathematical logic and formal learning theory . This paper presents the model-theoretic foundations of parametric logic and some of their connections to formal learning theory and topology.

1 INTRODUCTION

1.1 Objective

Since the early developments of formal logic, the question has been addressed whether *logical* is an absolute notion. What should be the object of investigation? *logic* or *logics*? From the perspective of artificial intelligence, the debate seems to be closed: there are various modes of reasoning that can only be formalized in distinct and often incompatible settings. Deduction is monotonic whereas induction is not, hence it is not possible for a single logical framework to account for both. In philosophy of science, there is no definite point of view, but induction is still usually opposed to deduction; deduction and induction are not conceived as complementary aspects of a more general logical concept. We present a framework where logical inferences that are usually considered to be incompatible—in particular, deductive and inductive inferences—can gracefully coexist. This framework also reconciles the *one* versus *many* views on formal logic by defining a *generic logic* with *parameters*. Setting the parameters to particular values amounts to defining instantiations of the generic logic or, alternatively, to defining particular logics.

1.2 The Tarskian Concept of Logical Consequence

It is instructive to get back to the sources, and more specifically, to Tarski's famous account of the concept of logical consequence. In [36], Tarski first criticizes the proof-theoretic notion of logical consequence, based on "[...] *few rules of inference that logicians thought exhausted the content of the concept of consequence.*" ([36], p. 410) Then Tarski reminds the reader of Gödel's incompleteness theorem, and infers: "[...] *this conjecture* [*that we can finally succeed in grasping the full intuitive content of the concept of consequence by the method sketched above, i.e. by supplementing the rules of inference used in the construction of deductive theories*] *is untenable.*"([36], p. 412) Tarski concludes that an alternative notion of logical consequence is needed: "*In order to obtain the proper concept of consequence, which is close in essentials to the common concept, we must resort to quite different methods and apply quite different conceptual apparatus in defining it.*" ([36], p. 413) The "*proper concept of consequence*" is the model-theoretic notion we are familiar with:

> (∗) "*The sentence* φ follows logically *from the sentences of the class K if and only if every model of the class K is also a model of the sentence* φ." ([36], p. 417)[1]

In view of Gödel's completeness theorem, proved in 1930, two points are in order here.

1. (∗) is by no means revolutionary. As Feferman points out: "*These* [*the semantical notions*] *had been used quite confidently* [...] *by a number of Tarski's contemporaries, including Skolem and Gödel.*" ([12], p. 79) The very statement of the completeness theorem would be meaningless without a notion of logical consequence based on (∗) or on a close variant of (∗). Tarski himself admits that"*the proposed treatment of the concept of consequence makes no very high claim to complete originality.*" ([36], p. 414)
2. In the particular case of first-order languages, the "*proper notion of logical consequence*" defined as (∗) is equivalent to the proof-theoretic notion.

What is genuinely new in (∗) is that the concept of logical consequence is not based on a particular language: the sentences mentioned in (∗) are not assumed to be first-order. A proof-theoretic notion of logical consequence, based on a well defined set of inference rules, presupposes the prior commitment to a particular language, whereas it is possible to apply (∗) to a whole range of logics that differ in their choice of language, or, in Tarski's words: "*The concept of satisfaction—like other semantical concepts—must always be relativized to some particular language.*"

Barwise's abstract model theory [4] is the most successful attempt to apply (∗) to a class of languages and "*achieve the necessary balance between strength and manageability.*" ([4], p. 222) Clearly, not every formal language qualifies as a logical formal language. Barwise defines a logic as "*an operation which assigns to each set L of symbols a syntax and a semantics such that*:

[1] What we denote by φ is denoted by X in Tarski's paper.

(1) *elementary syntactical operations (like relativizing and renaming symbols) are performable,*
(2) *isomorphic structures satisfy the same sentences.*" ([4], p. 224)

The "*elementary syntactic operations*" are defined so that the class of languages that Barwise has in mind, including, in particular, some subsets of the infinitary language $\mathcal{L}_{\infty\omega}$, can be encompassed in the resulting framework.[2]

1.3 Intended Interpretations

So both Tarski and Barwise envisioned different possible formal languages, but the notion of *model of a sentence* or *model of a set of sentences* is fixed: it is the usual notion from classical model theory. Restrictions to subclasses of the class of all structures have been proposed, that resulted in new semantics. Of particular relevance to our work is Leblanc's substitutional semantics [23]. In substitutional semantics, interpretations range over the set of Henkin structures, i.e., structures all of whose individuals interpret a closed term.[3] Such structures often exhaust the class of *legitimate* interpretations. For instance, a number theorist who has no inclination towards nonstandard models of the natural numbers works with a unique interpretation in mind, namely, the unique Henkin model of Peano arithmetic. Most applications in artificial intelligence, once formalized into logical theories T, have only Henkin interpretations as legitimate models of T, because every object in that part of reality being modeled has a name in the formalizing language. The same remark applies to most physical theories. More generally, it seems natural that various sets of structures be considered in the logical foundations of many disciplines, ranging from philosophy of science to artificial intelligence. There is also a need for various logical languages in these disciplines, and listing each and every formal logical language that has been developed would prove to be a difficult task. Contrary to Barwise's abstract model theory, which unifies a whole range of different languages, no framework in artificial intelligence or epistemology unifies various sets of structures. Parametric logic unifies sets of languages *and* sets of structures in a common framework. But let us first examine what could be the form of such a unification.

1. A class $\mathcal{C}_1$ of possible sets of sentences would be defined.
2. A class $\mathcal{C}_2$ of possible sets of structures would be defined.

[2] A formula in $\mathcal{L}_{\infty\omega}$ can contain conjunctions and disjunctions over sets of arbitrary cardinality, but it cannot contain an infinite sequence of consecutive quantifiers.

[3] A closed term is a term that does not contain any occurrence of a variable.

3. Given a sentence φ in a selected member $\mathcal{L}$ of $\mathcal{C}_1$ and given a structure $\mathfrak{M}$ in a selected member $\mathcal{W}$ of $\mathcal{C}_2$, the notion '$\mathfrak{M}$ is a model of φ' would be defined.
4. The Tarskian notion of logical consequence would be applied on the basis of $\mathcal{L}$ and $\mathcal{W}$. This would reinterpret $(*)$ as: a sentence φ in $\mathcal{L}$ follows logically from a set K of sentences in $\mathcal{L}$ if and only if every model of K in $\mathcal{W}$ is a model of φ.

1.4 Intended Models

Two questions would arise: (1) What should be the definition of $\mathcal{C}_1$? Should it be the collection of languages formalized in Barwise's abstract model theory? (2) What should be the definition of $\mathcal{C}_2$? Should it be the collection of the sets of structures considered in some existing frameworks? Before even thinking of addressing these issues, a third question should take precedence: (3) Would the proposed unification be of any relevance to a significant number of important logic-based disciplines? The answer to (3) is negative, for reasons that are essentially the same for many of the logic-based disciplines we might choose to focus on. In this paper, we will closely analyze the main concepts of formal learning theory (or inductive inference) in order to justify in detail why the answer to (3) is negative, but the point we want to make can be intuitively explained on the basis of many other disciplines. Take for instance logic programming. The elementary part of this discipline takes for $\mathcal{C}_1$ the class of definite clauses,[4] and for $\mathcal{C}_2$ the set of Herbrand structures, i.e., the set of particular Henkin structures all whose individuals interpret a unique closed term. It is accepted that an *intended* model of a subset K of $\mathcal{C}_1$—a definite logic program—is not any member of $\mathcal{C}_2$ that makes all members of K true, but only the minimal Herbrand model of K [10]. For another example, take the branch of artificial intelligence known as nonmonotonic reasoning [25]. This discipline is developed on the basis of semantic notions like preferential models [33] or aggregative models [22], where an *intended* model of a set K of sentences taken from the set selected from $\mathcal{C}_1$ is not any model of K in the set $\mathcal{W}$ of structures selected from $\mathcal{C}_2$, but a model $\mathfrak{M}$ of K in $\mathcal{W}$ such that no other model of K in $\mathcal{W}$ is "preferred" to $\mathfrak{M}$. Logic programming, nonmonotonic reasoning and, as will be seen in detail in this paper, inductive inference, share a notion of *minimality* that is essential in order to define the set of *intended models* of a set of sentences. The generic notion of logical consequence of parametric logic is based on a generic minimality principle, aimed at providing the extra generality that is

[4] A definite clause is the universal closure of a formula of the form $\psi \rightarrow \alpha$ where α is an atomic formula, and ψ is a conjunction of atomic formulas.

essential to so many logic-based disciplines. We develop a modified version of the previous plan that is characterized by the following key features.

1. A class $\mathcal{C}_1$ of possible sets of sentences is defined.
2. A class $\mathcal{C}_2$ of possible sets of structures is defined.
3. Given a sentence φ in a selected member $\mathcal{L}$ of $\mathcal{C}_1$ and given a structure $\mathfrak{M}$ in a selected member $\mathcal{W}$ of $\mathcal{C}_2$, the notion '$\mathfrak{M}$ is a model of φ' is defined.
4. Given a set K of sentences in $\mathcal{L}$, the notion of intended model of K in $\mathcal{W}$ is defined.
5. The notion of logical consequence is generalized from $(*)$ as follows. A sentence φ in $\mathcal{L}$ follows logically from a set K of sentences in $\mathcal{L}$ if and only if every intended model of K in $\mathcal{W}$ is a model of φ.

The notion of intended model of a set of sentences that we choose is a particular case of preferential entailment that generalizes the notion of minimal Herbrand model: whereas the latter applies a minimality principle to the set of all atomic sentences, the former will apply the same minimality principle to a set $\mathcal{D}$ of sentences, chosen from the set $\mathcal{L}$ of selected sentences. The counterpart to $\mathcal{D}$ in the theory of preferential models is a binary relation over the set $\mathcal{W}$ of selected structures.

1.5 Parameters

Two questions should be addressed: (1) How should $\mathcal{C}_1$ be defined? (2) How should $\mathcal{C}_2$ be defined? There is no better guide than Barwise's remark that "*the necessary balance between strength and manageability* [should be] *achieved.*" The formal developments of a theory should shape the basic notions so that natural, elegant and fruitful structures can be revealed. In parametric logic, $\mathcal{C}_1$ and $\mathcal{C}_2$ can be described as follows. Prior to the definition of $\mathcal{C}_1$ and $\mathcal{C}_2$, a parameter $\mathcal{V}$ is introduced, that denotes an arbitrary countable vocabulary (i.e., a countable set of predicate and function symbols). Then $\mathcal{C}_2$ is defined as the class of all nonempty sets of $\mathcal{V}$-structures. Sets from three classes of members of $\mathcal{C}_2$ play a fundamental role in our framework, for a given choice of $\mathcal{V}$:[5]

[5] The requirement that members of $\mathcal{C}_2$ be sets and not proper classes is meant to simplify the definition of further concepts. For example, we define a topology over the members of $\mathcal{C}_2$; if these members could be proper classes, then we would have to modify the usual treatment of topological spaces over sets. That would be an unworthy complication, because having to assume that the members of $\mathcal{C}_2$ are sets is not really restrictive. In the particular case of classical first-order logic, the Löwenheim-Skolem theorem enables us to consider a set of countable structures rather than the class of all structures.

- sets S of $\mathcal{V}$-structures such that every countable $\mathcal{V}$-structure is isomorphic to some member of S (this corresponds without loss of generality to classical first order logic);
- sets of Henkin $\mathcal{V}$-structures (the set of all Henkin structures is the set selected in substitutional semantics);
- sets of Herbrand $\mathcal{V}$-structures (the set of all Herbrand structures is the set selected in logic programming).

Choosing a member of $\mathcal{C}_2$ is achieved by giving the desired value to some parameter. The definition of $\mathcal{C}_1$ also depends on the parameter $\mathcal{V}$. Let $\mathcal{L}^{\mathcal{V}}_{\omega\omega}$ denote the set of first-order $\mathcal{V}$-formulas, and let $\mathcal{L}^{\mathcal{V}}_{\omega_1\omega}$ denote the extension of $\mathcal{L}^{\mathcal{V}}_{\omega\omega}$ that accepts disjunctions and conjunctions over countable sets of formulas. A *fragment* of $\mathcal{L}^{\mathcal{V}}_{\omega_1\omega}$ is a subset of $\mathcal{L}^{\mathcal{V}}_{\omega_1\omega}$ that is closed under subformulas, finite boolean operators, quantification, and substitution of variables by terms [26].[6] In parametric logic, the class $\mathcal{C}_1$ consists of the sets of closed members of the countable fragments of $\mathcal{L}^{\mathcal{V}}_{\omega_1\omega}$. Choosing a member of $\mathcal{C}_1$ is also achieved by giving the desired value to some parameter. So, infinitary languages play a key role in our framework, as they do in abstract model theory. Clearly, the set of closed members of $\mathcal{L}^{\mathcal{V}}_{\omega\omega}$ is the weaker language that can be selected from $\mathcal{C}_1$, and most notions can also be motivated on the basis of this familiar language.

1.6 Departing from Tarski: Distinguishing Logical and Deductive

Every choice of possible values for the parameters of parametric logic determines a generic notion of logical consequence, in accordance with the generalization of the Tarskian concept of logical consequence that 5. in Section 1.4 suggests (at this stage, it is not necessary to specify which parameters are involved in the generic notion of logical consequence of parametric logic). Suppose that values have been assigned to the parameters, with $\mathcal{L}$ denoting the language (the selected member of $\mathcal{C}_1$), and let $\mathbf{Cn}(K)$ denote the set of members of $\mathcal{L}$ that are logical consequences of a subset K of $\mathcal{L}$ w.r.t. the generic notion of logical consequence determined by the values of the parameters. In [35], Tarski claims that "*we have no alternative but to regard the concept of consequence as primitive*," before introducing "*four axioms* (*Ax.* 1–4) *which express certain elementary properties of the primitive concepts and are satisfied in all known formalized disciplines*." ([35], p. 63). Using the previous notation, these axioms are the following.[7]

[6] This definition will be recalled again more formally later, when needed.

[7] We use $\subseteq$ for inclusion between classes and $\subset$ for strict inclusion.

Ax. 1. $\mathcal{V}$ is countable.
Ax. 2. For all $K \subseteq \mathcal{L}$, $K \subseteq \mathbf{Cn}(K) \subseteq \mathcal{L}$.
Ax. 3. For all $K \subseteq \mathcal{L}$, $\mathbf{Cn}(\mathbf{Cn}(K)) = \mathbf{Cn}(K)$.
Ax. 4. For all $K \subseteq \mathcal{L}$, $\mathbf{Cn}(K) = \cup\{\mathbf{Cn}(D) \mid \text{finite } D \subseteq K\}$.

Are those axioms satisfied in parametric logic? Ax. 1 is satisfied precisely because parametric logic accepts any countable vocabulary, but no noncountable vocabulary, as possible value of $\mathcal{V}$. Ax. 2 will be trivially satisfied. A variant of Ax. 3 will be satisfied, namely: for every subset K of $\mathcal{L}$ that is a *possible knowledge base*, $\mathbf{Cn}(\mathbf{Cn}(K)) = \mathbf{Cn}(K)$. Observe that in the particular case of classical first-order logic, Ax. 3 can be replaced by Ax. 3': for all *consistent* $K \subseteq \mathcal{L}$, $\mathbf{Cn}(\mathbf{Cn}(K)) = \mathbf{Cn}(K)$. The possible knowledge bases of parametric logic will be the counterpart to the consistent theories of classical first-order logic. As such they will be in parametric logic the only legitimate starting points for logical investigation, in the same way that in classical first-order logic, the consistent theories are the only legitimate starting points for logical investigation. Most interesting is Ax. 4, which states that **Cn** is compact. Here is how Tarski justifies the axiom: "*Finally, it should be noted that, in concrete disciplines, the rules of inference with the help of which the consequences of a set of sentences are formed are in practice always operations which can be carried out only on a finite number of sentences (usually even on one or two sentences).*" ([35], p. 64) The title of the paper mentions "*deductive sciences,*" whereas the passages that have been quoted mention "*all known formalized disciplines*" and "*concrete disciplines.*" The paper makes no distinction between "deductive sciences," "all known formalized disciplines" and "concrete disciplines" and it seems that Tarski considers that the three expressions are interchangeable. Are they really interchangeable? Also, in [36], Tarski discusses the "*formalized concept of consequence*" in relation to "*formalized deductive theories.*" Can formalized deductive theories be characterized as frameworks that formalize the concept of consequence? Parametric logic does not identify *logical consequence* and *deductive consequence*, but defines *deductive consequence* as a particular case of *logical consequence*, w.r.t. the generic notion of logical consequence, i.e., irrespectively of the setting of the parameters.

To clarify the issue, consider a set K of sentences in $\mathcal{L}$ and a sentence φ in $\mathcal{L}$ such that φ follows logically from K, *i.e.*, $\varphi \in \mathbf{Cn}(K)$ where **Cn** is the specific notion of logical consequence determined by the values of the parameters, in particular, by the notion of intended model of K. By 5. in Section 1.4, every intended model of K is a model of φ. In other words, $\varphi \in \mathbf{Cn}(K)$ is meant to capture that:

if all members of K are true, then φ is true.

The previous statement expresses a relation between structures (abstractions of possible realities) and languages. It says nothing of *how* the truth of φ could be discovered from the truth of K, nor even *what* is meant by discovering the truth of φ from the truth of K. Deducing φ from K, on the other hand, answers the *how* and *what* questions.

What: come to the conclusion, with certainty, that φ is true, under the assumption that all members of K are true.

How: by carrying out finite operations on finitely many members of K, in a finite number of steps.

Model-theoretically, the above can arguably be formalized as: φ is a deductive consequence of K iff φ is a compact consequence of K. With this view, Ax. 4 is not a characterization of the notion of logical consequence, but it is a characterization of the notion of deductive consequence. Logical consequence and deductive consequence are two equivalent notions iff the notion of logical consequence itself is compact. Otherwise, the deductive consequences of K should be precisely the logical consequences of K that happen to be compact consequences of K. In classical first-order logic, where, for a given choice of $\mathcal{V}$, the selected member of $\mathcal{C}_2$ is assumed to be the class of all $\mathcal{V}$-structures[8] and the selected member of $\mathcal{C}_1$ is assumed to be the set of all first-order $\mathcal{V}$-sentences, the notion of logical consequence is compact. Hence it is possible, *in this particular case*, to identify logical consequence with deductive consequence. But the compactness of the predicate calculus is a property that follows from choosing particular members of $\mathcal{C}_1$ and $\mathcal{C}_2$. So the usual claim that deduction is captured by the predicate calculus amounts to the claim that deduction is captured by:

- the Tarskian notion of logical consequence;
- a particular selection of language and class of structures for $\mathcal{C}_1$ and $\mathcal{C}_2$, respectively.

We disagree with this claim. Choosing particular classes of languages and structures is unrelated to the nature of deduction. When the selected member of $\mathcal{C}_2$ is the class of all $\mathcal{V}$-structures and when the selected member $\mathcal{L}$ of $\mathcal{C}_1$ is the set of closed members of some countable fragment of $\mathcal{L}^{\mathcal{V}}_{\omega_1\omega}$, the

[8] As emphasized in Footnote 5, the selected member of $\mathcal{C}_2$ should actually be a set rather than a proper class, and in the particular case of classical first-order logic, this can be done without loss of generality thanks to the Löwenheim-Skolem theorem. In this paper, we prefer not to clutter the text with inessential technicalities.

compactness theorem usually fails. Still, such a choice for $\mathcal{L}$ is perfectly justified both with respect to the Tarskian notion of logical consequence and with respect to abstract model theory. Moreover, it encompasses classical first-order logic, in case all members of the set K of premises as well as the conclusion φ are finite sentences. In parametric logic, the notion of deductive consequence is reinterpreted and subsumed under a generalized notion of logical consequence, leaving room for more complex logical inferences, *i.e.*, logical consequences that are not deductive, or compact, consequences of the set K of premises.

1.7 Compactness and Weak Compactness

The previous considerations are relevant to epistemology. Ideally, and very simplistically, physics can be described as a discipline whose aim is the discovery of *true* laws, from a set of premises K consisting of theoretical statements and sentences that record observations or experimental results, assumed to be true. Here, what is meant by discovering the truth of law φ from the truth of premises K receives an answer based on the notion of inductive consequence, as opposed to deductive consequence. Whereas the latter expects a *certain* conclusion of φ from K, the former can only expect a *plausible* conclusion of φ from K. What should be meant by *plausible* will be discussed at length, and will result in a formal notion of inductive consequence. But we want to stress first that we advocate an approach where deductive consequence and inductive consequence are complementary rather than incompatible notions. Identifying deductive consequence with logical consequence forbids any logical approach to induction, or at least, any unified logical approach to deduction and induction. We claim that in the same way that deductive consequence is a particular case of logical consequence, inductive consequence is another particular case of logical consequence (where both deductive and inductive consequences of a set K of premises can conveniently be assumed to be the same in case the logical consequences of K are exhausted by the deductive consequences of K). Parametric logic defines inductive consequences as a particular case of generalized logical consequences. Like deduction, induction offers a particular answer to the questions of *what* is meant by discovering the truth of φ from the truth of K and of *how* the truth of φ could be discovered from the truth of K. Such answers make sense for any choice of parameters, and are provided uniformly in these parameters.

The compactness property expressed in Ax. 4 is the hallmark of deduction; that does not imply that every generalized logical consequence of K that is not a compact consequence of K is an inductive consequence of K. The *how* that characterizes inductive processes is the same as the *how* that characterizes deductive processes: finite operations are carried out on finitely many members of K, in a finite number of steps, to deduce φ from K, or to induce φ from K. But both processes draw conclusions of distinct qualities: the conclusion that φ can be deduced from K is certain, whereas the conclusion that φ can be induced from K is just plausible. What should be meant by plausible? We resort to a variant of Popper's *refutation principle* [31]: φ can be induced from K when it is certain that for all legitimate sets of premises[9] K' that contain a well chosen finite subset of K, *if* φ does not logically follow from K' (in the general sense), then it should be possible to refute φ from K'. We interpret *refute* φ *from* K' as *deduce the negation of* φ *from* K'. In terms of the idealized and very simplistic view of physics we have used before, being certain to be able to refute a hypothetical law means that finitely many theoretical statements, observations and results of experiments (all members of K') will allow to conclude with certainty that the law is incorrect, if such is indeed the case. Hence φ is a plausible consequence of K if:

1. φ is indeed a generalized logical consequence of K, and
2. it is possible, on the basis of a finite subset D of K, to conclude with certainty that for all legitimate sets of premises K' that contain D, either φ is a generalized logical consequence of K', or $\neg\varphi$ can be deduced from K'.

The previous pair of statements can be seen as a property of *weak compactness*. In case φ is an inductive but not a deductive consequence of K, φ cannot be derived with certainty from a finite subset of K; but φ can still be derived with confidence from a finite subset of K, where confidence is also defined in terms of finite sets of sentences. That is, induction does not demand that inference rules be used with infinitely many premises, since this would indeed, to use Tarki's words, be at odds with *concrete disciplines*. But induction demands giving up certainty for confident belief.

Usually, deductive and inductive inferences of a set K of sentences do not exhaust all logical consequences of K. In this paper, we shall examine carefully how such inferences can be systematically defined and classified. We shall show how these inferences are closely related to concepts from formal learning theory.

[9] A legitimate set of premises will be technically defined as a possible knowledge base—the counterpart in parametric logic of a consistent theory in first-order logic.

2 OVERVIEW OF PARAMETRIC LOGIC

2.1 The Parameters

Parametric logic is a family of (*logical*) *paradigms*, defined as sequences $\mathcal{P}$ of five *parameters*.[10] We will devote a lot of time defining and justifying the raison d'être of these parameters, but for the impatient reader, we summarize here some of the key aspects of parametric logic, many of which will be discussed in this paper. The parameters are:

- a *vocabulary* $\mathcal{V}$ (a countable set of predicate and function symbols, possibly with equality);
- a set $\mathcal{W}$ of *possible worlds* (a set of $\mathcal{V}$-structures);[11]
- a *language* $\mathcal{L}$ (the set of closed members of a countable fragment of $\mathcal{L}^{\mathcal{V}}_{\omega_1\omega}$, with $\mathcal{L}$ equal to the set of first-order $\mathcal{V}$-sentences as simplest case);[12]
- a set $\mathcal{D}$ of sentences (a subset of $\mathcal{L}$) called set of *possible data*;
- a set $\mathcal{A}$ of sentences (another subset of $\mathcal{L}$), disjoint from $\mathcal{D}$, called set of *possible assumptions*.

Intuitively, a possible datum (a member of $\mathcal{D}$) will represent observations or results of experiments. Given a legitimate set X of premises, a possible datum φ will be either false in all intended models of X, or true in all intended models of X. In the second case, φ will have to be a member of X: it will be one of the premises and it will not need any logical investigation to be discovered to be true. Even though a possible datum φ that is true in all intended models of a legitimate set X of premises has to be a member of X, we can think of φ as an axiom that will eventually be discovered to be true thanks to a 'nonlogical' or 'physical' enumeration procedure that generates a stream of observations or results of experiments.[13] In contrast, a possible

[10] In full generality, there is a sixth parameter, namely, a countable nonnull ordinal κ, that represents the number of levels above the object level at which reasoning is being formalized. In this paper, we assume for simplicity that κ takes the value 1.

[11] When the countable nonnull ordinal κ is included as a sixth parameter, a notion of κ-multistructure is defined, and $\mathcal{W}$ is replaced by a particular set of particular κ-multistructures.

[12] The notion of fragment of $\mathcal{L}^{\mathcal{V}}_{\omega_1\omega}$ will be precisely defined later. When the countable nonnull ordinal κ is included as a sixth parameter, a notion of infinitary κ-statement is defined, and $\mathcal{L}$ denotes the set of closed members of a countable fragment of the set of infinitary κ-statements.

[13] In classical logic, if a set X of axioms is not recursive, and especially if it is recursively enumerable, then it is also legitimate to conceive of a member of X as a formula that will

assumption (a member of $\mathcal{A}$) will represent an axiom or a member of some background knowledge. A possible assumption will possibly be false in some intended models of a legitimate set of premises, and true in others. A possible assumption will sometimes be true in all intended models of a legitimate set of premises without being one of the premises, in which case it could still be discovered to be true following some logical investigation. Parametric logic does not impose any syntactic condition on possible data and possible assumptions: any formula can be chosen as a possible datum or as a possible assumption (but not both).

First-order logic is a particular paradigm of parametric logic, with the parameters taking the following values:

- $\mathcal{V}$ is arbitrary;
- $\mathcal{W}$ is the class of all $\mathcal{V}$-structures;
- $\mathcal{L}$ is the set of closed members of $\mathcal{L}^{\mathcal{V}}_{\omega\omega}$;
- $\mathcal{D}$ is empty;
- $\mathcal{A}$ is $\mathcal{L}$.

Classical logic sets to $\mathcal{A}$ to $\mathcal{L}$ because every sentence is a potential assumption: every sentence can be used as a premise, every sentence can be a member of a set of axioms. As a consequence, since $\mathcal{A}$ and $\mathcal{D}$ are disjoint, $\mathcal{D}$ has to be set to $\emptyset$, which is consistent with the fact that classical logic incorporates no notion of observation or result of experiment.

2.2 The Generic Notion of Logical Consequence

A generic model-theoretic notion of *logical consequence in* $\mathcal{P}$ is defined. When $\mathcal{P}$ is the paradigm of first-order logic, then the notion of logical consequence in $\mathcal{P}$ is the classical notion of logical consequence. The latter is *compact*. This means that for all theories T and sentences φ:

> φ is a logical consequence of T iff there exists a finite subset D of T such that φ is a logical consequence of D.

When T is consistent, the former condition is equivalent to:

> φ is a logical consequence of T iff there exists a finite subset D of T such that for all consistent theories T', if $D \subseteq T'$ then φ is a logical consequence of T'.

In first-order logic, the theories of interest are those that are *consistent*, but the notion of consistent theory turns out to be too weak for other paradigms of

eventually be discovered to be true thanks to a 'nonlogical' enumeration procedure, rather than as a formula that is given prior to logical investigation.

parametric logic. The generalization of consistent theories to other paradigms will be the *possible knowledge bases*, derived from the parameters $\mathcal{W}$, $\mathcal{D}$ and $\mathcal{A}$. In an arbitrary paradigm $\mathcal{P}$, the possible knowledge bases play the role of the consistent theories of first-order logic (the possible knowledge bases *are* the consistent theories when $\mathcal{P}$ is the paradigm of first-order logic), in the sense that they are the legitimate starting points for logical investigation. So more generally, assuming that we have formalized the notion of logical consequence in $\mathcal{P}$, we say that this notion has the compactness property just in case for all possible knowledge bases T and sentences φ:

> φ is a logical consequence of T in $\mathcal{P}$ iff there exists a finite subset D of T such that for all possible knowledge bases T', if $D \subseteq T'$ then φ is a logical consequence of T' in $\mathcal{P}$.

Usually, for most paradigms $\mathcal{P}$, the notion of logical consequence in $\mathcal{P}$ is not compact. Let T be a possible knowledge base. When $\mathcal{P}$ is the paradigm of first-order logic, the notion of logical consequence in $\mathcal{P}$ is the classical notion of logical consequence; hence the following three clauses define the same set.

- The sentences that are compact logical consequences of T in $\mathcal{P}$.
- The sentences that are logical consequences of T in $\mathcal{P}$.
- The sentences that are logical consequences of T.

2.3 The Logical Hierarchies

When the notion of logical consequence in $\mathcal{P}$ is not compact, the set of sentences that are compact logical consequences of T in $\mathcal{P}$ can be viewed as the set of 'easy' logical consequences of T in $\mathcal{P}$. The main purpose of this framework is to study the set of sentences that are "not easy" logical consequences of T in $\mathcal{P}$ and to assess their degree of complexity (i.e., "not easiness"). The bulk of the logical consequences of T in $\mathcal{P}$ can be captured by a hierarchy $\mathcal{H}$ built over T, with sentences higher in $\mathcal{H}$ being of higher complexity than sentences lower in $\mathcal{H}$.[14] The hierarchy $\mathcal{H}$ consists of *levels* $\mathcal{H}_1$, $\mathcal{H}_2, \ldots$ The first level $\mathcal{H}_1$ of $\mathcal{H}$ is built using a *refutation principle* that generalizes the compactness property to a property referred to as *β-weak compactness*, where β is an ordinal. So $\mathcal{H}_1$ is made of *layers* $\mathcal{H}_{1,\beta}$, consisting of sentences that are defined as β-weak compact consequences of T in $\mathcal{P}$. Since the notion of 0-weak compact consequence in $\mathcal{P}$ turns out

[14] Under reasonable assumptions, $\mathcal{H}$ will be complete, i.e., every logical consequence of T in $\mathcal{P}$ will belong to $\mathcal{H}$. However, in general, $\mathcal{H}$ will not be complete. Even so, all inferences that may be of interest will occur in the lower part of the hierarchy.

to be nothing but a reformulation of the notion of compact consequence in $\mathcal{P}$,[15] the first layer of the first level of $\mathcal{H}$, i.e., $\mathcal{H}_{1,0}$, consists precisely of the compact consequences of T in $\mathcal{P}$. As a result, when $\mathcal{P}$ is the paradigm of first-order logic , $\mathcal{H} = \mathcal{H}_{1,0}$. In general, we view $\mathcal{H}_{1,0}$ as the set of *deductive consequences of* T *in* $\mathcal{P}$, and $\mathcal{H}_{1,1}$ as the set of *inductive consequences of* T *in* $\mathcal{P}$.[16] The higher levels of $\mathcal{H}$ are obtained by β-weak compactness applied to T and sentences in the lower levels of $\mathcal{H}$. So $\mathcal{H}_2$ is made of layers $\mathcal{H}_{2,\beta}$, consisting of sentences that are defined as β-weak compact consequences of T and $\mathcal{H}_1$ in $\mathcal{P}$. The process is iterated: for all nonnull ordinals α and for all ordinals β, we inductively define the set $\mathcal{H}_{\alpha,\beta}$ of sentences that are β-weak compact logical consequences of T and $\cup_{0<\gamma<\alpha}\mathcal{H}_\gamma$ in $\mathcal{P}$. So the levels of $\mathcal{H}$ are indexed by a nonnull ordinal α, representing iterations of the property of β-weak compactness that determines layer β of level α of $\mathcal{H}$, for any ordinal β. The higher a sentence φ occurs in the hierarchy $\mathcal{H}$, i.e., the larger the values of α and β that characterize the first levels and layers in $\mathcal{H}$ where φ occurs, the more complex the model-theoretic fact that φ is a logical consequence of T in $\mathcal{P}$. Hence we have a notion of *logical complexity*.

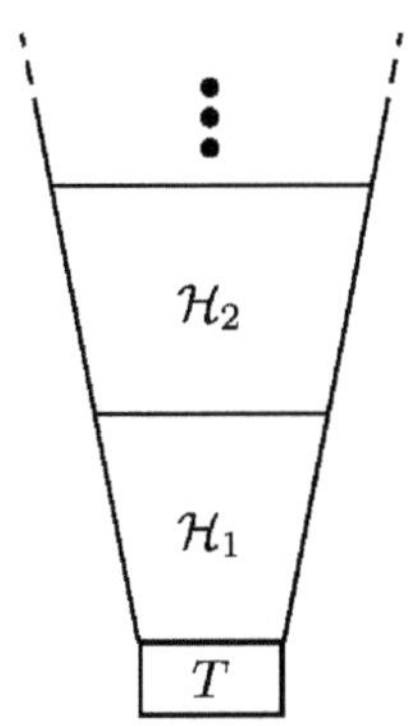

Figure 3.1. The levels of the logical hierarchies

[15] The larger the value of β, the weaker the condition of β-weak compactness. The intuitive meaning of '0-weak compactness' is 'no weak compactness,' i.e., 'compactness.'

[16] This is because the property of β-weak compactness with $\beta = 1$ is a reformulation of the property of weak compactness discussed in Section 1.7.

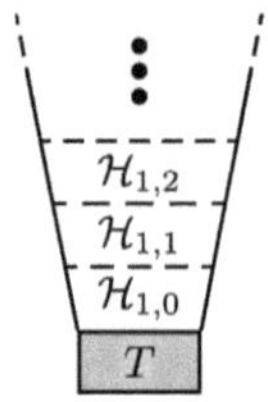

Figure 3.2. The layers of the first level of the logical hierarchies

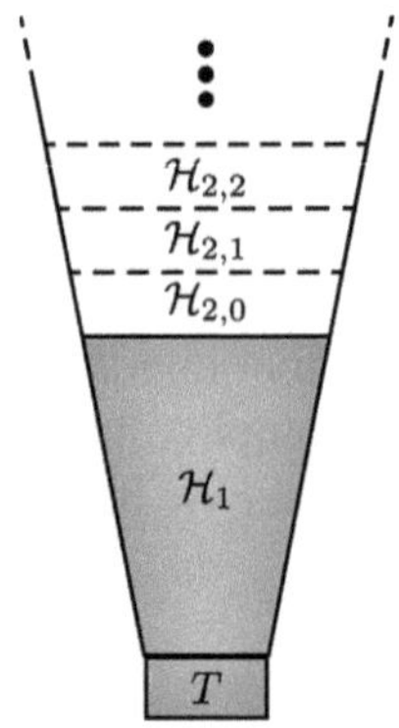

Figure 3.3. The layers of the second level of the logical hierarchies

2.4 The Other Notions of Complexity

The logical notion of complexity that emerges from the hierarchies of logical consequences in $\mathcal{P}$ of a possible knowledge base have fundamental relationships, that can even take the form of equivalences under some reasonable assumptions, with the following notions of complexity.

- A notion of topological complexity, given by the Borel hierarchy and the difference hierarchies over some topological space defined from $\mathcal{W}$ and $\mathcal{D}$.
- A notion of syntactic complexity, given by a normal form of sentences.
- A notion of learning-theoretic complexity, derived from the concepts of classification in the limit, or classification with a bounded number of mind changes.

A proof theory can be defined and completeness theorems can be obtained, under some assumptions, from the relationships between the various notions of complexity. Our aim in this paper is to introduce the main learning-theoretic notions and use learning scenarios to both motivate and illustrate the main concepts of parametric logic. We will devote special attention to the notions of deductive and inductive consequence in paradigm $\mathcal{P}$.

We focus on conceptual issues, and include only the arguments that are essential for understanding these issues. Technical proofs are omitted, and can be found in [29], with a preliminary version of some results in [28].

3 THE RELATIONSHIP BETWEEN LOGIC AND LEARNING

3.1 The Need for Parameters

Consider the following expressions.

(A) φ is a logical consequence of T.
(B) φ is a deductive consequence of T.
(C) φ is an inductive consequence of T.

In the realm of first-order logic, (A) and (B) are synonymous whereas (C) does not have any meaning. In other words, if first-order logic is the framework that correctly formalizes the notion of logical/deductive consequence, then 'inductive consequence' is not a logical concept, unless it is formalized in a framework incompatible with first-order logic. More specialized arguments lead to the same conclusion, e.g., the formal notion of deductive consequence is monotonic, whereas induction falls in the field of nonmonotonic reasoning. Despite such arguments, we would still like each of (A), (B) and (C) to make sense in a common framework. What should be the relation between (A), (B) and (C)? The three statements can be paraphrased as below.

(A') If all members of T are true, it can be concluded that φ is true.
(B') If all members of T are true, it can be deduced that φ is true.
(C') If all members of T are true, it can be induced that φ is true.

(A') is about inference of φ from T; (B') and (C') are special cases—deduction and induction—of such an inference. Of these two mechanisms, let us accept the view that deduction is more demanding than induction because the attributes of the former are conclusiveness and certainty, whereas the attributes of the latter are refutability—weaker than conclusiveness—and believability—weaker than certainty. We can then claim that the relationship between (A), (B) and (C) could be represented as follows.[17]

[17] Actually, postulating that deductive consequences are particular cases of inductive consequences would be mainly for convenience, though φ would be a 'genuine' inductive

(B) φ is a deductive consequence of T

$\Downarrow$

(C) φ is an inductive consequence of T

$\Downarrow$

(A) φ is a logical consequence of T

The relation between (A), (B) and (C) as depicted above asks for a redefinition of the classical notion of logical/deductive consequence into two distinct notions, without postulating that the latter is equivalent to the former. Though this would satisfy the desideratum that one framework provides a formalization to each of (A), (B) and (C), it raises the issue that the notions of logical and deductive consequences would be formalized differently in incompatible frameworks, namely, in the new framework and in first-order logic. We certainly have no intention to suggest that the way first-order logic formalizes the notions of logical and deductive consequences, postulating their equivalence, is wrong. The dilemma is easily solved by creating a framework equipped with a list $\mathcal{P}$ of parameters, and defining the notions

(A") φ is a deductive consequence of T in $\mathcal{P}$,
(B") φ is an inductive consequence of T in $\mathcal{P}$, and
(C") φ is a logical consequence of T in $\mathcal{P}$.

First-order logic would be a particular case of this framework when the members of $\mathcal{P}$ take some particular values. With respect to those values, (A"), (B") and (C") would just be equivalent notions, equivalent to the classical notion of logical consequence.[18] We call $\mathcal{P}$ a (*logical*) *paradigm* and define the constituent parameters of $\mathcal{P}$ in the sequel.

3.2 Possible Worlds

Let us first adopt a convention to be used throughout the paper. If we say that a vocabulary contains $\overline{0}$ and s, then $\overline{0}$ denotes a constant and s a unary function symbol. Moreover, given a nonnull member n of the set $\mathbb{N}$ of natural numbers, $\overline{n}$ is used as an abbreviation for the term obtained from

consequence of T in case it is an inductive consequence of T that is not a deductive consequence of T. For further arguments for supporting the view that (C) implies (A), see [7].

[18] More precisely, the equivalence between (A") and (B") in the particular case where $\mathcal{P}$ represents first-order logic requires a shift from 'the notion of inductive consequence is meaningless'—the classical view—to "the notion of inductive consequence is degenerate"—the view of parametric logic. Both points of view are technically indistinguishable.

$\overline{0}$ by n applications of s to $\overline{0}$. ($\{\overline{n} \mid n \in \mathbb{N}\}$ is the set of numerals.) Consider a vocabulary that contains $\overline{0}$ and s. The following is a typical example of what could reasonably be presented as an inductive inference in $\mathcal{P}$, hence as a logical inference in $\mathcal{P}$, for some choice of $\mathcal{P}$.[19]

$$(\odot) \quad \frac{P(\overline{0}) \quad P(\overline{1}) \quad P(\overline{2}) \quad P(\overline{3}) \quad P(\overline{4}) \qquad \ldots}{\forall x\, P(x)}$$

Obviously, $\forall x\, P(x)$ is not a logical consequence of $\{P(\overline{n}) \mid n \in \mathbb{N}\}$ in the classical sense. What could justify the claim that, *in some sense*, $\forall x\, P(x)$ is a logical consequence of $\{P(\overline{n}) \mid n \in \mathbb{N}\}$?

- First, the assumption that the underlying vocabulary contains no function symbol except $\overline{0}$ and s, hence that the numerals exhaust all closed terms.
- Second, the assumption that the structures under consideration are such that each of their individuals interprets a closed term; every individual has a name, there is no nonstandard element.[20]

In the context of group theory, the class of intended interpretations encompasses both standard and nonstandard structures. But in other contexts, in particular in artificial intelligence, it is legitimate to assume that any reasonable abstraction of that part of reality we are interested in consists exclusively of structures all of whose individuals have a name. These are the only *intended* interpretations.[21] The class of intended interpretations might as well consist exclusively of finite structures, etc. An intended interpretation is not a fixed notion, but depends on the context of formalization, which suggests rephrasing (C") as:

(C*) every *intended* model of T is a model of φ.

The formal notion of logical consequence of first-order logic then appears as a cautious formalization because it does not restrict whatsoever the class of intended models of a set of sentences; it is the limiting case of an extended notion of logical consequence that can accommodate various assumptions

[19] This is an instance of the *ω-rule*. See [18] for an interesting discussion on whether it should be considered as a logical rule. In parametric logic, whether it is logical or not depends on the values of the parameters.

[20] Indeed, $(\odot)$ is not sound in the classical sense due to the existence of a model $\mathfrak{M}$ of $\{P(\overline{n}) \mid n \in \mathbb{N}\}$ whose domain contains some element x that does not have property P; if the vocabulary contains no function symbol except $\overline{0}$ and s then x has to be nonstandard. Of course, both assumptions discussed here are only sufficient (i.e., they are not necessary) to support the view that *in some sense*, $\forall x\, P(x)$ is a logical consequence of $\{P(\overline{n}) \mid n \in \mathbb{N}\}$.

[21] See [7] for more detailed arguments of this claim.

on the nature of the intended models of a theory. We have justified the introduction of the first two parameters in $\mathcal{P}$:

- a countable *vocabulary* $\mathcal{V}$, *i.e.*, a countable set of function symbols and predicate symbols,[22] with or without equality;[23]
- a set $\mathcal{W}$ of $\mathcal{V}$-structures, referred to more simply as structures, of possible interpretations, called *possible worlds*.

Let us introduce some terminology. A *Henkin* structure is a structure all of whose individuals are nonempty sets of closed $\mathcal{V}$-terms, referred to more simply as terms, consisting precisely of those terms that they interpret. A *Herbrand* structure is a Henkin structure all of whose individuals are singletons. Hence every individual of a Herbrand structure interprets a unique closed term, and is usually identified with it. We consider Herbrand structures in case $\mathcal{V}$ is equality free, whereas we usually consider Henkin structures otherwise. We say that a structure $\mathfrak{M}$ is *standard* if either $\mathcal{V}$ contains equality and $\mathfrak{M}$ is Henkin, or $\mathcal{V}$ is equality free and $\mathfrak{M}$ is Herbrand.

3.3 Numerical Learning Paradigms

If $\mathcal{V}$ and $\mathcal{W}$ were the only parameters in $\mathcal{P}$, an *intended* model of a set T of $\mathcal{V}$-sentences[24]—referred to more simply as sentences—would be a model of T that belongs to $\mathcal{W}$. To examine further whether "member of $\mathcal{W}$" is good enough as a formal definition of "intended model," we now introduce some concepts from formal learning theory, or inductive inference. This discipline offers a formal treatment of induction by defining various learning paradigms. Most learning paradigms are expressed in a numerical setting [19], while others are expressed in a logical setting [21], [27]. The distinction, however, is not very relevant as virtually all numerical learning paradigms can be formulated as logical learning paradigms. It is howeveruseful to first understand learning paradigms in the numerical

22 In relation to (⊙) above, the choice of function symbols was crucial whereas the choice of predicate symbols was irrelevant, but some developments of parametric logic are only possible by choosing a specific set of predicate symbols.

23 We want to be able to consider both languages with equality and languages without equality, and hence depart from the usual practice of including only nonlogical symbols in vocabularies.

24 The notion of "$\mathcal{V}$-sentence" is intentionally left formally unspecified here. It is meant to be a closed $\mathcal{V}$-formula, which we do not define now. A definition will be given after we have addressed the question: "what should be a formula in parametric logic?" At this stage, it is not clear why this question *has* to be addressed. Before the notion is fully formalized, the reader can think of "first-order sentence" when reading "sentence."

setting. The next section will present the logical counterpart to the notions discussed here.

A *language* is a recursively enumerable (r.e.) set of natural numbers. Such a set can be conceived of as an abstract model of a natural language. Indeed, assume that all finite strings of words over some alphabet (the Latin alphabet for instance) are coded as numbers. At some abstraction level, a natural language (English for instance) can be reduced to the set L of strings of words that are grammatically correct, hence to some set of natural numbers (those which code the grammatically correct strings) that can legitimately be assumed to be recursive. A *grammar* in the theory of formal languages is a procedure that *generates* a set of strings, hence a correct grammar for L would generate precisely the members of L. The fact that the property of being recursively enumerable captures the notion of effective generation (by an unrestricted grammar) explains why languages are assumed to be r.e. rather than recursive. One of the parameters of a numerical learning paradigm is a set $\mathfrak{L}$ of languages, that could be called the set of *possible languages*. Intuitively, $\mathfrak{L}$ represents a set of coded natural languages that a child might be genetically programmed to learn, when the social context puts the child in contact with one of these languages. For an example, we will consider the learning paradigm where $\mathfrak{L} = \{\mathbb{N}, \mathbb{N} \setminus \{0\}, \mathbb{N} \setminus \{0, 1\}, \ldots, \emptyset\}$, i.e., $\mathfrak{L}$ is the set of final segments of $\mathbb{N}$.

Let $\sharp$ be a symbol whose intuitive meaning is 'no information,' or 'no datum.' Given a language L in $\mathfrak{L}$, a (*numerical*) *text for* L is an ω-sequence of members of $L \cup \{\sharp\}$ in which each member of L occurs at least once. For instance, $(4, 3, 2, \sharp, 6, 5, 4, \sharp, 8, 7, 6, \sharp, \ldots)$ is a numerical text for $\mathbb{N} \setminus \{0, 1\}$. Intuitively, a text for a language is an enumeration of all (codes of) grammatically correct sentences of this language, that represents a very idealized learning scenario where a child would be presented with nothing but grammatically correct sentences, and all grammatically correct sentences if no time restriction were imposed. Note that repetitions of data are allowed, and that $\sharp$ is necessary to define the (only) text for the empty language, assuming that $\emptyset$ is indeed a member of $\mathfrak{L}$. Two kinds of learning scenarios will be considered: identification and classification. In formal learning theory, paradigms of identification are much more common subject of study than paradigms of classification.[25] But in an uncomputable setting, identification is a particular case of classification, as will be seen, and classification is more closely related to parametric logic than identification,

[25] The seminal work on identification in the limit for numerical paradigms is [17]. Identification in the limit was extended to logical paradigms in [16]. For a study of classification in numerical paradigms, see [15], [34].

hence we will discuss both identification and classification, starting with the former. The following notation is needed. Given a set X, a sequence e of members of X, and a $k \in \mathbb{N}$ that, in case e is finite, is at most equal to the length of e, we denote by $e_{|k}$ the initial segment of e of length k. Given a set X, let $X^\star$ denote the set of finite sequences of members of X. So $(\mathbb{N} \cup \{\sharp\})^\star$ is the set of finite sequences of members of $\mathbb{N} \cup \{\sharp\}$.

Assume that an acceptable enumeration $(\varphi_i)_{i \in \mathbb{N}}$ of all partial recursive functions has been fixed, and for all $i \in \mathbb{N}$, denote by W_i the domain of φ_i (i is said to be an r.e. index for W_i). A (*numerical*) *identifier* is a partial function from $(\mathbb{N} \cup \{\sharp\})^\star$ into $\mathbb{N}$—where an output $i \in \mathbb{N}$ is meant to represent the language W_i.[26] So an identifier is fed with longer and longer finite initial segments of a text e for a language L in $\mathfrak{L}$, in response to which it possibly outputs a *hypothesis*, a guess on what it thinks L might be. The aim of the identifier is to *discover* or *learn* the particular L that generates e. What is meant by 'discover' or 'learn' is formalized by a *learning criterion*. The most common criterion is that of *identification in the limit*, that expects the identifier to eventually converge to an r.e. index of the correct language. Formally:[27]

Definition 3.1 *A numerical identifier f is said to* identify $\mathfrak{L}$ in the limit from texts *iff for all $L \in \mathfrak{L}$ and texts e for L, $f(e_{|k}) = f(e_{|k+1})$ and $W_{f(e_{|k})} = L$ for cofinitely many $k \in \mathbb{N}$.*

$\mathfrak{L}$ is said to be identifiable in the limit from texts *iff some numerical identifier identifies $\mathfrak{L}$ in the limit from texts.*[28]

The criterion of identification in the limit from texts is a very simplified representation of children who learn their native language when they eventually find out which language they get output from, and stick to this correct hypothesis in the face of new data. The key point is that identifiers are usually unable to know that they have hit the convergence point. Consider the example for $\mathfrak{L}$ given above. Let f be the numerical identifier that outputs a fixed r.e. index for L in response to a member σ of $(\mathbb{N} \cup \{\sharp\})^\star$, where $L = \emptyset$ if no natural number occurs in σ, and where $L = \{n, n+1, \ldots\}$ if some natural

[26] Most learning paradigms require that identifiers are computable, and possibly satisfy some extra conditions.

[27] Let two sets X, Y and a partial function g from X into Y be given. Given two members x, x' of X, we write $g(x) = g(x')$ when both $g(x)$ and $g(x')$ are defined and equal; we write $g(x) \neq g(x')$ otherwise.

[28] This criterion is known in the literature as EX-identification (explanatory identification), as opposed to BC-identification (behaviorally correct identification). In an uncomputable setting, the distinction is irrelevant. The kind of data that is dealt with will play a crucial role, hence we emphasize its importance in the terminology.

number occurs in σ and n is the least such number. Obviously, f identifies $\mathfrak{L}$ in the limit from texts.

We now turn to classifiers. A (*numerical*) *classifier* is a partial function from $(\mathbb{N} \cup \{\sharp\})^\star$ into $\{0, 1\}$. So a classifier is also fed with longer and longer finite initial segments of a text e for a language L in $\mathfrak{L}$, in response to which it possibly guesses whether L does or does not have some given *property*. The aim of the classifier is to *discover* whether the particular L that generates e has that property. Here again, what is meant by "discover" is subject to different formalizations. An important criterion is that of *positive classification in the limit*. That is, the classifier eventually converges to 1 if and only if the language the classifier gets data from has the property of interest. The property of interest can be identified with a subset X of the set of possible languages, and having property X means being a member of X. Formally:

Definition 3.2 *Let a subset X of $\mathfrak{L}$ be given.*

A numerical classifier f is said to positively classify $\mathfrak{L}$ in the limit, from texts and following X *iff for all $L \in \mathfrak{L}$ and texts e for L, $L \in X$ iff $f(e_{|k}) = 1$ for cofinitely many $k \in \mathbb{N}$.*

$\mathfrak{L}$ is said to be positively classifiable in the limit, from texts and following X *iff some numerical classifier positively classifies $\mathfrak{L}$ in the limit, from texts and following X.*

For instance, if $X = \{\mathbb{N}, \mathbb{N} \setminus \{0, 1\}, \mathbb{N} \setminus \{0, 1, 2, 3\}, \ldots\}$ then the property of interest is: the language is nonempty and its least element is even. Clearly, $\mathfrak{L}$ is positively classifiable in the limit, from texts and following X.

3.4 Logical Learning Paradigms

To represent a language (an r.e. subset of $\mathbb{N}$) in a logical setting, we consider a vocabulary $\mathcal{V}$ containing $\overline{0}$, s and a unary predicate symbol P, possibly plus equality and some extra predicate and function symbols. We say that a structure *represents* a language L if it is the unique standard structure $\mathfrak{M}_L$ such that:

- each of $\mathfrak{M}_L$'s individuals interprets a unique numeral (i.e., the domain of $\mathfrak{M}_L$ can be thought of as being $\mathbb{N}$);
- the interpretation in $\mathfrak{M}_L$ of all function and predicate symbols in $\mathcal{V}$ except P is fixed by some predicates and functions over $\mathbb{N}$;
- for all $n \in \mathbb{N}$, $P(\overline{n})$ is true in $\mathfrak{M}_L$ iff n is a member of L.

Note that $\mathfrak{M}_L$ is a Henkin structure. If $\mathcal{V}$ is equality free and contains no function symbol except $\overline{0}$ and s, then $\mathfrak{M}_L$ is actually a Herbrand structure. Clearly, $\mathfrak{M}_L$ is the counterpart in a logical setting to the language L of the numerical setting. To represent in the logical setting the class $\mathfrak{L}$ of possible languages of the numerical setting, we define the set $\mathcal{W}$ of possible worlds to be $\{\mathfrak{M}_L \mid L \in \mathfrak{L}\}$. To every numerical text for L corresponds a (*logical*) *text for* $\mathfrak{M}_L$ where the sentence $P(\overline{n})$ replaces all occurrences of n, for all $n \in \mathbb{N}$. So the logical text associated with the previous example of numerical text is:

$$(P(\overline{4}), P(\overline{3}), P(\overline{2}), \sharp, P(\overline{6}), P(\overline{5}), P(\overline{4}), \sharp, P(\overline{8}), P(\overline{7}), P(\overline{6}), \sharp, \ldots).$$

Note that $\sharp$ can occur both in a numerical and in a logical text. As a counterpart to the numerical identifiers, we then define a (*logical*) *identifier* to be any partial function from the set of finite sequences of sentences and $\sharp$'s into the set of sentences that have a unique standard model. A sentence φ in the codomain of a logical identifier plays the role of a natural number i in the codomain of a numerical identifier: φ refers to a standard structure via the notion of model, whereas i refers to a language L via the notion of r.e. index. As a counterpart to the numerical classifiers, we define a (*logical*) *classifier* to be any partial function from the set of finite sequences of sentences and $\sharp$'s into $\{0, 1\}$.[29] For the logical analogue of Definition 3.1, we have to assume that the property of being $\mathfrak{M}$, for any possible world $\mathfrak{M}$, is expressible by a sentence in the logical language, which parallels the representation of a language (an r.e. subset of $\mathbb{N}$) by an r.e. index.

Definition 3.3 *A logical identifier f is said to* identify $\mathcal{W}$ in the limit from texts *iff there exists a family $(\varphi_{\mathfrak{M}})_{\mathfrak{M}\in\mathcal{W}}$ of sentences such that the following holds for all possible worlds $\mathfrak{M}$.*

1. *For all possible worlds $\mathfrak{N}$, $\mathfrak{N} \models \varphi_{\mathfrak{M}}$ iff $\mathfrak{N} = \mathfrak{M}$.*
2. *For all texts e for $\mathfrak{M}$, $f(e_{|k}) = \varphi_{\mathfrak{M}}$ for cofinitely many $k \in \mathbb{N}$.*[30]

$\mathcal{W}$ is said to be identifiable in the limit from texts *iff some logical identifier identifies $\mathcal{W}$ in the limit from texts.*

[29] Note that logical identifiers and classifiers are more general than their numerical counterparts, because they are functions from a set that is more inclusive than $(\{P(\overline{n}) \mid n \in \mathbb{N}\} \cup \{\sharp\})^\star$. This will turn out to be convenient when we generalize the notion of text to a notion of environment, allowing for infinitely many variations on the kind of data presented to logical identifiers and classifiers. This will include in particular the case where $\{P(\overline{n}), \neg P(\overline{n}) \mid n \in \mathbb{N}\}$ is the set of data, which is the counterpart to the notion of informant of the numerical framework.

[30] Recall that $e_{|k}$ denotes the sequence of the first k elements of e.

For instance, with $\mathfrak{L} = \{\mathbb{N}, \mathbb{N} \setminus \{0\}, \mathbb{N} \setminus \{0, 1\}, \ldots, \emptyset\}$, each of the possible languages to be identified in the limit is adequately represented by one and only one member of

$$\{\forall x(P(x) \leftrightarrow \overline{n} \leq x) \mid n \in \mathbb{N}\} \cup \{\forall x \neg P(x)\}.$$

So if $\mathcal{V}$ contains $\leq$, interpreted as expected, there is a logical learner that does the same job as the numerical learner described in the previous section, and identifies $\mathcal{W}$ in the limit. When working from a logical text for, say, $\mathbb{N} \setminus \{0, 1\}$, such a successful logical learner outputs cofinitely many times the sentence $\forall x(P(x) \leftrightarrow \overline{2} \leq x)$, or another sentence that represents $\mathfrak{M}_{\mathbb{N} \setminus \{0,1\}}$ properly.

We define as follows the logical counterpart to Definition 3.2.

Definition 3.4 *Let a subset X of $\mathcal{W}$ be given.*

A logical classifier f is said to positively classify $\mathcal{W}$ in the limit, from texts and following X *iff for all $\mathfrak{M} \in \mathcal{W}$ and texts e for $\mathfrak{M}$, $\mathfrak{M} \in X$ iff $f(e_{|k}) = 1$ for cofinitely many $k \in \mathbb{N}$.*

$\mathcal{W}$ is said to be positively classifiable in the limit, from texts and following X *iff some logical classifier positively classifies $\mathcal{W}$ in the limit, from texts and following X.*

Given a sentence φ, if X is the set of models of φ in $\mathcal{W}$, then we say 'following φ' instead of 'following X.''

As suggested in the last clause of the definition, we will be interested in positive classification in the limit following *definable* subsets of $\mathcal{W}$. Which subsets of $\mathcal{W}$ (which properties of possible worlds) are definable depends on the members of $\mathcal{V}$ except $\overline{0}$, s and P, and their interpretation; it also depends on how we define the notion of a sentence. For instance, consider again $X = \{\mathbb{N}, \mathbb{N} \setminus \{0, 1\}, \mathbb{N} \setminus \{0, 1, 2, 3\}, \ldots\}$. If $\mathcal{V}$ consists just of $\overline{0}$, s, and P, then $\{\mathfrak{M}_L \mid L \in X\}$ is not definable by a first-order sentence. On the other hand, if $\mathcal{V}$ also contains $\leq$ and $\times$ (interpreted as expected), then $\{\mathfrak{M}_L \mid L \in X\}$ can be defined by

$$\exists y \forall x(P(x) \leftrightarrow x \geq \overline{2} \times y).$$

As for formal learning theory in the numerical setting, we could impose that logical identifiers and classifiers are computable. For logical classifiers, we would then have to decide if we wanted convergence to the same output on all texts for a given world (explanatory learning) or if we accepted convergence to different representations of some world on different texts for this world (behaviorally correct learning). Such considerations are irrelevant here. In particular, we do not want to impose any computability condition since we are introducing the model-theoretic part of parametric logic. We want to proceed as we do in first-order logic, where a notion of truth is introduced, and it turns out to be computable (in the sense that the set of all valid sentences is r.e.). Computability is not introduced at the model-theoretic level.[31] Finally, we note that due to the fact that Definitions 3.3 and 3.4 impose no computability condition, positive classification in the limit is more general than identification in the limit, provided that the set of possible worlds is countable—an assumption that holds for all logical paradigms that represent numerical learning paradigms. Indeed, a straightforward enumeration technique enables us to show the following.

Remark 3.5 *Suppose that $\mathcal{W}$ is countable. Let $(\varphi_{\mathfrak{M}})_{\mathfrak{M}\in\mathcal{W}}$ be a family of sentences such that for all $\mathfrak{M}$, $\mathfrak{N} \in \mathcal{W}$, $\mathfrak{N} \models \varphi_{\mathfrak{M}}$ iff $\mathfrak{N} = \mathfrak{M}$. Then $\mathcal{W}$ is identifiable in the limit from texts iff for all $\mathfrak{M} \in \mathcal{W}$, $\mathcal{W}$ is positively classifiable in the limit, from texts and following $\varphi_{\mathfrak{M}}$.*

Some learning paradigms do not consider texts, but *informants*, i.e., sequences of positive or negative data, whereas texts may contain positive data only. An informant for a language L is an enumeration of the graph of the characteristic function of L (hence an enumeration of pairs of the form either $(n, 1)$ with $n \in L$ or $(n, 0)$ with $n \in \mathbb{N} \setminus L$). For instance, one of the numerical informants for the language $\mathbb{N} \setminus \{0, 1\}$ starts with

$$((4, 1), (3, 1), (2, 1), (1, 0), \sharp, (6, 1), (5, 1), (4, 1), (0, 0), \sharp).$$

In the associated logical informant, all occurrences of $(n, 1)$ are replaced by $P(\overline{n})$, and all occurrences of $(n, 0)$ are replaced by $\neg P(\overline{n})$, for all natural numbers n. So the logical informant that corresponds to the previous numerical informant starts with

$$(P(\overline{4}), P(\overline{3}), P(\overline{2}), \neg P(\overline{1}), \sharp, P(\overline{6}), P(\overline{5}), P(\overline{4}), \neg P(\overline{0}), \sharp).$$

Whereas logical texts are sets of closed atoms (closed atomic formulas), logical informants are sets of closed literals (closed atoms or negations of closed atoms). The (numerical or logical) notion of identification in the limit from informants, as well as the (numerical or logical) notion of positive classification from informants, are modeled on the definitions above, replacing texts by informants. We omit the formal details, since they are straightforward.

[31] Under reasonable assumptions, parametric logic also enjoys a computable proof-theoretic characterization of its model-theoretic notions. The proof-theoretic aspects of parametric logic will not be discussed in this paper.

3.5 Possible Data

We have explained how learning paradigms can be defined in a logical setting. But what we are after is a deeper connection between formal learning theory and logic. We want to examine the relation between learning and some notion of logical consequence, by looking at the following issue. Let a sentence φ, an $\mathfrak{M} \in \mathcal{W}$, and $\llbracket$ a text | an informant $\rrbracket$ e for $\mathfrak{M}$ be given. Let T be the set of sentences that occur in e.

- Let a logical identifier f be given. Assume that f identifies $\mathcal{W}$ in the limit from $\llbracket$ texts | informants $\rrbracket$. If $f(e_{|k}) = \varphi$ for cofinitely many $k \in \mathbb{N}$, is every intended interpretation of T a model of φ?
- Let a logical classifier f be given. Assume that f positively classifies $\mathcal{W}$ in the limit, from $\llbracket$ texts | informants $\rrbracket$ and following φ. Is every intended interpretation of T a model of φ iff $f(e_{|k}) = 1$ for cofinitely many $k \in \mathbb{N}$?

In other words, we want to know whether the sentence output by a logical identifier that correctly converges on $\llbracket$ a text | an informant $\rrbracket$ e for T is a logical consequence of T, in some sense of 'logical consequence'. We want to know whether a logical classifier correctly converges on $\llbracket$ a text | an informant $\rrbracket$ e for T just in case the sentence that defines the task of positive classification is a logical consequence of T, in some sense of "logical consequence". If we asked the classifier to output either φ or $\neg\varphi$ instead of 1 or 0, respectively, the similarity between both issues would become even more apparent: is it the case that φ is logical consequence of T, for some sense of "logical consequence", iff a successful identifier or a successful (modified) classifier 'correctly' outputs φ (in the limit) when it processes T (working on one of the $\llbracket$ texts | informants $\rrbracket$ generated by T)?

If 'intended model of T' means "model of T that belongs to $\mathcal{W}$", the answer to these questions is clearly yes if e is an informant, but no if e is a text. For instance, consider $\varphi = \forall x(P(x) \leftrightarrow x \geq \overline{2})$, and let $\mathfrak{L}$ be defined as in the previous sections. Since φ uniquely describes $\mathfrak{M}_{\mathbb{N}\setminus\{0,1\}}$ among the set of structures $\mathfrak{M}_L$, $L \in \mathfrak{L}$, any identifier that identifies $\mathcal{W}$ in the limit from texts will converge in the limit to φ on any text for $\mathfrak{M}_{\mathbb{N}\setminus\{0,1\}}$. And any classifier that positively classifies $\mathcal{W}$ in the limit, from texts and following φ will converge in the limit to 1 on any text for $\mathfrak{M}_{\mathbb{N}\setminus\{0,1\}}$. The set of sentences occurring in e is $T = \{P(\overline{2}), P(\overline{3}), P(\overline{4}), \ldots\}$. Obviously, $\mathfrak{M}_{\mathbb{N}}$ and $\mathfrak{M}_{\mathbb{N}\setminus\{0\}}$ are models of T in $\mathcal{W}$, but they are not models of φ.

The problem is that $T = \{P(\overline{2}), P(\overline{3}), P(\overline{4}), \ldots\}$ is not a faithful translation of a logical text for $\mathfrak{M}_{\mathbb{N}\setminus\{0,1\}}$, or equivalently, of a numerical text for the language $L = \mathbb{N} \setminus \{0, 1\}$. A text for L contains occurrences of *all* members of L, which means that since 0 and 1 do not occur in a text for L, then

0 and 1 do not belong to L. This implicit condition on texts for L has no counterpart in T when we consider the models of T in $\mathcal{W}$. Defining the intended models of T as the models of T that belong to $\mathcal{W}$ is too crude, because such models can freely make $P(\overline{0})$ and $P(\overline{1})$ either true or false instead of *implicitly* constraining both $P(\overline{0})$ and $P(\overline{1})$ to be false. A natural approach is to include a new parameter in the definition of paradigm $\mathcal{P}$, called the *set of possible data*, denoted by $\mathcal{D}$. With the current example and the notion of text, we would give $\mathcal{D}$ the value $\{P(\overline{n}) \mid n \in \mathbb{N}\}$. With the current example and the notion of informant, $\mathcal{D}$ would take the value $\{P(\overline{n}), \neg P(\overline{n}) \mid n \in \mathbb{N}\}$. More generally, $\mathcal{D}$ could be any set of sentences that, intuitively, represent potential observations, or results of experiments. If the possible datum $P(\overline{1})$ does not occur in T, it indicates that the intended model of T—in our example $\mathfrak{M}_{\mathbb{N}\setminus\{0,1\}}$—makes $P(\overline{1})$ false. This is an instance of the *closed world assumption* [32], [24]. Intuitively, the closed world assumption expresses that a fact can be inferred to be false if it cannot be logically derived from an underlying set of hypotheses. Of course, this idea cannot be applied carelessly; for instance, neither $P(\overline{0})$ nor $P(\overline{1})$ can be logically derived from the hypothesis that $P(\overline{0}) \vee P(\overline{1})$ is true, but it would be inconsistent to conclude from that hypothesis and an incorrect interpretation of the closed world assumption that both $P(\overline{0})$ and $P(\overline{1})$ are false. Theories of a specific syntactic form can avoid this inconsistency issue and are good candidates to apply a closed world assumption: for instance a set T of *Horn clauses*, e.g., a set of disjunctions of atomic formulas or their negations with at most one positive (atomic) disjunct, has a model $\mathfrak{M}$ where all closed atomic sentences not logically implied by T are false in $\mathfrak{M}$. Leaving technical issues aside, the previous discussion leads to a closed world assumption in the form of the convention that if an observation or experimental result can be obtained from a possible world $\mathfrak{M}$ abstracting the reality under consideration, then such an observation or experimental result should occur in any theory related to $\mathfrak{M}$. Hence if a theory T does not contain some possible datum ψ, no intended interpretation of T should be a model of ψ.

3.6 Logical Consequence in $\mathcal{P}$

Armed with the parameters $\mathcal{W}$ and $\mathcal{D}$, we can define the notion of logical consequence in $\mathcal{P}$. First we have to introduce some notation. Given a structure $\mathfrak{M}$, we use $Diag_{\mathcal{D}}(\mathfrak{M})$ to represent the *$\mathcal{D}$-diagram* of $\mathfrak{M}$, that is, the set of all members of $\mathcal{D}$ true in $\mathfrak{M}$. Given a set of sentences T, we denote by $Mod(T)$ the class of all models of T, and by $Mod_{\mathcal{W}}(T)$ the class of all models of T in $\mathcal{W}$ ($Mod_{\mathcal{W}}(T) = Mod(T) \cap \mathcal{W}$).

Definition 3.6 *Let a set T of sentences and a structure $\mathfrak{M}$ be given. We say that $\mathfrak{M}$ is a $\mathcal{D}$-*minimal model of T in $\mathcal{W}$* iff $\mathfrak{M} \in Mod_{\mathcal{W}}(T)$ and for all $\mathfrak{N} \in Mod_{\mathcal{W}}(T)$, $Diag_{\mathcal{D}}(\mathfrak{N}) \not\subset Diag_{\mathcal{D}}(\mathfrak{M})$.*

Note that if $\mathcal{W}$ is the set of all Herbrand structures, $\mathcal{D}$ the set of all atomic sentences, and T a set of definite clauses, then a $\mathcal{D}$-minimal model of T in $\mathcal{W}$ is what is known in the literature on logic programming as a minimal Herbrand model of T. Given a set of sentences T, let us denote by $Mod^{\mathcal{D}}_{\mathcal{W}}(T)$ the class of all $\mathcal{D}$-minimal models of T in $\mathcal{W}$. By the discussion above, the class of intended models of T is not $Mod_{\mathcal{W}}(T)$, but $Mod^{\mathcal{D}}_{\mathcal{W}}(T)$. Note that $Mod^{\mathcal{D}}_{\mathcal{W}}(T)$ and $Mod_{\mathcal{W}}(T)$ are equal when $\mathcal{D}$ is semantically closed under negation, which covers the case $\mathcal{D} = \emptyset$. Consider for instance a theory T that has four models in $\mathcal{W}$, say $\mathfrak{M}_1$, $\mathfrak{M}_2$, $\mathfrak{M}_3$ and $\mathfrak{M}_4$, and for all $i \in \{1, 2, 3, 4\}$, denote by Z_i the set of all sentences that are true in $\mathfrak{M}_i$. Figure 3.4 is a possible representation of Z_1, Z_2, Z_3, Z_4 and $\mathcal{D}$. Here the $\mathcal{D}$-minimal models of T in $\mathcal{W}$ are $\mathfrak{M}_2$ and $\mathfrak{M}_4$; their $\mathcal{D}$-diagrams are represented by the dashed areas. The structure $\mathfrak{M}_2$ is a $\mathcal{D}$-minimal model of T in $\mathcal{W}$ because $Z_2 \cap \mathcal{D}$ does not strictly contain $Z_1 \cap \mathcal{D}$, $Z_3 \cap \mathcal{D}$ or $Z_4 \cap \mathcal{D}$. Similarly, $\mathfrak{M}_4$ is a $\mathcal{D}$-minimal model of T in $\mathcal{W}$ because $Z_4 \cap \mathcal{D}$ does not strictly contain $Z_1 \cap \mathcal{D}$, $Z_2 \cap \mathcal{D}$ or $Z_3 \cap \mathcal{D}$. On the other hand, since $Z_2 \cap \mathcal{D}$ is strictly included in both $Z_1 \cap \mathcal{D}$ and $Z_3 \cap \mathcal{D}$, neither $\mathfrak{M}_1$ nor $\mathfrak{M}_3$ is a $\mathcal{D}$-minimal model of T in $\mathcal{W}$.

Since $Mod^{\mathcal{D}}_{\mathcal{W}}(T)$ is the class of intended models of T, the next definition captures the notion of logical consequence in $\mathcal{P}$ introduced as (C") in Section 3.1 and rephrased as (C*) in Section 3.2.

Definition 3.7 *Given a set of sentences T and a sentence φ, we say that φ is a* logical consequence *of T in $\mathcal{P}$ iff $Mod^{\mathcal{D}}_{\mathcal{W}}(T) \subseteq Mod(\varphi)$.*

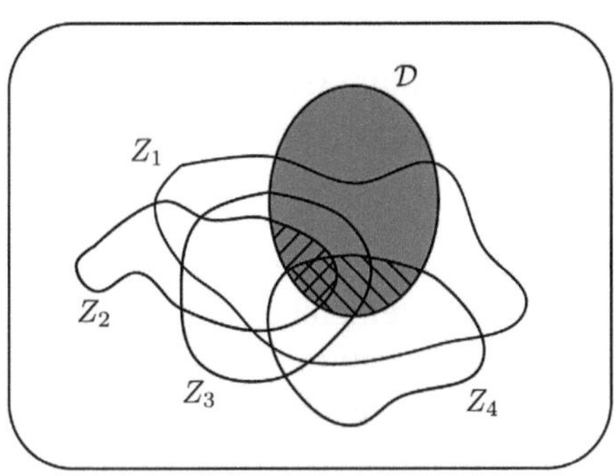

Figure 3.4. Two $\mathcal{D}$-minimal models out of four models

3.7 Possible Assumptions

A logical paradigm that represents a learning paradigm usually takes for $\mathcal{W}$ a countable set of standard structures, and the theories that should be the starting point for logical investigation—because they are associated with a text or with an informant for a member of $\mathcal{W}$—are the $\mathcal{D}$-diagrams of the possible worlds. Such theories contain "data" only. Of course, they form an extremely restrictive class of theories, more flexibility is needed. First-order logic is purely axiomatic in the sense that it accommodates no notion of observations or results of experiments. This means that first-order logic sets $\mathcal{D}$ to the empty set. On the other hand, in first-order logic, any (consistent) sentence can be part of a theory that would be the starting point for logical investigation, any (consistent) sentence can be used as an axiom.

This justifies the addition of a fourth parameter to $\mathcal{P}$, called set of *possible assumptions*, denoted by $\mathcal{A}$. Like $\mathcal{D}$, $\mathcal{A}$ can be any set of sentences, provided that it is disjoint from $\mathcal{D}$. Technically, no real difference would result from assuming that $\mathcal{D}$ is included in $\mathcal{A}$, or that no *a priori* relation between $\mathcal{D}$ and $\mathcal{A}$ exists. We require that $\mathcal{A}$ and $\mathcal{D}$ are disjoint because this sometimes allows to slightly simplify some formal arguments. With this observation in mind, the paradigms of formal learning theory we have examined before are such that $\mathcal{A}$ is empty. With $\mathcal{W}$, $\mathcal{D}$ and $\mathcal{A}$ in hand, what is the counterpart to the classical notion of a consistent theory? Since the intended interpretations of a theory T are the $\mathcal{D}$-minimal models of T in $\mathcal{W}$ and since we assume that a starting point for logical investigation consists of possible data and possible assumptions only, a quick answer could be: any set X of sentences such that $X \subseteq \mathcal{D} \cup \mathcal{A}$ and X has at least one $\mathcal{D}$-minimal model in $\mathcal{W}$. But among the sets X of sentences that satisfy this condition, some have very bad properties, as the next proposition shows.

Proposition 3.8 *Suppose that $\mathcal{V}$ consists of infinitely many constants and a unary predicate symbol. There exists a set W of standard structures such that if $\mathcal{W}=W$, $\mathcal{D}$ is the set of atomic sentences and X is the set of first-order sentences that are logical consequences of $\emptyset$ in $\mathcal{P}$, then $Mod^{\mathcal{D}}_{\mathcal{W}}(\emptyset)$ is a strict subset of $Mod^{\mathcal{D}}_{\mathcal{W}}(X)$.*

In the previous proposition, we would rather expect and prefer to have $Mod^{\mathcal{D}}_{\mathcal{W}}(\emptyset) = Mod^{\mathcal{D}}_{\mathcal{W}}(X)$. Also, remember that we justified the introduction of $\mathcal{D}$ claiming that a possible datum ψ represents an observation or an experimental result obtained from a possible world $\mathfrak{M}$ abstracting the reality under consideration, and that the absence of ψ from any theory T related to $\mathfrak{M}$ should mean that all intended interpretations of T make ψ false. In other words, for any 'reasonable' theory T and for any intended interpretation $\mathfrak{M}$

of T, the intersection of T with $\mathcal{D}$ should be equal to the $\mathcal{D}$-diagram of $\mathfrak{M}$. What makes $Mod_{\mathcal{W}}^{\mathcal{D}}(\emptyset)$ a strict subset of $Mod_{\mathcal{W}}^{\mathcal{D}}(X)$ in Proposition 3.8 is that, in particular, no possible world has $\emptyset$ as $\mathcal{D}$-diagram. These considerations suggest the following definition.

Definition 3.9 A possible knowledge base *is a set of the form* $Diag_{\mathcal{D}}(\mathfrak{M}) \cup A$ *where* $\mathfrak{M}$ *is a member of* $\mathcal{W}$ *and* A *a subset of* $\mathcal{A}$ *all of whose members are true in* $\mathfrak{M}$.

We denote by $\mathcal{B}$ the set of all possible knowledge bases. Hence:

$$\mathcal{B} = \{Diag_{\mathcal{D}}(\mathfrak{M}) \cup A \mid \mathfrak{M} \in \mathcal{W},\ A \subseteq \mathcal{A},\ \mathfrak{M} \models A\}.$$

Let a possible world $\mathfrak{M}$ be given, and let Z be the set of sentences that are true in $\mathfrak{M}$. Figure 3.5 represents three possible knowledge bases of the form $Diag_{\mathcal{D}}(\mathfrak{M}) \cup X$ with $X \subseteq \{\varphi \in \mathcal{A} \mid \mathfrak{M} \models \varphi\}$. The leftmost theory T_1 is the smallest knowledge base of this type: X is empty, and T_1 is just the $\mathcal{D}$-diagram of $\mathfrak{M}$. The rightmost theory T_3 is the largest knowledge base of this type: T_3 is the set of all members of $\mathcal{D} \cup \mathcal{A}$ that are true in $\mathfrak{M}$. The middle theory T_2 is an intermediate case, that includes only some of the members of $\mathcal{A}$ that are true in $\mathfrak{M}$, besides the $\mathcal{D}$-diagram of $\mathfrak{M}$.

Since $\mathcal{B}$ is fully determined by the parameters in $\mathcal{P}$, $\mathcal{B}$ is not conceived of as a primitive parameter of parametric logic, but as a derived one; the primitive parameters are the members of $\mathcal{P}$. First-order logic takes for $\mathcal{A}$ the set of all first-order sentences, which, together with the fact that $\mathcal{D}$ is set to $\emptyset$, implies that the set of possible knowledge bases is the set of all consistent theories. So when casting first-order logic and paradigms of formal learning theory into parametric logic, we obtain two limiting cases of logical paradigms: $\mathcal{D}$ is equal to $\emptyset$ for the former, whereas $\mathcal{A}$ is equal to $\emptyset$ for the latter.

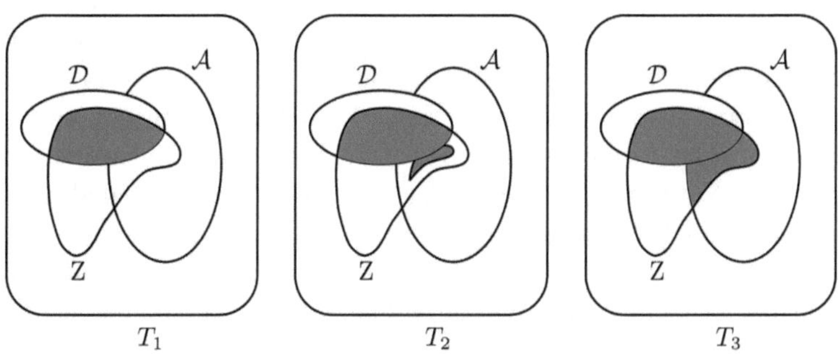

Figure 3.5. Three possible knowledge bases

3.8 Deductive Consequence in *P*

Let us get back to (A"), (B") and (C") as well as the inference

$$(\odot)\quad \frac{P(\overline{0})\;\; P(\overline{1})\;\; P(\overline{2})\;\; P(\overline{3})\;\; P(\overline{4})\qquad \ldots}{\forall x\, P(x)}$$

introduced in Sections 3.1 and 3.2. We have formalized (C") as Definition 3.7, but (A") and (B") have still not been formally addressed. We can now justify why ($\odot$) is a logical inference in $\mathcal{P}$ for natural choices of $\mathcal{P}$. Indeed, assume that the logical paradigm $\mathcal{P}$ we are working with is such that:

(i) $\mathcal{V}$ contains no function symbol except $\overline{0}$ and s;
(ii) $\{P(\overline{n}) \mid n \in \mathbb{N}\} \subseteq \mathcal{D} \subseteq \{P(\overline{n}), \neg P(\overline{n}) \mid n \in \mathbb{N}\}$;
(iii) $\mathcal{W}$ is a set of standard structures, and the $\mathcal{D}$-diagram of some member of $\mathcal{W}$ is equal to $\{P(\overline{n}) \mid n \in \mathbb{N}\}$.

Then $\{P(\overline{n}) \mid n \in \mathbb{N}\}$ is a possible knowledge base and every $\mathcal{D}$-minimal model of $\{P(\overline{n}) \mid n \in \mathbb{N}\}$ in $\mathcal{W}$ is a model of $\forall x\, P(x)$, which shows that $\forall x\, P(x)$ is a logical consequence of $\{P(\overline{n}) \mid n \in \mathbb{N}\}$ in $\mathcal{P}$. The assumptions about $\mathcal{P}$ above are natural, and they are tacitly accepted when ($\odot$) is viewed as some form of logical inference. But we have not examined why ($\odot$) can moreover be conceived of as an inductive inference in $\mathcal{P}$, for some choice of $\mathcal{P}$. Conditions (i)–(iii) make $\forall x\, P(x)$ a logical consequence of $\{P(\overline{n}) \mid n \in \mathbb{N}\}$ in $\mathcal{P}$. Do they also make $\forall x\, P(x)$ an inductive consequence of $\{P(\overline{n}) \mid n \in \mathbb{N}\}$ in $\mathcal{P}$, in a precise sense that still has to be formalized?

Before tackling (B"), we first examine how to formalize (A"). Note that a logical consequence of T in the classical sense is a logical consequence of T in $\mathcal{P}$, since every $\mathcal{D}$-minimal model of T in $\mathcal{W}$ is a model of T. Shall we define the deductive consequences of T in $\mathcal{P}$ as the logical/deductive consequences of T in the classical sense? It would be unnatural, having defined a generic notion of logical consequence that depends, in particular, on a set of possible worlds $\mathcal{W}$ possibly distinct from the class of all structures, to have a model-theoretic interpretation of "deductive consequence in $\mathcal{P}$" that does not refer to the member $\mathcal{W}$ of $\mathcal{P}$, but invariantly to the class of all structures. We reminded the reader that the attributes of deduction are certainty and conclusiveness. Why does an inference such as ($\odot$) not have these attributes? Because no finite subset of the set of premises enables us to conclude that $\forall x\, P(x)$ is true. We suggest taking into account the fact that we are not considering arbitrary sets of sentences, but possible knowledge bases only, and accept the converse of the implication expressed in the previous statement as the hallmark of deduction. Formally:

Definition 3.10 *Given $T \in \mathcal{B}$ and a sentence φ, we say that φ is a* deductive consequence *of T in $\mathcal{P}$ iff there exists a finite subset D of T such that for all $T' \in \mathcal{B}$:*

if T' includes D then φ is a logical consequence of T' in $\mathcal{P}$.

In the case of first-order logic where the set of possible knowledge bases is the set of consistent theories, the condition expressed in Definition 3.10 is equivalent to the usual compactness property. This provides an illuminating reason for the equivalence between logical and deductive consequences in first-order logic: the compactness theorem. In the general case, it is easy to verify that for all possible knowledge bases T and sentences φ, φ is a deductive consequence of T in $\mathcal{P}$ iff there exists a finite subset D of T such that $Mod_{\mathcal{W}}(D) \subseteq Mod(\varphi)$. Hence the only difference between the compactness property of first-order logic and the compactness property expressed in Definition 3.10 is that all structures are involved in the former, and only the members of $\mathcal{W}$ in the latter.

Example 3.11 *Assume that $\mathcal{W} = \{\mathfrak{M}_L \mid L \in \mathfrak{L}\}$ where $\mathfrak{L}$ is the set of final segments of $\mathbb{N}$. Set $\mathcal{D} = \{P(\overline{n}) \mid n \in \mathbb{N}\}$, $\mathcal{A} = \emptyset$, and let $T = \{P(\overline{n}) \mid n \geq 2\}$. Then $\varphi = \forall x P(s(s(s(x))))$ is a deductive consequence of T in $\mathcal{P}$. Indeed, T contains $P(\overline{3})$, and every model of $P(\overline{3})$ in $\mathcal{W}$ is a model of φ. Note that $T \not\models \varphi$.*

3.9 Inductive Consequence in *P*

Let us consider inference ($\odot$) again and examine under which conditions it could be viewed as some form of induction. Because we want every inductive consequence of a possible knowledge base T in $\mathcal{P}$ to be a logical consequence of T in $\mathcal{P}$ (as explained in Section 3.1), we assume that conditions (i)–(iii) above are satisfied: they are the natural hypotheses on the basis of which we can infer that $\forall x\, P(x)$ is a logical consequence of $\{P(\overline{n}) \mid n \in \mathbb{N}\}$ in $\mathcal{P}$, as defined in Definition 3.7.

- First assume that $\mathcal{D}$ is equal to $\{P(\overline{n}), \neg P(\overline{n}) \mid n \in \mathbb{N}\}$. Working from $\{P(\overline{n}) \mid n \in \mathbb{N}\}$, can we believe in $\forall x\, P(x)$, supporting this belief on a notion of refutability—the other attribute of induction? Yes because either $\forall x\, P(x)$ is true, or a conclusive refutation of $\forall x\, P(x)$ is guaranteed. More formally, if T is any possible knowledge base then we have:
  - either $\forall x\, P(x)$ is a logical consequence of T in $\mathcal{P}$ (in which case T can only be $\{P(\overline{n}) \mid n \in \mathbb{N}\}$), or
  - $\exists x \neg P(x)$ is a deductive consequence of T in $\mathcal{P}$ (because T contains $\neg P(\overline{n})$ for some $n \in \mathbb{N}$).

Then assume that $\mathcal{D}$ is equal to $\{P(\overline{n}) \mid n \in \mathbb{N}\}$. Still working from $\{P(\overline{n}) \mid n \in \mathbb{N}\}$, can we also believe in $\forall x\, P(x)$, supporting this belief on a notion of refutability? No because if we were *actually* working from $\{P(\overline{n}) \mid n \geq 2\}$, then we could not refute the hypothesis that $\forall x\, P(x)$ is not a logical consequence of $\{P(\overline{n}) \mid n \geq 2\}$ in $\mathcal{P}$: $\exists x \neg P(x)$ is not a deductive consequence of $\{P(\overline{n}) \mid n \geq 2\}$ in $\mathcal{P}$. So for this choice of $\mathcal{D}$, though $\forall x\, P(x)$ is a logical consequence of $\{P(\overline{n}) \mid n \in \mathbb{N}\}$ in $\mathcal{P}$, we do not view $\forall x\, P(x)$ as an inductive consequence of $\{P(\overline{n}) \mid n \in \mathbb{N}\}$ in $\mathcal{P}$.
Finally, assume that $\mathcal{D} = \{P(\overline{n}) \mid n \in \mathbb{N}\} \cup \{\neg P(\overline{n}) \mid n \geq 2\}$. Once again working from $\{P(\overline{n}) \mid n \in \mathbb{N}\}$, can we believe in $\forall x\, P(x)$, supporting this belief on a notion of refutability? Though the argument used for the first choice of $\mathcal{D}$ cannot be used for this third choice of $\mathcal{D}$, the answer is yes. Indeed, once we have $P(\overline{0})$ and $P(\overline{1})$ in hand, *then* it becomes possible to believe in $\forall x\, P(x)$: if T is any possible knowledge base that contains $\{P(\overline{0}), P(\overline{1})\}$ then we have:

- either $\forall x\, P(x)$ is a logical consequence of T in $\mathcal{P}$ (in which case T can only be $\{P(\overline{n}) \mid n \in \mathbb{N}\}$), or
- $\exists x \neg P(x)$ is a deductive consequence of T in $\mathcal{P}$ (because T contains $\neg P(\overline{n})$ for some $n \geq 2$).

Remember that the first choice of $\mathcal{D}$ is related to a positive classification in the limit from informants, whereas the second choice of $\mathcal{D}$ is related to a positive classification in the limit from texts. What the first two cases express is that $\forall x\, P(x)$ can be refuted—hence can be viewed as an inductive consequence in $\mathcal{P}$ of every possible knowledge base from which it is logically implied in $\mathcal{P}$—by classifiers that positively classify $\mathcal{W}$ in the limit, from informants and following φ, but not by classifiers that positively classify $\mathcal{W}$ in the limit, from texts and following φ. The third choice of $\mathcal{D}$ indicates that the notion of refutability described for the first choice of $\mathcal{D}$ can be made more general. This more general notion turns out to be the right one. Formally, we can give the following definition.

Definition 3.12 *Given $T \in \mathcal{B}$ and a sentence φ, we say that φ is an* inductive consequence *of T in $\mathcal{P}$ iff it is a logical consequence of T in $\mathcal{P}$ and there exists a finite subset D of T such that for all $T' \in \mathcal{B}$:*

if T' includes D and $\neg\varphi$ is not a deductive consequence of T' in $\mathcal{P}$, then φ is a logical consequence of T' in $\mathcal{P}$.

Note that the condition in Definition 3.12 weakens the compactness condition expressed in Definition 3.10: it can be seen as a property of *weak compactness*. At this stage, the notions (A"), (B") and (C") of Section 3.1 have all been formalized.

Example 3.13 *Assume that $\mathcal{V}$ contains $\leq$, and let $\mathcal{W}$, $\mathcal{D}$, $\mathcal{A}$ and T be defined as in Example 3.11. Then $\varphi = \forall x(\overline{2} \leq x \leftrightarrow P(x))$ is not a deductive consequence of T in $\mathcal{P}$ because for all finite $D \subseteq T$, some model of D in $\mathcal{W}$ is not a model of φ. But φ is an inductive consequence of T in $\mathcal{P}$. Indeed, T contains $P(\overline{2})$, and every $T' \in \mathcal{B}$ that contains $P(\overline{2})$ but does not logically imply φ in $\mathcal{P}$ is such that $\neg\varphi$ is a deductive consequence of T' in $\mathcal{P}$, since it contains $P(\overline{1})$.*

3.10 Classification Generalized

Consider again the logical learning paradigm described in Sections 3.4 and 3.5 where $\mathcal{W}$ takes the value $\{\mathfrak{M}_L \mid L \in \mathfrak{L}\}$, $\mathcal{D}$ is set to $\{P(\overline{n}) \mid n \in \mathbb{N}\}$, and $\mathcal{A}$ is empty.[32] We have seen in Section 3.9 that $\forall x\, P(x)$ is a logical consequence of $\{P(\overline{n}) \mid n \in \mathbb{N}\}$ in $\mathcal{P}$, but it is not an inductive consequence of $\{P(\overline{n}) \mid n \in \mathbb{N}\}$ in $\mathcal{P}$. Still, some identifiers identify $\mathcal{W}$ in the limit from texts, hence converge in the limit to $\forall x\, P(x)$ on all texts for $\mathfrak{M}_{\mathbb{N}}$, and some classifiers positively classify $\mathcal{W}$ in the limit, from texts and following $\forall x\, P(x)$, hence converge in the limit to 1 on all texts for $\mathfrak{M}_{\mathbb{N}}$, and no others. This shows that the scope of formal learning theory goes beyond what we have described as induction, if we accept that the informal notion of induction is well captured by Definition 3.12. To investigate more deeply the relationship between formal learning theory and the notion of logical consequence in $\mathcal{P}$, we first need to generalize the concept of positive classification formalized in Definition 3.4, and then we must introduce the concept of *mind change bound*.

Most learning paradigms, like the one described in Section 3.3, are cast into logical paradigms where every possible world is uniquely determined by its $\mathcal{D}$-diagram. This property is obviously satisfied whenever $\mathcal{W}$ is a set of standard structures and $\mathcal{D}$ contains all atomic sentences that do not have a fixed interpretation in all members of $\mathcal{W}$. When a possible world $\mathfrak{M}$ is uniquely determined by its $\mathcal{D}$-diagram, the conditions '$\mathfrak{M}$ is a model of φ' and 'φ is a logical consequence of $Diag_{\mathcal{D}}(\mathfrak{M})$ in $\mathcal{P}$' are equivalent. They imply that for every possible knowledge base T and sentence φ, either φ or $\neg\varphi$ is a logical consequence of T in $\mathcal{P}$: $\mathcal{P}$ has *complete* possible knowledge bases w.r.t. the notion of logical consequence in $\mathcal{P}$. Paradigms whose possible knowledge bases are complete in this sense play an important role in our framework.

Definition 3.14 *$\mathcal{P}$ is said to* have complete knowledge bases *iff for all $T \in \mathcal{B}$*

[32] This logical paradigm was derived from the numerical learning paradigm described in Section 3.3—with a set $\mathfrak{L}$ of possible languages equal the set of final segments of $\mathbb{N}$.

and sentences φ, either φ or $\neg\varphi$ is a logical consequence of T in $\mathcal{P}$.

Though most learning paradigms are represented by paradigms of parametric logic having complete knowledge bases, having complete knowledge bases is not fundamentally related to the notion of positive classification. It is possible to generalize Definition 3.4 to any paradigm in a natural way. The first step is to generalize the notions of text and informant. Texts and informants correspond to particular choices of $\mathcal{D}$ (closed atoms—atomic formulas—for the former, closed literals—atomic formulas or their negations—for the latter). But $\mathcal{D}$ can be any set of sentences, corresponding to other kinds of data. Moreover, if a possible knowledge base contains a possible assumption (a member of $\mathcal{A}$), there is no reason not to make it accessible to the identifiers and to the classifiers. We have seen that the distinction between a member of $\mathcal{D}$ and a member of $\mathcal{A}$ is essential in order to define the set $\mathcal{B}$ of possible knowledge bases, but it can be ignored for the purpose of generalizing the notions of text and informant, which is done in the next definition.

Definition 3.15 *Given a possible knowledge base T, we call an* environment for T (in $\mathcal{P}$) *an ω-sequence of members of $T \cup \{\sharp\}$ in which every member of T occurs at least once.*

Suppose that $\mathcal{A}$ is empty. Depending on the choice of $\mathcal{D}$, the enumeration of a possible knowledge base (with $\sharp$ possibly occurring in the enumeration), i.e., the enumeration of the $\mathcal{D}$-diagram of a possible world, can correspond to a text, or to an informant, or have no counterpart in inductive inference. Generalizing further, we also consider enumerations of knowledge bases in case $\mathcal{A}$ is not empty. We use *environment* as a general term for an enumeration of a possible knowledge base.

So environments are texts or informants when $\mathcal{A} = \emptyset$ and $\mathcal{D}$ is set to the right value. To generalize Definition 3.4 to arbitrary paradigms (including paradigms that do not have complete possible knowledge bases), we must exclude the unwanted situation where a classifier would converge to 0 on an environment for a possible knowledge base that does not logically imply in $\mathcal{P}$ the sentence that expresses the property that the classifier has to discover:

Definition 3.16 *Let a sentence φ be given.*

A logical classifier f is said to positively classify $\mathcal{B}$ in the limit following φ (in $\mathcal{P}$) *iff for all $T \in \mathcal{B}$ and environments e for T:*

- *φ is a logical consequence of T in $\mathcal{P}$ iff $\{k \in \mathbb{N} \mid f(e_{|k}) = 1\}$ is cofinite.*
- *If $\{k \in \mathbb{N} \mid f(e_{|k}) = 0\}$ is cofinite then $\neg\varphi$ is a logical consequence of*

T in $\mathcal{P}$.

$\mathcal{B}$ is said to be positively classifiable in the limit following φ *(in $\mathcal{P}$) iff some logical classifier positively classifies $\mathcal{B}$ in the limit following φ.*

3.11 Mind Change Bounds

The notion of mind change bound provides a measure of complexity for functions defined on a set of finite sequences, hence it can be applied to both numerical and logical identifiers, and to both numerical and logical classifiers. Let us focus on logical classifiers. We first introduce some terminology. Recall that a binary relation R on a class X is well-founded iff for every nonempty set $Y \subseteq X$, Y contains an element x such that no member y of Y satisfies $R(y, x)$. Suppose that R is well-founded. We then denote by ρ_R the unique function from X into the class of ordinals such that for all $x \in X$:

$$\rho_R(x) = \sup\{\rho_R(y) + 1 \mid y \in X,\ \ R(y, x)\}.$$

The length of R is the least ordinal not in the range of ρ_R. Note that the length of R is equal to 0 iff $X = \emptyset$. For example, Figure 3.6 depicts a finite binary relation R of length 5. In this diagram, an arrow joins a point y to a point x iff $R(y, x)$ holds. For all points x in the field of R, the value of $\rho_R(x)$ is indicated. The four points with no predecessor are mapped to 0 by ρ_R. All other points x are mapped by ρ_R to the smallest $n \in \mathbb{N}$ such that one of x's predecessor is mapped to $n - 1$ and none of x's predecessors is mapped to an integer at least equal to n. Since 4 is the largest number in the range of ρ_R, the length of ρ_R is 5.

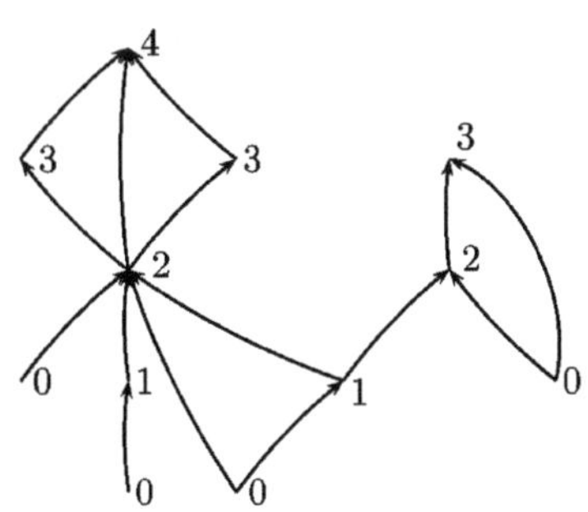

Figure 3.6. A finite binary relation of length 5

Definition 3.17 *Let a nonnull ordinal β, a sentence φ, and a logical classifier f be given.*

Let X be the set of all $\sigma \in (\mathcal{D} \cup \mathcal{A} \cup \{\sharp\})^\star$ such that the set of formulas in σ has a model in $\mathcal{W}$ and $f(\tau) = \downarrow$ for some initial segment τ of σ. We denote by R_f the binary relation over X such that for all $\sigma, \tau \in X$, $R_f(\sigma, \tau)$ holds iff $\tau \subset \sigma$ and $f(\sigma) \neq f(\tau)$.[33]

We say that f positively classifies $\mathcal{B}$ with fewer than β mind changes following φ (in $\mathcal{P}$) *iff the length of R_f is defined and smaller than or equal to β, and f positively classifies $\mathcal{B}$ in the limit following φ.*

We say that $\mathcal{B}$ is positively classifiable with fewer than β mind changes following φ (in $\mathcal{P}$) *iff some classifier positively classifies $\mathcal{B}$ with fewer than β mind changes following φ.*

Rather than $\llbracket$ 'fewer than $\beta + 1$ mind changes' | 'fewer than 1 mind change' $\rrbracket$, we say $\llbracket$ 'at most β mind changes' | 'no mind change' $\rrbracket$.

An equivalent definition is to equip f with an *ordinal counter*. Let us refer to the outputs of f as hypotheses (on whether φ does or not logically follow in $\mathcal{P}$ from the underlying knowledge base). When f first outputs a hypothesis, the ordinal counter is set to an ordinal γ. Then every time f switches from a hypothesis to another, or does not output any hypothesis, the ordinal counter is decreased. Saying that the ordinal complexity of f is smaller than an ordinal β is equivalent to saying that the behavior just described is possible, and that the ordinal counter only uses ordinals smaller than β.

Straightforward modifications to the previous definition enable us to define the notion of identification with fewer than β mind changes. In the literature, the notion of identification with at most β mind changes is more often found [1], [2]. Note that in case β is a limit ordinal, identification with at most β mind changes means either identification with fewer than β mind changes or identification with fewer than $\beta + 1$ mind changes, so the "fewer than" formulation is a refinement of the "at most" one. Getting back again to our favorite example, the numerical identifier described in Section 3.3 identifies $\mathfrak{L}$ with fewer than $\omega + 1$ mind changes, but not with fewer than ω mind changes. Indeed, as long as this identifier only gets $\sharp$'s, it outputs the hypothesis $\emptyset$ and the ordinal counter is set to ω. Otherwise, the identifier outputs the hypothesis $\{n, n+1, \ldots\}$ where n is the least number that has appeared so far, and the ordinal counter is set to n. It is easy to verify that no numerical identifier can do better, as far as mind change bounds are

[33] When a partial function f is not defined on an argument x, we write $f(x) = \uparrow$; otherwise, we write $f(x) = \downarrow$. We use $\llbracket \subseteq | \subset \rrbracket$ to denote that a finite sequence is a $\llbracket$ subsequence | strict subsequence $\rrbracket$ of another sequence, either finite or infinite.

concerned. Of course, some classes of languages are identifiable in the limit, but with no mind change bound, e.g., the class of finite languages.

Thanks to the concept of mind change bound, we can perfectly characterize the notions of deductive and inductive consequence in $\mathcal{P}$. Let a sentence φ be given. Let a classifier f that positively classifies $\mathcal{B}$ with no mind change following φ be given. Presented with members of an underlying possible knowledge base T, f waits until enough sentences from T have appeared that enable it to output φ with absolute confidence. If no such finite set of sentences exists, then T does not logically imply φ in $\mathcal{P}$, and f never outputs any hypothesis. This observation shows that we have the following property.

Remark 3.18 For all sentences φ, the following are equivalent.

- For all possible knowledge bases T that logically imply φ in $\mathcal{P}$, φ is a deductive consequence of T in $\mathcal{P}$.
- $\mathcal{B}$ is positively classifiable with no mind change following φ.

Similarly, it is easily verified that inductive consequences in $\mathcal{P}$ enjoy the following characterization.

Remark 3.19 For all sentences φ, the following are equivalent.

- For all possible knowledge bases T that logically imply φ in $\mathcal{P}$, φ is an inductive consequence of T in $\mathcal{P}$.
- $\mathcal{B}$ is positively classifiable with at most one mind change following φ.

3.12 Finite Telltales

We have observed in Section 3.10 that positive classification in the limit in $\mathcal{P}$ goes beyond deductive and inductive consequences in $\mathcal{P}$. Thanks to Properties 3.18 and 3.19, we know exactly by how much. To continue our investigation of the relationship between formal learning theory and parametric logic, we turn to the characterization, in the numerical setting, of the notion of identification in the limit from texts. It is know as the *finite telltale* condition, and is expressed as follows [3]:

Proposition 3.20 *A (countable) set $\mathfrak{L}$ of languages is identifiable in the limit from texts iff there exists a sequence $(E_L)_{L \in \mathfrak{L}}$ of finite subsets of $\mathbb{N}$ such that for all $L \in \mathfrak{L}$, $E_L \subseteq L$ and no $L' \in \mathfrak{L}$ satisfies $E_L \subseteq L' \subset L$.*

It is easy to generalize Proposition 3.20 to positive classification in the limit in $\mathcal{P}$, under the assumption that the set of possible knowledge bases is countable (the counterpart to the hypothesis that the set of possible languages is countable), and obtain the following result.

Proposition 3.21 *Suppose that $\mathcal{B}$ is countable. Let a sentence φ be given. The following are equivalent.*

- *$\mathcal{B}$ is positively classifiable in the limit following φ.*
- *For all possible knowledge bases T that logically imply φ in $\mathcal{P}$, there exists a finite subset E of T such that for all $T' \in \mathcal{B}$, if $E \subseteq T' \subseteq T$ then T' logically implies φ in $\mathcal{P}$.*

To see that Proposition 3.21 generalizes Proposition 3.20, assume that $\mathcal{V}$ satisfies the conditions stated in Section 3.4, and recall how a language L is mapped to a structure $\mathfrak{M}_L$ that represents L. Let $\mathcal{W}$ be set to $\{\mathfrak{M}_L \mid L \in \mathfrak{L}\}$, and let $\mathcal{D}$ take the value $\{P(\overline{n}) \mid n \in \mathbb{N}\}$. Assume that $\mathcal{A} = \emptyset$. Finally, suppose that for all $L \in \mathfrak{L}$, there exists a sentence φ_L such that for all $L' \in \mathfrak{L}$, $\mathfrak{M}_{L'} \models \varphi_L$ iff $L = L'$. So $\mathcal{P}$ is a logical paradigm that faithfully represents the numerical learning paradigm of identification in the limit where $\mathfrak{L}$ is the set of possible languages. Recall from Property 3.5 that since $\mathcal{W}$ is countable, identifying $\mathcal{W}$ in the limit from texts is equivalent to positively classifying $\mathcal{B}$ in the limit, from texts and following φ_L, for all $L \in \mathfrak{L}$. Together with Proposition 3.21, this shows that $\mathcal{W}$ is identifiable in the limit from texts iff for all $L \in \mathfrak{L}$, there exists a finite subset E_L of $Diag_{\mathcal{D}}(\mathfrak{M}_L)$ such that for all $L' \in \mathfrak{L}$, if $E_L \subseteq Diag_{\mathcal{D}}(\mathfrak{M}_{L'}) \subseteq Diag_{\mathcal{D}}(\mathfrak{M}_L)$, then $Diag_{\mathcal{D}}(\mathfrak{M}_{L'})$ logically implies φ_L in $\mathcal{P}$. We infer immediately that $\mathcal{W}$ is identifiable in the limit from texts iff for all $L \in \mathfrak{L}$, there exists a finite subset E_L of $Diag_{\mathcal{D}}(\mathfrak{M}_L)$ such that no $L' \in \mathfrak{L}$ satisfies $E_L \subseteq Diag_{\mathcal{D}}(\mathfrak{M}_{L'}) \subset Diag_{\mathcal{D}}(\mathfrak{M}_L)$. Obviously, by the choice of $\mathcal{D}$, this is equivalent to the finite telltale condition in Proposition 3.20.

It is worth looking more carefully at the second clause in Proposition 3.21 under the assumption that $\mathcal{A} = \emptyset$. Let T be a possible knowledge base that logically implies φ in $\mathcal{P}$. Let E be a finite subset of T such that for all $T' \in \mathcal{B}$:

$$(*)\ \text{if } E \subseteq T' \subseteq T \text{ then } T' \text{ logically implies } \varphi \text{ in } \mathcal{P}.$$

Condition $(*)$ says that in order to discover that φ holds in every intended model of T, a classifier can proceed in two steps:

- First induce $\bigwedge\{\neg\psi \mid \psi \in \mathcal{D} \setminus T\}$.
- Then 'deduce' φ from E and $\bigwedge\{\neg\psi \mid \psi \in \mathcal{D} \setminus T\}$.

Indeed, $\bigwedge\{\neg\psi \mid \psi \in \mathcal{D} \setminus T\}$ is an inductive consequence of T in $\mathcal{P}$, as defined in Definition 3.12 where we can choose D to be empty: it is a logical consequence of T in $\mathcal{P}$, and if another possible knowledge base T' does not logically imply $\bigwedge\{\neg\psi \mid \psi \in \mathcal{D} \setminus T\}$ in $\mathcal{P}$, then T' has to logically imply in $\mathcal{P}$ some member ψ of $\mathcal{D} \setminus T$, and ψ is obviously a deductive consequence of T' in $\mathcal{P}$ since it has to belong to T'. Moreover, since $\mathcal{A} = \emptyset$, any member

T' of $\mathcal{B}$ logically implies $\wedge\{\neg\psi \mid \psi \in \mathcal{D} \setminus T\}$ in $\mathcal{P}$ iff it is included in T. It then follows from condition $(*)$ that if a possible knowledge base T' contains E and logically implies $\wedge\{\neg\psi \mid \psi \in \mathcal{D} \setminus T\}$ in $\mathcal{P}$, then we can conclusively infer that φ is a logical consequence of T in $\mathcal{P}$. Admittedly, φ is probably not a deductive consequence of T in $\mathcal{P}$, as defined in Definition 3.10: there is no reason why T would contain a finite subset D with the property that φ is true in all intended models of any possible knowledge base that contains D. But φ is obtained by a *second level of deduction* in $\mathcal{P}$ where the sentences involved are both members of the underlying theory T (the members of E) and sentences generated from T by induction (only one such sentence actually, namely, $\wedge\{\neg\psi \mid \psi \in \mathcal{D} \setminus T\}$): it is possible to conclusively infer that φ is true from the *finite* set $E \cup \{\wedge\{\neg\psi \mid \psi \in \mathcal{D} \setminus T\}\}$ of formulas all of whose members belong either to level 0—the members of the subset E of T—or to level 1—the inductive consequence $\wedge\{\neg\psi \mid \psi \in \mathcal{D} \setminus T\}$ of T in $\mathcal{P}$. This suggests:

positive classification = induction step, followed by deduction.

The meaning of "deducing" here is not captured by Definition 3.10. It is still informal and refers to what we claim is the hallmark of deduction: the ability to conclude with certainty on the basis of a finite set of sentences. Finally, let us point out that the hypotheses that $\mathcal{A} = \emptyset$ and $\mathcal{B}$ is countable are by no means essential in the previous argument; we just used them for simplicity. With no assumption on $\mathcal{W}$, $\mathcal{D}$ and $\mathcal{A}$, it is still possible to develop the same kind of reasoning (despite the fact that Proposition 3.21 cannot be used since $\mathcal{B}$ might be uncountable) and argue that a positive classification in the limit can be decomposed as an induction followed by a deduction. We will get back to this point later.

3.13 Languages

There is a slight problem with the argument developed in the previous section. We said that is was possible to induce $\wedge\{\neg\psi \mid \psi \in \mathcal{D} \setminus T\}$. But $\wedge\{\neg\psi \mid \psi \in \mathcal{D} \setminus T\}$ is not a first-order sentence. In some cases, there might be a first-order sentence φ to which $\wedge\{\neg\psi \mid \psi \in \mathcal{D} \setminus T\}$ is logically equivalent, or at least, such that for all possible knowledge bases T', T' logically implies φ in $\mathcal{P}$ iff T' logically implies $\wedge\{\neg\psi \mid \psi \in \mathcal{D} \setminus T\}$ in $\mathcal{P}$. This is not true in general, and is one of the many reasons why we have to include in $\mathcal{P}$ a last parameter: a (possible) *language*.

Denote by $\mathcal{L}^{\mathcal{V}}_{\omega\omega}$ the set of first-order formulas, i.e., expressions built from $\mathcal{V}$ and a countable set of first-order variables using negation, disjunction and conjunction of (possibly empty) finite sets of formulas, and quantification. So for all finite $D \subseteq \mathcal{L}^{\mathcal{V}}_{\omega\omega}$, the disjunction of all members of D, written $\vee D$, and the conjunction of all members of D, written $\wedge D$, both belong to $\mathcal{L}^{\mathcal{V}}_{\omega\omega}$. Denote by $\mathcal{L}^{\mathcal{V}}_{\omega_1\omega}$ the extension of $\mathcal{L}^{\mathcal{V}}_{\omega\omega}$ that accepts conjunctions and disjunctions of countable sets of formulas. So for all countable $X \subseteq \mathcal{L}^{\mathcal{V}}_{\omega_1\omega}$, the disjunction of all members of X, written $\vee X$, and the conjunction of all members of X, written $\wedge X$, both belong to $\mathcal{L}^{\mathcal{V}}_{\omega_1\omega}$.[34] It turns out that the right choice of parameter for languages is the set of closed members of a *countable fragment* of $\mathcal{L}^{\mathcal{V}}_{\omega_1\omega}$. Note that $\mathcal{L}^{\mathcal{V}}_{\omega_1\omega}$ is uncountable. A fragment of $\mathcal{L}^{\mathcal{V}}_{\omega_1\omega}$ is a subset L of $\mathcal{L}^{\mathcal{V}}_{\omega_1\omega}$ with the following properties.[35]

- All members of $\mathcal{L}^{\mathcal{V}}_{\omega\omega}$ are in L.
- All subformulas of the members of L are in L.
- For all $\varphi \in L$, variables x and terms t, $\varphi[t/x]$ is in L.
- For all $\varphi \in L$, $\neg\varphi$ is in L.
- For all finite $D \subseteq L$, $\vee D$ and $\wedge D$ are in L.
- For all $\varphi \in L$ and variables x, $\exists x\varphi$ and $\forall x\varphi$ are in L.

(In the fourth clause above, $\varphi[t/x]$ denotes a formula obtained from φ by substituting all free occurrences of x in φ by t.) Clearly, $\mathcal{L}^{\mathcal{V}}_{\omega\omega}$ is the smallest fragment of $\mathcal{L}^{\mathcal{V}}_{\omega_1\omega}$. It is easy to verify that given any countable subset X of $\mathcal{L}^{\mathcal{V}}_{\omega_1\omega}$, there exists a smallest fragment of $\mathcal{L}^{\mathcal{V}}_{\omega_1\omega}$ which contains X; it is countable and is called the *fragment of* $\mathcal{L}^{\mathcal{V}}_{\omega_1\omega}$ *generated by* X. A (possible) language of parametric logic is the set of closed members of a countable fragment of $\mathcal{L}^{\mathcal{V}}_{\omega_1\omega}$. We now have all the parameters in hand that make up our logical paradigm $\mathcal{P}$. Let us put them together.

Definition 3.22 A logical paradigm *is a quintuple* $\mathcal{P}$ *consisting of:*

- *a countable vocabulary* $\mathcal{V}$*;*
- *the set* $\mathcal{L}$ *of closed members of a countable fragment of* $\mathcal{L}^{\mathcal{V}}_{\omega_1\omega}$*;*
- *a set of* $\mathcal{V}$*-structures (possible worlds);*
- *a subset* $\mathcal{D}$ *of* $\mathcal{L}$ *(possible data);*
- *a subset* $\mathcal{A}$ *of* $\mathcal{L}$ *disjoint from* $\mathcal{D}$ *(possible assumptions).*

[34] $\mathcal{L}^{\mathcal{V}}_{\omega_1\omega}$ is a more natural extension of $\mathcal{L}^{\mathcal{V}}_{\omega\omega}$ when disjunction and conjunction in $\mathcal{L}^{\mathcal{V}}_{\omega\omega}$ are defined as unary operators whose arguments are finite sets, rather than being defined as operators taking two statements as arguments. It also simplifies many definitions and proofs.

[35] It is sometimes assumed that a fragment of $\mathcal{L}^{\mathcal{V}}_{\omega_1\omega}$ is also closed under a syntactic operator $\sim$ [5].

Retrospectively as well as from now on, "sentence" means "member of $\mathcal{L}$." For instance, Definitions 3.4 and 3.6 apply to the members of $\mathcal{L}$, etc. Let a possible knowledge base T be given. We denote by $Cn_{\mathcal{W}}^{\mathcal{D}}(T)$ the set of sentences (members of $\mathcal{L}$) that are logical consequences of T in $\mathcal{P}$ (as defined in Definition 3.7).

4 LOGICAL AND TOPOLOGICAL HIERARCHIES

4.1 Introduction to the Logical Hierarchies

We have proposed model-theoretic counterparts to the notion of positive classification with fewer than 2 mind changes, and a hint on how to define a model-theoretic counterpart to the notion of positive classification in the limit (an induction followed by a deduction). We now want to convert the hint into a formal definition, and find a model-theoretic counterpart to the notion of positive classification with fewer than β mind changes when β is an ordinal greater than 2. Remember that deduction in $\mathcal{P}$ has been characterized by a compactness property, whereas induction in $\mathcal{P}$ has been characterized by a property of weak compactness. It is actually possible to generalize and unify the compactness and weak compactness properties by defining a property of β-weak compactness with compactness being 0-weak compactness, and weak compactness being 1-weak compactness. Weaker and weaker forms of compactness are obtained with larger and larger values of β. The property of β-weak compactness will provide a model-theoretic counterpart to the notion of positive classification with fewer than β mind changes.

Let a possible knowledge base T be given. Consider the set X_1 of sentences obtained from T by β-weak compactness, for all ordinals β. It is then possible to apply the property of β-weak compactness from T *and* X_1 instead of T alone, and generate more sentences. The sentences obtained from T and X_1 by (0-weak) compactness will provide a model-theoretic counterpart to the notion of positive classification in the limit. More generally, for all ordinals α greater than 1, we can generate a set X_α of sentences by applying the property of β-weak compactness from T *and* $\bigcup_{0<\gamma<\alpha} X_\gamma$. This process will generate a hierarchy of sentences of the following kind. The hierarchy has levels indexed by a nonnull ordinal α, above T that sits on level 0. For every nonnull ordinal α, sentences that occur on level α of the hierarchy over T are further stratified into layers indexed by a second ordinal β. The sentences that occur on layer β of the first level will be obtained by β-weak compactness from sentences on level 0 (all members of T). The sentences that occur on layer β of the second level will be obtained by β-weak compactness from sentences on level 0 or on level 1. The sentences that occur on layer β of the third level will be obtained by β-weak compactness from sentences on level 0 or on level 1 or on level 2. Etc. A sentence that occurs on layer β of level α (where α, β are ordinals with $\alpha \neq 0$) of the hierarchy over T also occurs on layer β' of level α' of that hierarchy if $\alpha < \alpha'$, or if $\alpha = \alpha'$ and $\beta \leq \beta'$. It is actually possible to consider countable ordinals only, since there exists a least nonnull countable ordinal λ such that all sentences that belong to the hierarchy over T occur in that hierarchy below level λ, and below layer λ of any level that contains that sentence. The higher the first pair (level, layer) that indexes a sentence φ occurring in the hierarchy built over T, the less confidence we should place in a derivation of φ from T. Ideally, all members of $Cn^{\mathcal{D}}_{\mathcal{W}}(T)$ would occur in the hierarchy built over T but for some paradigms (for some values of the parameters in $\mathcal{P}$) and some possible knowledge bases T, some members of $Cn^{\mathcal{D}}_{\mathcal{W}}(T)$ might be left out of the hierarchy built over T.

4.2 Definition of the Logical Hierarchies

Let us now describe the hierarchies more precisely. Let a nonnull ordinal α and an ordinal β be given. Suppose that the hierarchies over all possible knowledge bases have been built up to below layer β of level α. Suppose that whenever a sentence ψ is a logical consequence in $\mathcal{P}$ of a possible knowledge base T' (T' can be equal to T or it can be distinct from T) that occurs below layer β of level α of the hierarchy built over T', we can discover that ψ is indeed a logical consequence of T' in $\mathcal{P}$. Consider a sentence φ.

(†) Let finite $K, H \subseteq \mathcal{L}$ be such that $K \subseteq T$ and all members of H occur in the hierarchy over T below level α. Assume that for every possible knowledge base T', if $K \subseteq T'$, $H \subseteq Cn^{\mathcal{D}}_{\mathcal{W}}(T')$ and $\varphi \notin Cn^{\mathcal{D}}_{\mathcal{W}}(T')$, then $\neg\varphi$ occurs in the hierarchy over T' below layer β of level α. In that case we can be sure, on the basis of K and H, that if φ were not a member of $Cn^{\mathcal{D}}_{\mathcal{W}}(T)$, then we could discover it. And we put φ on layer β of level α of the hierarchy over T.

In case $\beta = 0$, (†) can be rewritten as follows.

(‡) Let finite $K, H \subseteq \mathcal{L}$ be such that $K \subseteq T$ and all members of H occur in the hierarchy over T below level α. Assume that for every possible knowledge base T', if $K \subseteq T'$ and $H \subseteq Cn^{\mathcal{D}}_{\mathcal{W}}(T')$, then $\varphi \in Cn^{\mathcal{D}}_{\mathcal{W}}(T')$. In that case we can be sure, on the basis of K and H, that φ belongs to $Cn^{\mathcal{D}}_{\mathcal{W}}(T)$. And we put φ on layer 0 of level α of the hierarchy over T.

We can view (‡) as a compactness property: a finite amount of information—data and axioms (subset of the underlying theory T) and hypotheses (sentences that have been 'discovered' before to belong to $Cn^{\mathcal{D}}_{\mathcal{W}}(T)$)—enables us to conclude that φ is a logical consequence of T in $\mathcal{P}$. More generally, we can view (†) as the description of a β-weak compactness property: a finite amount of information—of the same kind as before—enables us to conclude that it is easier to discover that φ fails to be a logical consequence of T in $\mathcal{P}$, than it is to discover that φ is a logical consequence of T in $\mathcal{P}$. Intuitively, the property of β-weak compactness allows to believe in the correctness of φ with a degree of disbelief measured by β. The case $\beta = 0$ means no disbelief, or total confidence, in accordance with the fact that the compactness property allows to conclude with certainty that φ is correct. Of course, "certainty" or "disbelief" are not absolute notions, they are relative to that part of the hierarchy which lies below the pair (level, layer) which is currently built. Only the sentences on layer 0 of level 1 of the hierarchy over T can be discovered to be logical consequences of T in $\mathcal{P}$ with absolute certainty.

Given ordinals $\alpha, \beta, \alpha', \beta'$, we write $(\alpha, \beta) < (\alpha', \beta')$ iff (α, β) is lexicographically smaller than (α', β'): $\alpha < \alpha'$, or $\alpha = \alpha'$ and $\beta < \beta'$. Given a nonnull ordinal α, an ordinal β and a possible knowledge base T, denote by $\Lambda^{\mathcal{P}}_{\alpha,\beta}(T)$ the set of all sentences that occur on layer β of level α of the hierarchy built over T. In accordance with the description above, a sentence φ belongs to $\Lambda^{\mathcal{P}}_{\alpha,\beta}(T)$ iff it belongs to T or there exists a finite $K \subseteq T$ and a finite $H \subseteq \bigcup_{(\alpha',\beta')<(\alpha,0)} \Lambda^{\mathcal{P}}_{\alpha',\beta'}(T)$ such that *φ belongs to $\Lambda^{\mathcal{P}}_{\alpha,\beta}(T)$ on the basis of K and H*, meaning that:

(†) for all $T' \in \mathcal{B}$, if $K \subseteq T'$, $H \subseteq Cn^{\mathcal{D}}_{\mathcal{W}}(T')$ and $\varphi \notin Cn^{\mathcal{D}}_{\mathcal{W}}(T')$, then $\neg\varphi \in \bigcup_{\gamma<\beta} \Lambda^{\mathcal{P}}_{\alpha,\gamma}(T')$.

In case $\beta = 0$, (†) can be rewritten as follows:

(‡) for all $T' \in \mathcal{B}$, if $K \subseteq T'$ and $H \subseteq Cn^{\mathcal{D}}_{\mathcal{W}}(T')$, then $\varphi \in Cn^{\mathcal{D}}_{\mathcal{W}}(T')$.

Condition (‡) expresses that it is possible to infer conclusively, on the basis of K and H, that φ is a logical consequence of T in $\mathcal{P}$; such an inference can be viewed as a *deduction, at level α*. For $\alpha = 1$, (‡) is equivalent to the condition in Definition 3.10, which shows that $\Lambda^{\mathcal{P}}_{1,0}(T)$ consists precisely of the deductive consequences of T in $\mathcal{P}$. For $\alpha = 1$ and $\beta = 1$, (†) is equivalent to the condition in Definition 3.12, which shows that $\Lambda^{\mathcal{P}}_{1,1}(T)$ consists precisely of the inductive consequences of T in $\mathcal{P}$.

4.3 Logical Complexity

We can now introduce a model-theoretic notion of logical complexity. Let a sentence φ and nonnull ordinals α, β be such that φ occurs below layer β of level α of the hierarchy built over T, for every possible knowledge base T that logically implies φ in $\mathcal{P}$. Then (α, β) is a natural upper bound on the logical complexity of φ in $\mathcal{P}$. If (α, β) is the least pair of ordinals (for the lexicographic ordering on the class of pairs of nonnull ordinals) having that property, then (α, β) can be thought of as the logical complexity of φ in $\mathcal{P}$. For a reason similar to what has been discussed in relation to positive classification with a bounded number of mind changes, the condition "*below* layer β of level α" is preferable to the alternative condition "*on* layer β of level α" for the case where β is a limit ordinal. If a sentence φ does not occur in the hierarchy built over T for some possible knowledge base T that logically implies φ in $\mathcal{P}$, then we consider that the logical complexity of φ in $\mathcal{P}$ is undefined. In summary:

Definition 4.1 *Let a sentence φ and a nonnull ordinal α be given.*

For all nonnull ordinals β, we say that φ is $\Sigma^{\mathcal{P}}_{\alpha,\beta}$ iff for all possible knowledge bases T that logically imply φ in $\mathcal{P}$, $\varphi \in \bigcup_{\gamma<\beta} \Lambda^{\mathcal{P}}_{\alpha,\gamma}(T)$.

We say that φ is $\Sigma^{\mathcal{P}}_{\alpha}$ if it is $\Sigma^{\mathcal{P}}_{\alpha,1}$.

We will see shortly that there are good reasons for adopting the notation $\Sigma^{\mathcal{P}}_{\alpha,\beta}$. Remember that first-order logic is the particular paradigm where $\mathcal{L} \subseteq \mathcal{L}^{\mathcal{V}}_{\omega\omega}$ (equivalently, $\mathcal{L}$ is the set of closed members of $\mathcal{L}^{\mathcal{V}}_{\omega\omega}$), $\mathcal{W}$ is the class of all structures, $\mathcal{D} = \emptyset$, and $\mathcal{A} = \mathcal{L}$. It is immediately verified that first-order logic, or more generally, any instance of the weak generalization of first-order logic where $\mathcal{A}$ can be chosen arbitrarily, is a limiting case in the sense that all hierarchies of logical consequences collapse to the first layer of the first level. This is due to the compactness of first-order logic, and accounts to the fact that first-order logic is a purely deductive paradigm:

Remark 4.2 If $\mathcal{L} \subseteq \mathcal{L}^{\mathcal{V}}_{\omega\omega}$, $\mathcal{W}$ is the class of all structures and $\mathcal{D} = \emptyset$, then every sentence is $\Sigma^{\mathcal{P}}_{1}$.

Let us get back to the general case. It is possible to show that positive classification with a bounded number of mind changes is actually equivalent to the model-theoretic notion of logical complexity formalized in Definition 4.1, applied to $\alpha = 1$:

Proposition 4.3 *For all nonnull ordinals β and $\varphi \in \mathcal{L}$, φ is $\Sigma^{\mathcal{P}}_{1,\beta}$ iff $\mathcal{B}$ is positively classifiable with fewer than β mind changes following φ.*

Note that Properties 3.18 and 3.19 are the particular cases of Proposition 4.3 where $\alpha = 1$ and $\beta = 2$, respectively. It is easy to see that for all sentences φ, if φ is $\Sigma_2^{\mathcal{P}}$ then $\mathcal{B}$ is positively classifiable in the limit following φ. The converse holds provided that $\mathcal{L}$ is rich enough, which is captured by the next proposition.

Proposition 4.4 *For all countable subsets L of $\mathcal{L}^{\mathcal{V}}_{\omega_1\omega}$, there exists a countable fragment L' of $\mathcal{L}^{\mathcal{V}}_{\omega_1\omega}$ that extends L with the following property. Suppose that $\mathcal{L}$ is the set of closed members of L'. Then for all sentences φ, φ is $\Sigma_2^{\mathcal{P}}$ iff $\mathcal{B}$ is positively classifiable in the limit following φ.*

Proposition 4.4 still does not enable us to formally justify the equation:

positive classification = induction step, followed by deduction.

We will do this after we have introduced the notion of topological complexity, and stated a result stronger than Proposition 4.4.

4.4 Borel Hierarchies

The study of the hierarchies of logical consequences in $\mathcal{P}$ is greatly facilitated by their close relationship to the Borel hierarchy and the difference hierarchies over the topological space that we define next. Remember that a possible datum φ represents some elementary information, or elementary fact, about the underlying world. Hence the models in $\mathcal{W}$ of a possible datum represent an elementary subset of $\mathcal{W}$. It is then natural to consider the topology generated by the set X of these elementary subsets of $\mathcal{W}$, i.e., the topology whose open sets are the countable unions of the finite intersections of the members of X.

Definition 4.5 *We denote by $\mathbb{W}$ the topological space over $\mathcal{W}$ generated by* $\{\mathrm{Mod}_{\mathcal{W}}(\varphi) \mid \varphi \in \mathcal{D}\}$.

Recall that the Cantor space is the set $\{0, 1\}^{\mathbb{N}}$ of all ω-sequences of 0's and 1's. The usual topology over the Cantor space is the topology generated by the sets of members of $\{0, 1\}^{\mathbb{N}}$ of the form X_σ, where σ is a finite sequence of 0's and 1's and X_σ is the set of all ω-sequences of 0's and 1's that extend σ.[36] Then $\mathbb{W}$ is a generalization of the usual topology over the Cantor space. Indeed, suppose that $\mathcal{V}$ is equality free, and that there are infinitely many closed atoms. Fix a repetition free enumeration $(\varphi)_{i\in\mathbb{N}}$ of the

[36] More precisely, the sets X_σ, where σ ranges over the set of finite sequence of 0's and 1's, constitute a basis for this topology: an open set for this topology is a countable union of sets of the form X_σ.

set of closed atoms. Assume that $\mathcal{W}$ is the set of all standard structures and $\mathcal{D}$ is the set of all closed *literals*. Let $f : \{0,1\}^{\mathbb{N}} \rightarrow \mathcal{W}$ map a member $\vec{c}$ of $\{0,1\}^{\mathbb{N}}$ to the unique member $\mathfrak{M}$ of $\mathcal{W}$ such that for all $i \in \mathbb{N}$, $\mathfrak{M} \models \varphi_i$ iff $\vec{c}(i) = 1$. Obviously, f is a homeomorphism from $\mathbb{W}$ into $\{0,1\}^{\mathbb{N}}$. For instance, the set of all $\vec{c} \in \{0,1\}^{\mathbb{N}}$ such that $\vec{c}(0) = 0$, $\vec{c}(1) = 0$ and $\vec{c}(2) = 1$ (that is, the set of members of $\{0,1\}^{\mathbb{N}}$ that extend $(0,0,1)$, represented by the leftmost crisscrossed triangle in Figure 3.7), the set of all $\vec{c} \in \{0,1\}^{\mathbb{N}}$ such that $\vec{c}(0) = 0$ and $\vec{c}(1) = 1$ (that is, the set of members of $\{0,1\}^{\mathbb{N}}$ that extend $(0,1)$, represented by the other crisscrossed triangle in Figure 3.7), and the set of all $\vec{c} \in \{0,1\}^{\mathbb{N}}$ such that $\vec{c}(0) = 1$, $\vec{c}(1) = 1$ and $\vec{c}(2) = 1$ (that is, the set of members of $\{0,1\}^{\mathbb{N}}$ that extend $(1,1,1)$, represented by the shaded triangle in Figure 3.7), are all open sets of $\{0,1\}^{\mathbb{N}}$ that correspond, respectively, to the three open sets of $\mathbb{W}$ described as:

- $O_1 = \{\mathfrak{M} \in \mathcal{W} \mid \mathfrak{M} \not\models \varphi_0,\ \mathfrak{M} \not\models \varphi_1 \text{ and } \mathfrak{M} \models \varphi_2\}$,
- $O_2 = \{\mathfrak{M} \in \mathcal{W} \mid \mathfrak{M} \not\models \varphi_0 \text{ and } \mathfrak{M} \models \varphi_1\}$, and
- $O_3 = \{\mathfrak{M} \in \mathcal{W} \mid \mathfrak{M} \models \varphi_0,\ \mathfrak{M} \models \varphi_1 \text{ and } \mathfrak{M} \models \varphi_2\}$.

But if $\mathcal{D}$ is the set of all closed *atoms*, then only O_3 (corresponding to the shaded triangle in Figure 3.7), is an open set of $\mathbb{W}$; neither O_1 nor O_2 is open in $\mathbb{W}$. When $\mathcal{D}$ is neither the set of closed literals nor the set of closed atoms, then the relationship between $\mathbb{W}$ and the Cantor space can be much more complicated. The definition of $\mathbb{W}$ is quite general and makes no assumption on $\mathcal{V}$, $\mathcal{W}$ or $\mathcal{D}$.

The Borel hierarchy over $\mathbb{W}$ provides a notion of topological complexity for some subsets of $\mathcal{W}$. It is built as follows. We call the sets built from $\{Mod_{\mathcal{W}}(\varphi) \mid \varphi \in \mathcal{D}\}$ by finite unions and finite intersections the Π_0 Borel sets of $\mathbb{W}$. The sets built from the complements in $\mathcal{W}$ of the members of $\{Mod_{\mathcal{W}}(\varphi) \mid \varphi \in \mathcal{D}\}$ by finite unions and finite intersections are called the Σ_0 Borel sets of $\mathbb{W}$. The Σ_α and Π_α Borel sets of $\mathbb{W}$, where α is a nonnull

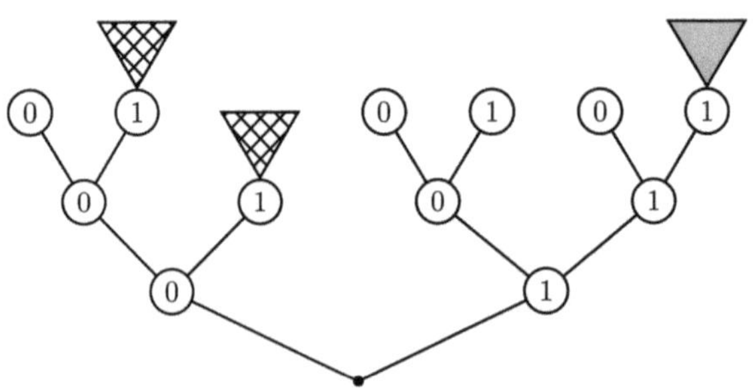

Figure 3.7. Some open sets

ordinal, are defined inductively as follows. A subset of $\mathcal{W}$ is Σ_α Borel in $\mathbb{W}$ iff it is built from the Π_β Borel subsets of $\mathcal{W}$, where β ranges over the class of ordinals smaller than α, by countable unions. A subset of $\mathcal{W}$ is Π_α Borel in $\mathbb{W}$ iff it is built from the Σ_β Borel subsets of $\mathcal{W}$, where β ranges over the class of the ordinals smaller than α, by countable intersections. Given an ordinal α, a subset of $\mathcal{W}$ is said to be Δ_α Borel in $\mathbb{W}$ iff it is both Σ_α and Π_α Borel in $\mathbb{W}$. Note that if a subset Z of $\mathcal{W}$ is Σ_0 Borel in $\mathbb{W}$ then Z is not necessarily Σ_1 Borel in $\mathbb{W}$, and if Z is Π_0 Borel in $\mathbb{W}$ then Z is not necessarily Π_1 Borel in $\mathbb{W}$. When $\mathcal{D}$ is semantically closed under negation, the Π_0 Borel sets of $\mathbb{W}$ are the Σ_0 Borel sets of $\mathbb{W}$, the Σ_0 Borel sets of $\mathbb{W}$ are Σ_1 Borel in $\mathbb{W}$, and the Π_0 Borel sets of $\mathbb{W}$ are Π_1 Borel in $\mathbb{W}$.

4.5 Difference Hierarchies

A refinement of the Borel hierarchy is given by the difference hierarchies. More precisely, every nonnull ordinal α determines a difference hierarchy, built from the class of Σ_α or Π_α Borel sets of $\mathbb{W}$, that consists of sets that are all $\Delta_{\alpha+1}$ Borel in $\mathbb{W}$.[37] Here is one way to define the difference hierarchies. Let a nonnull ordinal α and a subset Z of $\mathcal{W}$ be given. We say that Z is $\Sigma_{\alpha,1}$ Borel in $\mathbb{W}$ iff Z is Σ_α Borel in $\mathbb{W}$. We say that Z is $\Pi_{\alpha,1}$ Borel in $\mathbb{W}$ iff Z is Π_α Borel in $\mathbb{W}$. Let an ordinal β greater than 1 be given. We say that Z is $\Sigma_{\alpha,\beta}$ Borel in $\mathbb{W}$ iff there exists two families $(A_i)_{i\in\mathbb{N}}$ and $(Z_i)_{i\in\mathbb{N}}$ of subsets of X and a family $(\beta_i)_{i\in\mathbb{N}}$ of nonnull ordinals smaller than β such that the following holds.

(i) For all $i \in \mathbb{N}$, A_i is Σ_α Borel in $\mathbb{W}$ and Z_i is Π_{α,β_i} Borel in $\mathbb{W}$.
(ii) For all $i, j \in \mathbb{N}$ and $x \in A_i \cap A_j$, $x \in Z_i$ iff $x \in Z_j$.
(iii) $Z = \bigcup_{i\in\mathbb{N}}(A_i \cap Z_i)$

We say that Z is $\Pi_{\alpha,\beta}$ Borel in $\mathbb{W}$ iff there exists two families $(A_i)_{i\in\mathbb{N}}$ and $(Z_i)_{i\in\mathbb{N}}$ of subsets of X and a family $(\beta_i)_{i\in\mathbb{N}}$ of nonnull ordinals smaller than β such that the following holds.

(i) For all $i \in \mathbb{N}$, A_i is Π_α Borel in $\mathbb{W}$ and Z_i is Σ_{α,β_i} Borel in $\mathbb{W}$.
(ii) For all $i, j \in \mathbb{N}$ and $x \in A_i \cap A_j$, $x \in Z_i$ iff $x \in Z_j$.
(iii) $Z = \bigcap_{i\in\mathbb{N}}(A_i \cup Z_i)$.

[37] Since $\mathbb{W}$ is a Polish space—a separable completely metrizable space, see [20]—the class of sets that belong to the difference hierarchy built from the class of Σ_α or Π_α Borel sets of $\mathbb{W}$ is precisely equal to the class of sets that are $\Delta_{\alpha+1}$ Borel in $\mathbb{W}$.

Intuitively, when building level β for a nonnull ordinal β, the difference hierarchies are defined like the Borel hierarchies, but consider families of $[\![\Pi_{\alpha,\gamma} \mid \Sigma_{\alpha,\gamma}]\!]$, $\gamma < \beta$, sets that are 'quasi-separated' by $[\![\Sigma_\alpha \mid \Pi_\alpha]\!]$ sets instead of arbitrary families of $[\![\Pi_\gamma \mid \Sigma_\gamma]\!]$, $\gamma < \beta$, sets. Indeed, both conditions (ii) above are trivially satisfied when $(A_i)_{i\in\mathbb{N}}$ is a family of pairwise disjoint sets: for instance, if $(Z_i)_{i\in\mathbb{N}}$ is a family of Π_α Borel sets of $\mathbb{W}$ and if $(A_i)_{i\in\mathbb{N}}$ is a family of pairwise disjoint Σ_α Borel sets of $\mathbb{W}$ such that for all $i \in \mathbb{N}$, $Z_i \subseteq A_i$, then $\bigcup_{i\in\mathbb{N}} Z_i$ is not only $\Sigma_{\alpha+1}$ Borel in $\mathbb{W}$, but also $\Sigma_{\alpha,2}$ Borel in $\mathbb{W}$. Conditions (ii) above generalize disjointness to a condition of 'quasi-disjointness.' The definition of the difference hierarchies we have presented is well suited to our purposes, but it is not the standard one [20]. The next proposition expresses the equivalence between the standard definition and the notion we take as primitive. Recall that the parity of an ordinal α is either even or odd: it is is even if α is of the form $\lambda + 2n$ where λ is either 0 or a limit ordinal and n is a finite ordinal; it is odd otherwise.

Proposition 4.6 *For all nonnull countable ordinals α, β and subsets Z of $\mathcal{W}$, Z is $\Sigma_{\alpha,\beta}$ Borel in $\mathbb{W}$ iff there exists a $\subseteq$-increasing sequence $(Y_\gamma)_{\gamma<\beta}$ of Σ_α Borel subsets of $\mathcal{W}$ with the following property. For all $x \in X$, $x \in Z$ iff $x \in \bigcup_{\gamma<\beta} Y_\gamma$ and the parity of the least $\gamma < \beta$ such that $x \in Y_\gamma$ is opposite to the parity of β.*

The characterization offered by Proposition 4.6 is illustrated in Figure 3.8, which depicts a $\Sigma_{\alpha,4}$ and a $\Sigma_{\alpha,5}$ Borel set, that are both members of the difference hierarchy built from the set of subsets of $\mathcal{W}$ that are Σ_α or Π_α Borel in $\mathbb{W}$. The former is represented as the crosshatched part of the diagram on the left, that shows a $\subseteq$-increasing sequence of Σ_α Borel subsets of $\mathcal{W}$ of length 4; the latter is represented as the crosshatched part of the diagram on the right, that shows a $\subseteq$-increasing sequence of Σ_α Borel subsets of $\mathcal{W}$ of length 5.

Figure 3.8. $\Sigma_{\alpha,4}$ and $\Sigma_{\alpha,5}$ Borel sets

4.6 Topological Complexity in Parametric Logic

Now that we have the topological background, we can define the topological complexity of a sentence φ, in terms of the Borel subset Z of $\mathcal{W}$ that φ represents, in case there is indeed such a set Z. Given an ordinal α, a sentence φ certainly represents a Σ_α Borel set A of $\mathbb{W}$ if the set of models of φ in $\mathcal{W}$ is equal to A. But a stronger, hereditary, notion of representation is needed. Assume that $\alpha > 0$. The set A is defined as the union of a countable set X of subsets of $\mathcal{W}$ each of which is Π_β Borel for some ordinal $\beta < \alpha$. For at least one such set X, it should be possible to represent each member of X by some sentence. And so on. Alternatively, the following definition of a sentence being $\Sigma^{\mathcal{L}}_\alpha$ or $\Pi^{\mathcal{L}}_\alpha$ Borel in $\mathbb{W}$ parallels the definition of a set being Σ_α or Π_α Borel in $\mathbb{W}$, with countable $\llbracket$ disjunctions | conjunctions $\rrbracket$ replacing countable $\llbracket$ unions | intersections $\rrbracket$. But such countable disjunctions or conjunctions might not belong to $\mathcal{L}$, in which case $\mathcal{L}$ should contain at least one sentence having the same class of models in $\mathcal{W}$ as the countable disjunction or conjunction being considered:

Definition 4.7 *Let a sentence φ be given.*

We say that φ is $\llbracket$ $\Sigma^{\mathcal{L}}_0$ | $\Pi^{\mathcal{L}}_0$ $\rrbracket$ Borel in $\mathbb{W}$ *iff $Mod_{\mathcal{W}}(\varphi)$ is* $\llbracket$ Σ_0 | Π_0 $\rrbracket$ *Borel in* $\mathbb{W}$.

Let a nonnull ordinal α be given. We say that φ is $\llbracket$ $\Sigma^{\mathcal{L}}_\alpha$ | $\Pi^{\mathcal{L}}_\alpha$ $\rrbracket$ Borel in $\mathbb{W}$ *iff there exists a set X of sentences each of which is* $\llbracket$ $\Pi^{\mathcal{L}}_\beta$ | $\Sigma^{\mathcal{L}}_\beta$ $\rrbracket$ *Borel in* $\mathbb{W}$ *for some $\beta < \alpha$ such that* $\mathrm{Mod}_{\mathcal{W}}(\varphi)$ *is equal to*

$$\llbracket\ \mathrm{Mod}_{\mathcal{W}}(\vee X) \mid \mathrm{Mod}_{\mathcal{W}}(\bigwedge X)\ \rrbracket.$$

The previous definition is immediately generalized to a notion of hereditary representation of a $\Sigma_{\alpha,\beta}$ Borel set of $\mathbb{W}$, where α and β are nonnull ordinals.

Definition 4.8 *Let a nonnull ordinal α and a sentence φ be given.*

We say that φ is $\llbracket$ $\Sigma^{\mathcal{L}}_{\alpha,1}$ | $\Pi^{\mathcal{L}}_{\alpha,1}$ $\rrbracket$ Borel in $\mathbb{W}$ iff *φ is* $\llbracket$ $\Sigma^{\mathcal{L}}_\alpha$ | $\Pi^{\mathcal{L}}_\alpha$ $\rrbracket$ *Borel in* $\mathbb{W}$.

Given an ordinal β greater than 1, we say that φ is $\llbracket$ $\Sigma^{\mathcal{L}}_{\alpha,\beta}$ | $\Pi^{\mathcal{L}}_{\alpha,\beta}$ $\rrbracket$ *Borel in* $\mathbb{W}$ iff there exists two sequences $(\psi_i)_{i\in\mathbb{N}}$ and $(\varphi_i)_{i\in\mathbb{N}}$ of sentences and a family $(\beta_i)_{i\in\mathbb{N}}$ of nonnull ordinals smaller than β such that the following holds.

- For all $i \in \mathbb{N}$, ψ_i is $\llbracket$ $\Sigma^{\mathcal{L}}_\alpha$ | $\Pi^{\mathcal{L}}_\alpha$ $\rrbracket$ Borel in $\mathbb{W}$.
- For all $i \in \mathbb{N}$, φ_i is $\llbracket$ $\Pi^{\mathcal{L}}_{\alpha,\beta_i}$ | $\Sigma^{\mathcal{L}}_{\alpha,\beta_i}$ $\rrbracket$ Borel in $\mathbb{W}$.

- For all $i, j \in \mathbb{N}$, $Mod_{\mathcal{W}}(\psi_i \wedge \psi_j) \subseteq Mod_{\mathcal{W}}(\varphi_i \leftrightarrow \varphi_j)$.
- $Mod_{\mathcal{W}}(\varphi)$ is equal to

$$\llbracket Mod_{\mathcal{W}}(\vee_{i\in\mathbb{N}}(\psi_i \wedge \varphi_i)) \mid Mod_{\mathcal{W}}(\bigwedge_{i\in\mathbb{N}}(\psi_i \vee \varphi_i)) \rrbracket.$$

4.7 Relationships between Logical and Topological Complexities

In all logical paradigms, the logical complexity of a sentence is a lower bound of its logical complexity.

Proposition 4.9 *For all nonnull ordinals α, β and sentences φ, if φ is $\Sigma^{\mathcal{L}}_{\alpha,\beta}$ Borel in $\mathbb{W}$ then φ is $\Sigma^{\mathcal{P}}_{\alpha,\beta}$.*

For sentences that are *decided in $\mathcal{P}$* by every possible knowledge base, logical and topological complexity are actually equivalent notions.

Proposition 4.10 *Let a sentence φ be such that every possible knowledge base logically implies either φ or $\neg\varphi$ in $\mathcal{P}$. For all nonnull ordinals α, β, φ is $\Sigma^{\mathcal{P}}_{\alpha,\beta}$ iff φ is $\Sigma^{\mathcal{L}}_{\alpha,\beta}$ Borel in $\mathbb{W}$.*

Now we can state the result which expresses that provided the language is rich enough,

positive classification = induction step, followed by deduction.

Indeed, it is immediately verified that every sentence that is $\Pi^{\mathcal{L}}_1$ Borel in $\mathbb{W}$, is $\Sigma^{\mathcal{L}}_{1,2}$ Borel in $\mathbb{W}$. By Proposition 4.9, every sentence χ that is $\Sigma^{\mathcal{L}}_{1,2}$ Borel in $\mathbb{W}$, is an inductive consequence in $\mathcal{P}$ of every possible knowledge base that logically implies χ in $\mathcal{P}$ (which by Property 3.19, is equivalent to $\mathcal{B}$ being positively classifiable with at most one mind change following the sentence φ). The proposition below then expresses that $\mathcal{B}$ is positively classifiable in the limit following φ iff for every possible knowledge base T that logically implies φ in $\mathcal{P}$, φ can be discovered from T by a deductive step from a finite subset of T and a sentence that, being $\Pi^{\mathcal{L}}_1$ Borel in $\mathbb{W}$, can be induced from T in $\mathcal{P}$.

Proposition 4.11 *For all countable subsets L of $\mathfrak{L}^{\mathcal{V}}_{\omega_1\omega}$, there exists a countable fragment L' of $\mathfrak{L}^{\mathcal{V}}_{\omega_1\omega}$ that extends L with the following property. Suppose that $\mathcal{L}$ is the set of closed members of L'. For all $\varphi \in \mathcal{L}$, the following are equivalent.*

- *$\mathcal{B}$ is positively classifiable in the limit following φ.*

- *For all $T \in \mathcal{B}$ that logically imply φ in $\mathcal{P}$, there exists a finite subset K of T and $\chi \in Cn^{\mathcal{D}}_{\mathcal{W}}(T)$ such that χ is $\Pi^{\mathcal{L}}_1$ Borel in $\mathbb{W}$ and for all $T' \in \mathcal{B}$, if $K \subseteq T'$ and $\chi \in Cn^{\mathcal{D}}_{\mathcal{W}}(T')$ then $\varphi \in Cn^{\mathcal{D}}_{\mathcal{W}}(T')$.*

Example 4.12 *Suppose that $\mathcal{V} = \{=, \overline{0}, s, R\}$ where R is a binary predicate symbol. Let $\mathcal{W}$ be the set of Herbrand structures where R is interpreted as a total ordering. Let $\mathcal{D}$ be equal to the set of atomic sentences, and assume that $\mathcal{A} = \emptyset$. Set*

$$\varphi = \exists x \exists y (x \neq y \wedge R(x, y) \wedge \forall z (x = z \vee R(y, z))).$$

So φ expresses that the ordering has a first and a second element. It is easily verified that $\mathcal{B}$ is positively classifiable in the limit following φ, but not with any mind change bound. Let $\mathfrak{M} \in \mathcal{W}$ have a first element, namely $\overline{4}$, and a second element, namely $\overline{2}$. So $Diag_{\mathcal{D}}(\mathfrak{M})$ logically implies φ in $\mathcal{P}$. Note that

$$\chi = \neg R(\overline{0}, \overline{2}) \wedge \neg R(\overline{1}, \overline{2}) \wedge \neg R(\overline{3}, \overline{2}) \wedge \forall x \neg R(s(s(s(s(s(x))))), \overline{2})$$

is $\Pi^{\mathcal{L}}_1$ Borel in $\mathbb{W}$. Set $K = \{R(\overline{4}, \overline{2})\}$. Then $K \subseteq Diag_{\mathcal{D}}(\mathfrak{M})$ and for all $T \in \mathcal{B}$, if $K \subseteq T$ and $\chi \in Cn^{\mathcal{D}}_{\mathcal{W}}(T)$ then $\varphi \in Cn^{\mathcal{D}}_{\mathcal{W}}(T)$: φ can be discovered from $Diag_{\mathcal{D}}(\mathfrak{M})$ by first inducing χ, and then deducing φ from $R(\overline{4}, \overline{2})$ and χ.

Thanks to Propositions 4.4 and 4.11, we know that provided that the language is rich enough, being of logical complexity $\Sigma^{\mathcal{P}}_2$ is equivalent to $\mathcal{B}$ being positively classifiable in the limit following φ, which is equivalent to: the truth of φ can be discovered in two steps, a first step which involves an induction (positive classification with at most one mind change), and a second step which involves a second level of deduction (a conclusive inference on the basis of a finite set of premises consisting of members of the underlying theory plus the sentence induced at the first step). Thanks to Proposition 4.3, we know that being of logical complexity $\Sigma^{\mathcal{P}}_{1,\beta}$ is equivalent to $\mathcal{B}$ being positively classifiable with fewer than βmind changes following φ. Moreover, Propositions 4.9, 4.10 and 4.11 relate logical complexity and learning-theoretic complexity to topological complexity .[38] Corollaries of these results can be obtained that relate:

[38] Relationships between topological complexity and learning complexity have first been investigated in [11].

- φ being $\triangle^{\mathcal{P}}_{\alpha,\beta}$, meaning that both φ and $\neg\varphi$ are $\Sigma^{\mathcal{P}}_{\alpha,\beta}$;
- φ being $\triangle^{\mathcal{L}}_{\alpha,\beta}$ Borel in $\mathbb{W}$, meaning that both φ and $\neg\varphi$ are $\Sigma^{\mathcal{L}}_{\alpha,\beta}$ Borel in $\mathbb{W}$;
- $\mathcal{B}$ being *classifiable* in the limit following φ, with or without a bounded number of mind changes, where the task of a classifier is to discover whether a possible knowledge base logically implies φ *or* $\neg\varphi$ in $\mathcal{P}$.

5 CONCLUSION

The relation "every $\mathcal{D}$-minimal model of T in $\mathcal{W}$ is a model of φ," where $\mathcal{D}$ is a set of sentences and $\mathcal{W}$ a set of structures, is a generalization of the classical notion of logical consequence "every model of T is a model of φ" that expands the scope of the first-order logic (of deduction) to induction and learning. Strong connections are made explicit between a notion of logical complexity, the Borel and difference hierarchies over a topological space simply defined from $\mathcal{W}$ and $\mathcal{D}$. These are explicitly connected to concepts from formal learning theory, and other notions (including a notion of syntactic complexity not discussed in this paper) provide strong evidence that the framework is natural and powerful. In this paper we focused on the model-theoretic aspects of the framework, and stated model-theoretic results that pave the way to a proof theory and a declarative extension of Prolog to the realm of scientific discovery [30].

ACKNOWLEDGMENTS

The authors would like to thank Michèle Friend and Valentina Harizanov for their very helpful comments.

REFERENCES

[1] Ambainis, A., Freivalds, R. and Smith, C. (1999). "Inductive Inference with Procrastination: Back to Definitions", *Fundamenta Informaticae* 40, 1–16.

[2] Ambainis, A., Jain, S. and Sharma, A. (1999). "Ordinal Mind Change Complexity of Language Identification", *Journal of Theoretical Computer Science* 220, 323–343.

[3] Angluin, D. (1980). "Inductive Inference of Formal Languages from Positive Data", *Information and Control* 45, 117–135.

[4] Barwise, J. (1974). "Axioms for Abstract Model Theory", *Annals of Mathematical Logic* 7, 221–265.

[5] Barwise, J. (1975). *Admissible Sets and Structures*, Perspectives in Mathematical Logic, Berlin: Springer-Verlag.

[6] Barwise, J. (ed.) (1977). *Handbook of Mathematical Logic*, Studies in Logic and the Foundations of Mathematics 90, Amsterdam: North-Holland.
[7] Chopra, S. and Martin, E. (2002). "Generalized Logical Consequence: Making Room for Induction in the Logic of Science", *Journal of Philosophical Logic* 31, 245–280.
[8] Corcoran, J. (ed.) (1983). *Logic, Semantics, Metamathematics*, 2nd ed., translations by Woodger, J.H., Indianapolis (Ind.): Hacket.
[9] Detlefsen, M. (ed.) (1992). *Proof, Logic and Formalization*, London: Routledge.
[10] Doets, K. (1994). *From Logic to Logic Programming*, Cambridge (Mass.): MIT Press.
[11] Ershov, Yu.L. (1968–69). "A Hierarchy of Sets I, II, III", *Algebra and Logic*, 7, 47–74; 7,15–47; 9, 34–51.
[12] Feferman, S. (2004), "Tarski's Conceptual Analysis of Semantical Notions," *in Sémantique et Épistémologie,* Benmakhlouf, A. (ed.), Editions Le Fennec, Casablanca [distrib. J. Vrin, Paris], 79–108.
[13] Gabbay, D. and Guenthner, F. (eds.) (1983). *Handbook of Philosophical Logic* 1, Dordrecht: Reidel.
[14] Gallaire, J. and Minker, J. (eds.) (1978). *Logic and Data Bases*, New York: Plenum Press.
[15] Gasarch, W., Pleszkoch, M., Stephan, F. and Velauthapillai, M. (1998). "Classification Using Information", *Annals of Mathematics and Artificial Intelligence*, selected papers from ALT 1994 and AII 1994, 23, 147–168.
[16] Glymour, C. (1985). "Inductive Inference in the Limit", *Erkenntnis* 22, 23–31.
[17] Gold, E.M. (1967). "Language Identification in the Limit", *Information and Control* 10, 447–474.
[18] Isaacson, D. (1992). "Some Considerations of Arithmetical Truth and the ω-rule", in Detlefsen, M. [9], 94–138.
[19] Jain, S., Osherson, D., Royer, J.S. and Sharma, A. (1999). *Systems That Learn: An Introduction to Learning Theory*, 2nd ed., Cambridge (Mass.): MIT Press.
[20] Kechris, A. (1994). "Classical Descriptive Set Theory", *Graduate Texts in Mathematics* 156, New York: Springer-Verlag.
[21] Kelly, K. (1996). *The Logic of Reliable Inquiry*, Oxford: Oxford University Press.
[22] Kraus, S., Lehmann, D. and Magidor, M. (1990). "Nonmonotonic Reasoning, Preferential Models and Cumulative Logics", *Artificial Intelligence* 44, 167–207.
[23] Leblanc, H. (1983). "Alternative to Standard First-Order Semantics", in Gabbay, D. and Guenthner, F. [13], 189–274.
[24] Lifschitz, V. (1985). "Closed-World Databases and Circumscription", *Artificial Intelligence* 27, 229–235.
[25] Lukaszewicz, W. (1990). "Non-Monotonic Reasoning, Formalization of Commonsense Reasoning", *Computational Intelligence* 4, 1–16.
[26] Makkai, M. (1977). "Admissible Sets and Infinitary Logic", in Barwise, J. [6].
[27] Martin, E. and Osherson, D. (1998). *Elements of Scientific Inquiry*, Cambridge (Mass.): MIT Press.
[28] Martin, E., Sharma, A. and Stephan, F. (2001). "A General Theory of Deduction, Induction, and Learning", in *Proceedings of the 4th International Conference on Discovery Science*, Jantke, K.P. and Shinohara, A. (eds.), London: Springer-Verlag, 228–242.
[29] Martin, E., Sharma, A. and Stephan, F. (2006). "Unifying Logic, Topology and Learning in Parametric Logic", *Theoretical Computer Science* 350, 103–124.
[30] Martin, E., Nguyen, P., Sharma, A. and Stephan, F. (2002). "Learning in Logic with RichProlog", in *Logic Programming: 18th International Conference* (ICLP

2002), Lecture Notes in Computer Science 2401, Stuckey, P.J. (ed.), Springer-Verlag, 239–254.

[31] Popper, K. (1959). *The Logic of Scientific Discovery*, London: Hutchinson.

[32] Reiter, R. (1978). "On Closed-World Data Bases", in Gallaire, J. and Minker, J. [14], 55–76.

[33] Shoham, Y. (1988). *Reasoning About Change*, Cambridge (Mass.): MIT Press.

[34] Stephan, F. (2001). "On One-Sided versus Two-Sided Classification", *Archive for Mathematical Logic* 40, 489–513.

[35] Tarski, A. (1983). "Fundamental Concepts of the Methodology of the Deductive Sciences", in Corcoran [8], 60–109.

[36] Tarski, A. (1983). "On the Concept of Logical Consequence", in Corcoran, J. [8], 409–420.

4

HOW SIMPLICITY HELPS YOU FIND THE TRUTH WITHOUT POINTING AT IT

KEVIN T. KELLY

Department of Philosophy, Carnegie Mellon University, Pittsburg, PA 15213, U.S.A., kk3n@andrew.cmu.edu

Abstract: It seems that a fixed bias toward simplicity should help one find the truth, since scientific theorizing is guided by such a bias. But it also seems that a fixed bias toward simplicity cannot indicate or point at the truth, since an indicator has to be sensitive to what it indicates. I argue that both views are correct. It is demonstrated, for a broad range of cases, that the Ockham strategy of favoring the simplest hypothesis, together with the strategy of never dropping the simplest hypothesis until it is no longer simplest, uniquely minimizes reversals of opinion and the times at which the reversals occur prior to convergence to the truth. Thus, simplicity guides one down the straightest path to the truth, even though that path may involve twists and turns along the way. The proof does not appeal to prior probabilities biased toward simplicity. Instead, it is based upon minimization of worst-case cost bounds over complexity classes of possibilities.

1 THE SIMPLICITY PUZZLE

There are infinitely many alternative hypotheses consistent with any finite amount of experience, so how is one entitled to choose among them? Scientists boldly respond with appeals to "Ockham's razor", which selects the "simplest" hypothesis among them, where simplicity is a vague family of virtues including unity, testability, uniformity of nature, minimal causal entanglement, and minimal ontological commitment. The debate over "scientific realism" in the philosophy of science hinges on the

M. Friend, N.B. Goethe and V.S. Harizanov (eds.), Induction, Algorithmic Learning Theory, and Philosophy, 111–143.
© 2007 *Springer*.

propriety of this response. Scientific realists view simplicity as a legitimate reason for belief and anti-realists do not. More recently, the question has spread to computer science, where the widespread adoption of simplicity-biased learning and data-mining software makes it all the more unavoidable [15].

Scientific realists infer from the rhetorical force of simplicity arguments that the simpler theory is better "confirmed" and, hence, that belief in the simpler theory is better justified [5]. Anti-realists [21] concede the rhetorical force of simplicity arguments, but wonder why they should be so compelling.[1] Presumably, epistemic justification is supposed to direct one toward the truth and away from error. But how could simplicity do any such thing? If you already know that the truth is simple or probably simple, then Ockham's razor is unnecessary, and if you don't already know that the truth is simple or probably simple, then how could a fixed bias toward simplicity steer you toward the true theory? For a fixed bias can no more indicate the truth than a compass whose needle is stuck can indicate direction.

There are answers in the literature, but only irrelevant or circular ones. The most familiar and intuitive argument for realism is that it would be a "miracle" if a complex, disunified theory with many free parameters were true when a unified theory accounts for the same data. But the alleged miracle is only a miracle with respect to one's personal, prior probabilities. At the level of theories, one is urged to be even-handed, so that both the simple theory and its complex competitor carry non-zero prior probability. Then since the complex theory has more free parameters to tweak than the simple theory has, each particular setting of its parameters has lower prior probability than does each of the parameter settings of the simple theory. So the miracle argument amounts to an *a priori* bias in favor of simple parameter settings over complex parameter settings. But that is just how a Bayesian agent implements Ockham's razor; the question under consideration is why one should implement it, so far as finding the true theory is concerned [12].

Another standard argument is that simple explanations are "better" and that one is entitled, somehow, to infer the "best" explanation [7]. But even

[1] Van Fraassen focuses on the problem of theories that are not distinguished even by all the evidence that might ever be collected. There is no question of simplicity guiding you to the truth in such cases, since no method based only on observations possibly could. On the other hand, it is almost always the case that simple and complex theories that disagree about some future observations are compatible with the current data and the simpler one is preferred (e.g., in routine curve-fitting). I focus exclusively on this ubiquitous, local problem of simplicity rather than on the hopelessly global one.

assuming that the simplest explanation is best, that sounds like wishful thinking [21], for one may like strong explanations, but liking them doesn't make them true. The same objection applies to the view that simplicity is just one virtue among many [13]. An apparently more promising idea is that simple or unified theories compatible with the data are more severely tested or probed by the data and, hence, are better "corroborated" [16] or "confirmed" [5]. But if the truth isn't simple, then the truth is less testable than falsehood, so why should one presume that the truth is simple? Either considerations like testability and explanatory power are irrelevant to the question at hand or one must assume, circularly, that the world is simple in order to explain why one is entitled to prefer more testable theories.

Another idea [20] is that if a simple theory is false, future data will lead to its retraction, so a simplicity-biased, rational agent will converge to the truth in the limit of inquiry. But the question at hand is not merely how to overcome one's simplicity bias. If Ockham's razor is truly helpful, as opposed to merely being a defeasible impediment, it should facilitate truth-finding better than competing biases. But since other biases would also be over-ruled by experience eventually, mere convergence to the truth does not explain why simplicity is a better bias than any other, so this approach is irrelevant to the realism debate.

Perhaps the most interesting of the standard arguments in favor of simplicity is based upon the concept of "overfitting" [2]. The idea is that predicting the future by means of an equation with too many free parameters compared to the size of the sample is more likely to produce a prediction far from the true value. But that argument has more to do with the size of the sample than with the nature of reality, for the same argument against overfitting still favors use of a simple theory for prediction from small samples even when you know that the true theory is very complex. So although this argument is sound and compelling, so far as using an equation for predictive purposes is concerned, it is also irrelevant to the question at hand, which concerns finding the true theory rather than using a false theory for predictive purposes.

Taking stock of the standard answers, it appears that the anti-realist's objection is insuperable, for it can only be met by showing how a fixed simplicity bias helps one find the truth even when the truth is complex. That sounds hopeless, for in complex worlds simplicity points in the wrong direction. Nonetheless, it is demonstrated below that simplicity is the best possible advice for a truth-seeker to follow, in a certain sense, no matter how complex the truth might be.

2 THE FREEWAY TO THE TRUTH

It is no fault of simplicity that it fails to point out or indicate the true theory, since nothing possibly could. General theories or models can always be overturned in the future by the discovery of subtle effects missed earlier even by the most diligent probing. So science is not an uneventful voyage along a compass course to the truth. It is more like an impromptu road trip through the mountains, with numerous hairpin twists and detours along the way. Taking this more appropriate metaphor seriously is the key to the simplicity puzzle.

Suppose that, on your way to a distant city, you exit the freeway for a rest stop and become lost in the neighboring town. If you ask for directions, you will be told the shortest route back to the freeway entrance ramp even before you say which city you are headed to, because the freeway is the best route to anywhere a stranger might wish to go (Figure 4.1). That remains true even if the shortest route to the entrance ramp takes you west for a few miles when your ultimate destination is east.

Suppose that you disregard the local resident's advice. You find yourself on small dirt tracks headed nowhere and, after enough of this, you make a U-turn and head back toward the entrance ramp. Your hubris is rewarded by the addition of one gratuitous course reversal to your route before you even begin the real journey on the freeway, with all of its unavoidable curves through the mountains. So even if directions to the freeway take you directly away from your ultimate goal at first, you ought to follow them.

The journey to the truth likewise occasions reversals and detours: revolutions or revisions in which one theory is retracted and replaced by another and the textbooks are rewritten accordingly [13]. Some retractions are unavoidable in principle given that one finds the truth at all, since accepting a general theory always occasions a risk of being surprised by an unanticipated anomaly later. In that case, retracting the theory is not merely excusable but virtuous—the alternative would be dogmatic commitment to error for

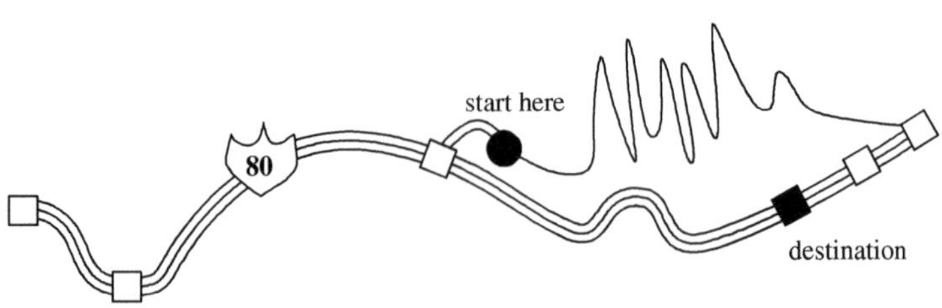

Figure 4.1. Entrance ramp

eternity, as Popper [16] emphasized. But gratuitous reversals in the course of inquiry are another matter entirely: it would be better to avoid them.[2]

Suppose that you violate Ockham's razor by selecting a theory more complex than experience requires. Then the simple experience up to now can be extended for eternity with equally uniform, simple experience, devoid of "effects" whose detection would indicate the need to postulate more causes or free parameters. If you refuse ever to retract to a simple hypothesis, you never arrive at the truth at all, so you have to take the bait, eventually, and fall back to the simplest theory. Now you are essentially where you would have been had you never violated Ockham's razor, except that you have already retracted once; and you are still subject to the future appearance of any number of subtle empirical effects that could not be detected at current sample sizes or using current instrumentation. Each such effect may occur sufficiently late to result in an unavoidable retraction. So you are stuck with an extra retraction at the outset added to all of these. Therefore, always presuming that the world is simple keeps you on the straightest path to the truth even though the truth may be arbitrarily complex! So both the realist and the anti-realist are right, since simplicity keeps one on the straightest path to the truth, but the straightest path may point in the wrong direction for the time being and for any finite number of times in the future as well, assuming that you converge to the truth at all.

3 ILLUSTRATION: COUNTING MARBLES

Suppose that you are studying a marble-emitting device that occasionally emits a marble (a new empirical effect). Your job is to determine how many marbles it will ever emit (how many free parameters the true theory has). You know nothing about when the marbles will be emitted (empirical effects may be arbitrarily small and hard to notice) but you do know on general grounds that at most finitely many marbles will be emitted (every theory under consideration has at most finitely many free parameters). Call the situation just described the *counting problem*.

In this simplistic setting, it seems that when exactly k marbles have been seen so far, k is the simplest answer compatible with experience. First,

[2] Retractions have been studied extensively in computational learning theory. For a review cf. [9]. The first version of the U-turn argument, albeit restricted to problems in which at most k marbles may be seen, is presented in [17]. An infinite ordinal version of the argument, based loosely on ideas in [3] is presented in [10], but that idea still can't handle the marble counting problem described below.

k posits the fewest entities among all answers compatible with experience, which accords with the standard formulation of Ockham's razor. Second, k is satisfied by the most uniform (i.e., eternally marble-free) course of future experience, for alternative answers involve discrete "kinks" in experience (i.e., each time another marble is seen). Third, k has the fewest free parameters (for answer $k + k'$ leaves the appearance time of each of the extra k' posited marbles unspecified). Fourth, k is the best explanation of the data, since $k + k'$ leaves each of the k' appearance times unexplained.[3] Fifth, k is most testable, for if k is false, it is refuted, eventually, but answer $k + k'$ is false but never strictly refuted if the truth is less than $k + k'$.

A *strategy* for solving the counting problem examines the current marble history at each stage and returns either a natural number k indicating the total number of marbles or the skeptical response "?", which indicates a refusal to guess. Such a strategy *solves* the counting problem in the limit if and only if it converges, on increasing data, to the true count k, no matter what the true k happens to be and no matter when the k marbles happen to appear.

Now suppose that you solve the counting problem in the convergent sense just defined. Suppose, further, that no marbles have appeared yet, so the Ockham answer is 0, but you violate Ockham's razor by guessing some k greater than 0 (Figure 4.2). Everything you have seen is consistent with the possibility of never seeing any marbles. Since you converge to the truth, it follows that if the truth is 0, you must eventually converge to 0, so you retract k and revise to 0 at some point. Now it is possible for you to see a marble followed by no more marbles. Since you converge to the truth, you retract 0 eventually and replace it with 1, and so forth. So each answer k is satisfied by a world compatible with the problem's background assumptions

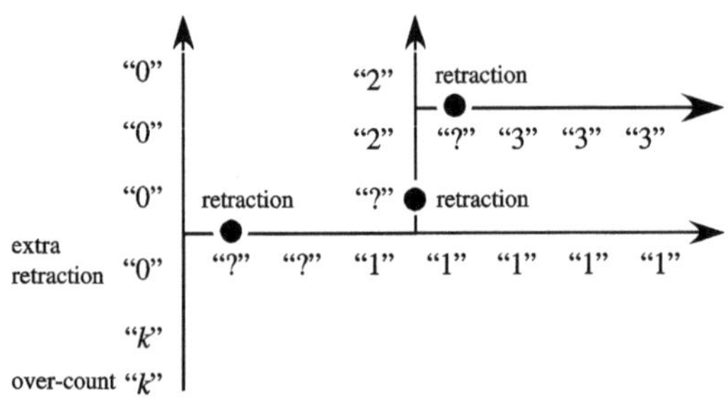

Figure 4.2. U-turn

3 One might object that if the k marbles have appeared at each stage so far, then one would expect them to continue appearing forever, but that violates the background assumption that they will stop appearing eventually.

in which you retract $k + 1$ times. But had you always produced the Ockham answer at each stage, you would have retracted at most k times in an arbitrary world satisfying answer k. So your worst-case retractions are worse than the Ockham strategy's over each answer. Your initial retraction is analogous to the initial U-turn back to the entrance ramp being added to all the course reversals encountered on one's journey home after getting on the freeway.

Another natural consequence of the U-turn argument is that, after having selected answer 0, you should never retract it until it ceases to be simplest. Call this property *stalwartness*. For suppose that no marbles have been seen and that you follow Ockham's advice by choosing answer 0. Suppose, later, that you retract this answer in spite of the fact that no marble has been observed (for general, skeptical reasons, perhaps). Then if you converge to the truth, this initial retraction gets added to all the others you perform, regardless of which answer is true. So your worst-case retraction bound in answer k is $k + 1$, whereas a stalwart Ockham strategy can converge to the truth with just k retractions in answer k.

As simple as it is, the preceding logic has applications to real scientific questions. For example, consider the case of finding the polynomial degree of the true law, assuming that the law is polynomial. It is plausible to assume that larger samples or improvements in instrumentation allow one to progressively narrow in on the true value of the dependent variable y for any specified, rational value of the independent variable x over some closed, bounded interval as time progresses. Any finite number of such observations for a linear law is compatible with the discovery of a small quadratic effect later. Then any finite amount of such data for a quadratic law is compatible with the discovery of a small cubic effect later, etc.[4] The occasional appearances of these arbitrarily small (i.e., arbitrarily late), higher-order effects are analogous to the occasional appearances of marbles and polynomial degree k is analogous to seeing exactly k marbles for eternity.

4 ITERATING THE ARGUMENT

To this point, the U-turn argument has been applied only in cases in which no marble (anomaly) has yet been detected. But suppose that a marble appears after you say 0 but you stubbornly retain the answer 0 (Figure 4.3). Suppose, further, that when the second marble appears you violate Ockham's

4 Popper [16] had a similar idea, except that he assumed exact measurements and counted the number of distinct measurements required to refute a given curve. In science, the observations are never exact and the logic is as I have described.

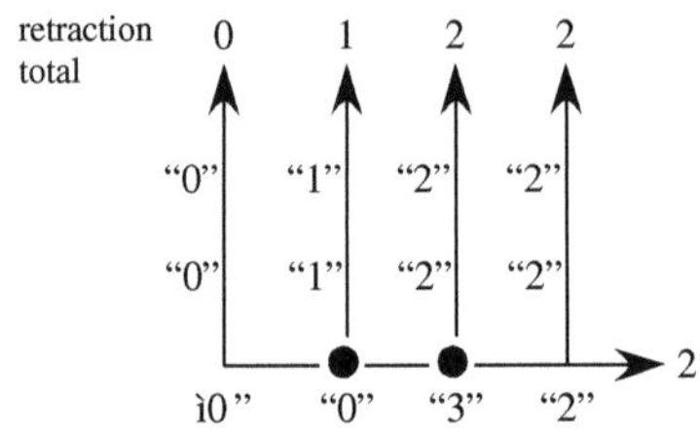

Figure 4.3. Ockham violator who is efficient *ex ante*

razor by producing 3. Thereafter, you follow Ockham's advice. The U-turn logic rehearsed above does not distinguish your performance from that of the natural strategy that just counts the current marbles, for although guessing 3 opens you to the risk of retracting back to 2 later, that extra retraction is concealed by the retraction you saved by not retracting 0 to 1 earlier. So you converge to the truth and match the Ockham strategy's performance in terms of overall, worst-case retractions within each answer.

The preceding analysis is carried out at the onset of inquiry (i.e., *ex ante*). The situation changes if your efficiency is assessed *ex post*, at the moment you first violate Ockham's razor by over-counting; e.g., by saying 3 upon seeing the last entry in input sequence $e = (e_0, \ldots, e_n)$ in which only two marbles are presented. At that very moment, the input data e are already fixed, as is the sequence $b = (B_0, \ldots, B_{n-1})$ of answers you chose at each stage along $e_{-} = (e_0, \ldots, e_{n-1})$. So only worlds that present e and only strategies that produce b along e_{-} should count when your efficiency is assessed at the moment e has been presented.

Now the U-turn argument rules out over-counting even after some marbles have been seen (Figure 4.4). For suppose that you over-count for the first time at the end of e. Consider the hybrid strategy σ that agrees with you along e_{-} and that returns the current count thereafter. Strategy σ

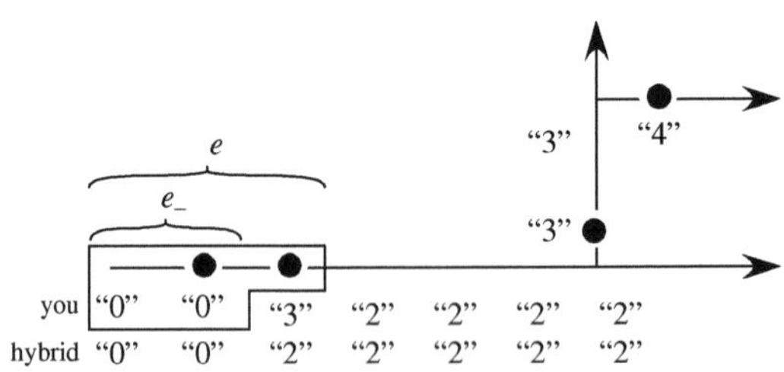

Figure 4.4. Inefficiency exposed *ex post*

converges to the right answer (by counting up to it). Like you, strategy σ saves a retraction by not noticing the first marble (which appears in e_-), but σ produces the current count k at the end of e rather than the over-count you produce. Moreover, σ never retracts again if the truth is k and, in general, retracts at most k' times after the end of e if the truth is $k + k'$. But if you converge to the truth, you eventually retract your over-count at e to k if the truth is k (the initial U-turn back to the freeway to the truth) and then retract k to $k + 1$ if another marble is presented thereafter, etc., so you retract $k' + 1$ times after the end of e in answer $k + k'$ (the initial U-turn gets added to the k inevitable hairpins along the freeway to destination $k + k'$). Since σ acts just like you along e_-, both you and σ retract the same number of times (say r) along e_-. Since e is your first over-count, you retract at e, so you retract $r + 1$ times along e, so your worst-case bound over answer $k + k'$ is $r + k' + 2$. Even if σ retracts at e, the worst-case retraction bound for σ over answer $k + k'$ is at most $r + k' + 1$. So if you over-count for the first time at e, then for each answer $k + k'$, your worst-case retraction bound over $k + k'$ exceeds that of σ. So you are *strongly beaten* by σ at e, in the sense that σ agrees with you along e_- and over each answer k compatible with e, your worst-case bound over worlds compatible with e in answer k is worse than that of σ. If σ does as well as you in each answer k and worse in some answer, then say that you are *weakly beaten* by σ at e.

The same argument works at each e at which you (a) fail to repeat the answer you produced at the immediately preceding stage e_- and (b) choose any answer other than the current count. For in that case you retract at e, do no better than the hybrid method along e_-, and do worse in the worst case after e (due to having to retract back to k if no more marbles are seen after e). Say that a *lagged* Ockham strategy is a strategy that only violates Ockham's razor by retaining the answer it selected at the preceding stage. So an arbitrary solution is strongly beaten at *each* violation of the lagged Ockham property.

By a similar argument, if you solve the problem then you are strongly beaten by the hybrid strategy σ at an arbitrary e at which you fail to be stalwart. For if you are not stalwart at e, you drop the answer B you selected at e_- even though B is Ockham at e, so your stalwart clone σ also produces B at e_- (because it is a clone) and does not drop B at e (by stalwartness). Then, as before, σ retracts no more than you after e in each answer compatible with e, so σ beats you at e.

Being strongly beaten is no sin if every solution suffers that fate. To clinch the U-turn argument, each stalwart, lagged Ockham solution σ (e.g., the strategy that always returns the current count) is *efficient* at each e in the sense that over *each* answer compatible with e, solution σ does as well in worst-case retraction performance as an *arbitrary* solution σ' agreeing with

σ along e_{-}. For let e be given and let σ' be just like σ along e_{-}. Then both σ and σ' retract the same number of times r along e_{-} and both produce the current count k at e_{-}. If σ retracts at e, then since σ is stalwart, it follows that a marble was presented at e and σ produces the current count $k+1$ at e. So if no more marbles are ever presented, σ' also has to retract to $k+1$ eventually in order to converge to the truth. So σ' achieves no better retraction bound than σ in answer $k+1$. Finally, σ retracts no more than k' times after e in answer $k+k'$ and σ' can be forced to retract at least k' times after e in answer $k+k'$ by presenting each of the remaining k marbles and waiting until σ' converges to the current count. So σ does at least as well as σ' in answer $k+k'$.

So the following has been shown.

Proposition 4.1 *Let σ solve the counting problem. Then for each finite input sequence e:*

1. *if σ violates either the lagged Ockham property or stalwartness at e then σ is strongly beaten in terms of retractions at e;*
2. *if σ satisfies stalwartness and the lagged Ockham property at e, then σ is efficient in terms of retractions at e.*

It is clear from the definitions that being strongly beaten implies being weakly beaten which implies inefficiency, so it follows that:

Corollary 4.2 *Let σ be a solution to the counting problem and let the cost be retractions. Then the following are equivalent:*

1. *σ is efficient at each e;*
2. *σ is weakly beaten at no e;*
3. *σ is strongly beaten at no e;*
4. *σ is stalwart and has the lagged Ockham property at each e.*

So the set of all solutions to the counting problem is neatly partitioned into the efficient solutions and the strongly beaten solutions, where the former are precisely the stalwart, lagged Ockham solutions. That is hardly obvious from the definitions of efficiency and beating, themselves. It reflects a substantive interaction between the criteria of evaluation and convergence to the truth.

Say that a *method* is a constraint on strategies, so the stalwart, lagged Ockham property is a method. Since violating this method results in being beaten at each violation, it follows that no matter what you did in the past, following the stalwart, lagged Ockham method will always look better at each stage (in terms of worst-case retractions) than violating it (given that you aim to converge to the truth). Thus, one may say that the stalwart, lagged

Ockham method is *stably retraction efficient* for agents who wish to converge to the truth in a retraction-efficient manner. Stability is crucial for explaining the history of science, for it has frequently occurred that a complex theory is selected because the simple theory has not yet been conceived or has been rejected on spurious grounds (e.g., Ptolemaic astronomy *vs.* Copernican astronomy or wave optics *vs.* Newtonian optics). If Ockham's razor is to explain the subsequent revision to the simpler theory, the rationale for preferring simpler theories must survive past violations.

The preceding results respond to an additional anti-realist challenge. Suppose that you have already seen $n - 1$ marbles at awkward, distant intervals and that after seeing each marble you came to believe, eventually, that you had seen all the marbles there are. The "negative induction" argument against realism [14] recommends the conclusion that one more marble will appear, since you were fooled each time before. But that policy would risk a gratuitous retraction, according to the preceding argument. So the realist wins, no matter how many times Ockham's razor led to disaster in the past![5]

5 TIMED RETRACTIONS

Retraction efficiency does not prohibit a solution from hanging onto its previous answer in spite of the appearance of new marbles, since no retraction is incurred thereby. Mere consistency with experience rules out under-counting, so consistency together with retraction efficiency entails that one never return a value other than the correct count. But that response is not sufficiently general, for suppose that the question is modified so that if the true number of marbles is even, all you have to say is "even".[6] When the first marble is seen, the right answer seems to be 1 rather than "even", but the lagged Ockham property together with consistency does not imply this conclusion, for "even" is consistent with any possible experience.

Here is a more general and unified explanation. Suppose that you hang on to answer "even" to save a retraction when the first marble is seen. Nature can withhold further marbles until you converge to answer 1. The obvious Ockham strategy would drop "even" immediately and would eventually gain enough confidence to say 1 later, so if the answer is 1, both you and the

[5] On the other hand, enough surprises might push one to rethink the problem by adding the answer "infinitely many marbles will appear". The U-turn argument concerns only the problem as presented, not other possible problems one might take one's self to be solving instead.

[6] Worst-case bounds must still be taken over total marble counts rather than over answer "even". The general theory of simplicity developed below works the same way.

Ockham strategy retract once, but you retract later than the Ockham strategy. That is worse, for one's state after the retraction is more enlightened than one's state prior to it (think of the Newtonians before and after they lost their faith that an ether drift would be detected) and needlessly delaying a retraction allows more subsidiary conclusions to accumulate that must be flushed when it finally occurs.

So instead of simply counting retractions, let the cost of inquiry in a given world w be represented by a possibly empty, finite sequence of ascending natural numbers $(r_1, \ldots, r_k)$ such that the strategy retracts exactly k times in w and for each i from 1 to k, the strategy retracts at moment r_i. It is necessary to rank such cost sequences. It would be unfortunate if Ockham's razor were to depend upon some fussy weighting of time against overall retractions so that, say, $(9) > (1, 2)$. Happily, it suffices in the following argument to restrict attention to weak Pareto dominance with respect to overall retractions and the times of occurrence thereof, which yields only a partial order over cost sequences. Accordingly, if c, c' are both cost vectors, let $c \leq c'$ if and only if there exists a sub-sequence d of c' whose length matches that of c such that the successive entries in d are at least as great as the corresponding entries in c. Then define $c < c'$ if and only if $c \leq c'$ but $c' \not\leq c$. For example:

$$(1, 3, 8) < (1, 5, 9) < (1, 2, 5, 9).$$

Refer to the cost concept just defined as *timed retractions*.

Next, consider bounds on sets of timed retraction cost sequences. Recall that ω is the least (infinite) ordinal upper bound on the natural numbers. A potential *timed retraction bound* is the result of substituting ω from some point onward in a cost sequence: e.g., $(1, 2, \omega, \omega)$. If S is a set of cost sequences and b is a potential bound, then b bounds S (written $S \leq g$) if and only if for each c in S, $c \leq b$. Thus, $(1, \omega)$ bounds the set of all sequences $(1, k)$ such that k is an arbitrary natural number.

Finally, say that a strategy is *Ockham* just in case it never chooses an answer other than the current count (or possibly '?'). Then one obtains the following, strengthened result.

Proposition 5.1 *Let a solution to the counting problem be given. Then:*

1. *if the solution violates either the Ockham property or stalwartness at e, then the solution is strongly beaten in terms of timed retractions at e;*
2. *if the solution satisfies stalwartness and the Ockham property at e, then the solution is efficient in terms of timed retractions at e.*

Proof. Suppose that you over or under count at e, which presents exactly k marbles. As before, let hybrid strategy σ be just like you along e_- and then always return the current count from e onward. Consider answer $k + k'$, where k' is an arbitrary natural number. Suppose that you retract at e if σ does. Then the cost sequence for σ along e is no worse than yours, which is, say, $(c_1, \ldots, c_r)$. Then since σ retracts at most once, for each of the additional marbles that appear after e in answer $k + k'$, the worst-case cost bound for σ over answer $k + k'$ is at most $(c_1, \ldots, c_r, \omega, \ldots, \omega)$, with k' repetitions of ω. Nature can withhold marbles after e until you eventually retract your answer (say, at stage i) in preparation for convergence to k. Furthermore, after you converge to k, nature can continue to withhold marbles until you say k an arbitrary number of times before presenting another marble. Eventually, you drop k in preparation for convergence to $k + 1$, etc. So your bound in answer $k + k'$ is at least $(c_1, \ldots, c_r, i, \omega, \ldots, \omega)$, with k' repetitions of ω. That is worse than the bound for σ because the bound for σ is a proper sub-sequence of your bound.

Now suppose that you don't retract at e but σ does. Then let your cost through e be $(c_1, \ldots, c_r)$, in which case the cost of σ through e is $(c_1, \ldots, c_r, i)$, where i is the length of e. Then since σ retracts at least once, arbitrarily late, for each of the additional marbles that appear after e in answer $k + k'$, the worst-case cost bound for σ over answer $k + k'$ is at most $(c_1, \ldots, c_r, i, \omega, \ldots, \omega)$, with k' repetitions of ω. But since you do not produce k at the end of e, nature can withhold marbles until (say, at stage $i' > i$) you retract your answer at e in preparation for convergence to k. Then nature can exact one retraction out of you, arbitrarily late, for each of the k' marbles that appears after e in answer $k + k'$. Hence, your worst case bound is at least $(c_1, \ldots, c_r, i', \omega, \ldots, \omega)$, where $i' > i$. So your bound is worse than that of σ. The beating argument for stalwartness and the efficiency argument for stalwart, Ockham solutions are similar.[7] ■

So when retraction delays are taken into account, every solution is either efficient, stalwart, and Ockham or strongly beaten. Again, there is no middle ground.

Corollary 5.2 *Let σ be a solution to the counting problem and let the cost be timed retractions. Then the following are equivalent:*[8]

1. *σ is efficient at each e;*
2. *σ is weakly beaten at no e;*

[7] In any event, more general arguments are provided in the appendix for propositions 12.2 and 12.4 below.

[8] Corollary 5.2 is an instance of corollary 12.5 below.

3. *σ is strongly beaten at no e;*
4. *σ is stalwart and Ockham at each e.*

6 GENERALIZING THE ARGUMENT

In order to argue, in general, that Ockham's razor is necessary for minimizing timed retractions, one must say, in general, what Ockham's razor amounts to. That may seem like a tall order compared to counting marbles. First, simplicity has such manifold characteristics—e.g., uniformity, unity, testability, and reduction of free parameters, causes, or ontological commitments—that one wonders if there is a single notion that underlies them all. Second, it seems that some aspects of simplicity are a mere matter of description. For example, if one describes inputs as marbles or non-marbles, then marble-free worlds are most uniform. But if an "n-ble" is a marble at each time other than n, when it is a non-marble, then uniformly marble-free experience is not uniformly n-ble free experience. Nor can one complain that the definition of n-ble is strange, since marbles are n-bles at each stage but n, when they are non-n-bles (Goodman [6]). So with respect to the syntactic complexity of definitions, the situation is entirely symmetrical. These sorts of observations have led to widespread skepticism about the prospects for a general, unified, objective account of simplicity . But the skepticism is premature, for in the marble counting problem, the question at hand concerns marbles rather than n-bles and simplicity may depend upon the structure of the problem one is trying to solve. Indeed, if simplicity is to have anything to do with efficiency, it must somehow reflect the structure of the problem one is trying to solve.

In the marble counting problem, answers positing more marbles are more complex. Presumably, then, worlds that present more marbles are more complex, assuming that simpler answers are answers satisfied by simpler worlds. One might plausibly say that each marble is an *anomaly* relative to the counting problem, since the previously simplest (best) explanation is no longer simplest after the marble appears. Some insight is gained into the nature of anomalies by characterizing the occurrence of a marble entirely in terms of the structure of the marble counting problem, itself.[9]

One structural feature of the marble counting problem is that, prior to seeing a third marble, nature can *force* an arbitrary solution to the problem to produce successive answers 2, 3, 4, . . . by presenting no marbles until the

[9] For a critique of this idea and a response, see [1] and [19].

solution converges to 2, one marble followed by no more until the method converges to 3, and so forth. But after seeing the third marble, nature can only force the solution to produce successive answers 3, 4, 5, ... So as a working hypothesis, it seems that an anomaly occurs when the sequence of answers nature can force is truncated (from the front). This might be expressed by saying that an anomaly occurs when nature uses up an opportunity to force the scientist to change her mind or, more colorfully, when nature leads the scientist one exit further down the freeway to the truth.

Nature may be capable of taking more than one step down the freeway at a time (e.g., modify the marble counting problem so that several marbles can be emitted at one time), in which case nature takes two steps down the forcible path (0, 1, 2, ...) when two marbles are presented at one time, for after these marbles are seen, only (2, 3, 4, ...) is forcible.

Also, there may be more than one freeway to the truth, in which case there may be several simplest answers to select among. For example (Figure 4.5), modify the counting problem so that marbles come in two colors, white and black, and you have to determine (i, j), where i is the total number of white marbles and j is the total number of black. If no marbles have been seen so far, then patterns of form ((0, 0), (1, 0), ...) and ((0, 0), (0, 1), ...) are forcible. Suppose you now hear a rumble in the machine, which guarantees that another marble is coming, but you don't see the color. Now (0, 0) is no longer forcible (the rumble can't be "taken back") so only patterns of form ((1, 0), ...) and ((0, 1), ...) are forcible. That is a step by nature down both possible paths, so the rumble constitutes an anomaly. Suppose that the announced marble is black. Now only patterns of form ((0, 1), ...) are forcible. No step is taken down path ((0, 1)), however, so seeing the black marble after hearing the noise is not an anomaly— intuitively, the anticipated marble has to have some color or other. The same is true if a black marble is seen. So no world in which just one more marble is seen presents any anomalies after the sound, so all such worlds are maximally simple in light of

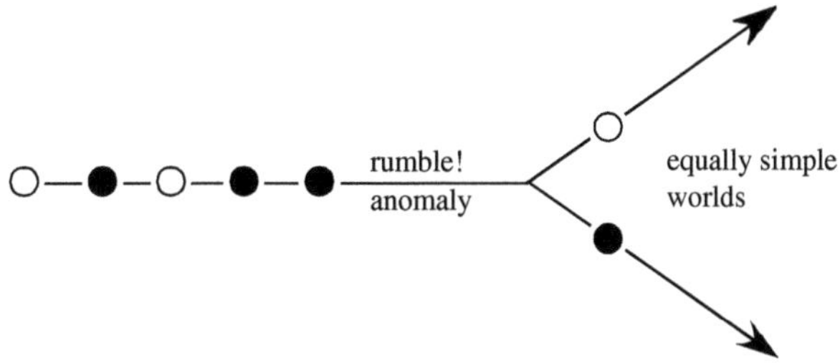

Figure 4.5. Nature chooses a path without stepping down it

the sound. Hence, answers $(1, 0)$ and $(0, 1)$ are both simplest after the sound, whereas answers that entail more than one marble are more complex than necessary. That is intuitive, since Ockham's razor seems to govern number rather than color in this example.[10]

As it is usually formulated, Ockham's razor requires that one never presume a more complex hypothesis than necessary, which allows for selection among simplest answers when the noise is heard: e.g., $(3, 3)$ over $(2, 4)$. Answers positing extra marbles—e.g., $(3, 4000)$ are plausibly ruled out. But there is still something odd about guessing one color rather than another before seeing what the color is: after the noise, it seems that one should simply wait to see what color the announced marble happens to be. Indeed, the problem's future structure is entirely symmetrical with respect to color, so there could be no efficiency advantage in favoring one color over another until one sees which color it is. Say that a method has the *symmetry* property at a given stage if it produces no answer other than the uniquely simplest answer at that stage.

In the counting problem, worst-case cost bounds were assessed over possible answers to the question. In the general theory presented below, worst-case bounds are assessed over complexity classes of worlds. One reason for this is to "break up" coarse answers sufficiently to recover the U-turn argument. For example, recall the problem in which you must count the marbles if the total count is odd and must return "even" if the total count is even. In this problem, Ockham violators are not necessarily strongly beaten because the retractions of an arbitrary solution are unbounded in answer "even", both in terms of retractions and in terms of timed retractions. In the general theory, the answer "even" is partitioned into anomaly complexity classes corresponding to each possible even count and retractions are bounded over these complexity classes so that the strong beating arguments rehearsed earlier for the counting problem can be lifted to this coarser problem. This agrees with standard practice in the theory of computational complexity, in which one examines an algorithm's worst-case resource consumption over sets of inputs of equal size [4].

[10] That is because color does not lead to unavoidable retractions in the example under discussion. If each white marble could spontaneously change color, just once, from white to black at an arbitrary time after being emitted, then white would be simpler than black. The same is true if a continuum of gray-tones between white and black is possible and marbles never get brighter. Then Ockham should say "presume no more darkness than necessary".

7 EMPIRICAL SIMPLICITY DEFINED

It remains to state the preceding ideas with mathematical precision. An *empirical problem* is a pair (K, Π), where K is a set of infinite sequences of inputs and Π partitions K. Elements of K are called *worlds* and cells in Π are called *potential answers*. A scientific *strategy* is a mapping from finite sequences of inputs to answers in Π (or to "?", signalling a refusal to choose). A *solution* is a strategy that converges to the true answer in each world in K. Let K_e denote the set of all elements of K that extend finite input sequence e and let Π_e denote the set of all answers A in Π such that A is compatible with e (i.e., such that K_e shares an element with A). Finally, say that e is *compatible* with K just in case some world in K extends e.

All of the following definitions are relative to a given problem (K, Π), which is suppressed to avoid clutter. Say that an *answer pattern* is a finite sequence of answers in which no answer occurs immediately after itself. Let g be an answer pattern. The *g-forcing game* given finite input sequence e compatible with K is played between the scientist and nature as follows. The scientist plays an answer (or "?"), nature plays an input, and so forth, forever.[11] In the limit, the two players produce an infinite play sequence p, of which p_N is the infinite subsequence played by nature and p_S is the infinite subsequence played by the scientist. Let i be the length of e and let $p_S - i$ denote the result of deleting the first i entries from the beginning of p_S. Then nature wins the game if and only if p_n is in K_e and either p_N does not converge to the answer true in p_N or g is a subsequence of $p_S - i$.

Strategies for the scientist have already been defined. A strategy for nature maps finite sequences of answers (or "?") to inputs. A strategy for the scientist paired with a strategy for nature determines a play sequence. A strategy is *winning* for a player if it wins against an arbitrary strategy for the other player. Say that g is *forcible* given e if and only if nature has a winning strategy in the g-forcing game given e. The g-forcing game is *determined* just in case one player or the other has a winning strategy. The assumption of determinacy for forcing games is so useful formally that I will restrict attention to such problems.

Restriction 1 (determinacy of forcing games) *The following results are restricted to problems such that for each pattern g, the g-forcing game is determined.*

[11] Cf. (Kechris 1991) for a general introduction to the pivotal role of infinite games in descriptive set theory.

The restriction turns out not to matter in typical applications, for D. Martin's Borel determinacy theorem (1975) has the following consequence:

Proposition 7.1 (determinacy of Borel forcing games) *If* (K, Π) *is solvable and if* K *is a Borel set and* e *is a finite input sequence, then for all answer patterns* g*, the* g*-forcing game in* (K, Π) *is determined given* e*.*

Since unsolvable problems are irrelevant to the results that follow, it suffices for determinacy of forcing games to assume that K is Borel. That is weaker than saying that K can be stated with some arbitrary number of quantifiers over observable predicates, which covers just about any empirical problem one might encounter in practice.[12] The antecedent of the proposition is not a necessary condition for the consequent, so the scope of the following results is broader still.

Say that answer pattern g is *backwards-maximally forcible* at e if and only if g is forcible given e and for each forcible answer pattern g' given e, if g is a sub-sequence of g' then g is an initial segment of g. Let $\triangle_e$ denote the set of all answer patterns that are backwards-maximally forcible at e. The backwards-maximality property is crucial to the results that follow. The point is to eliminate gaps from all the sequences in $\triangle_e$. For example, in the marble counting problem, if e presents no marbles, then $\triangle_e$ looks like:

$$()$$
$$(0)$$
$$(0, 1)$$
$$(0, 1, 2)$$
$$(0, 1, 2, 3)$$
$$\vdots$$

whereas the forcible sequences include all (gappy) sub-sequences of these, such as $(4, 7, 9)$.

It is not necessarily the case that each forcible pattern b at e can be extended to a backwards-maximally forcible pattering at e. For example, suppose that tomorrow you may see any number of marbles and that any of the marbles may disappear at any time thereafter. At the outset, each finite, descending sequence of marble counts is forcible, so each forcible pattern can be extended at the beginning to a forcible pattern. The following formal development is simplified by, frankly, ignoring such problems.

[12] A typical sort of K (e.g., for marble counting and for inferring polynomial degree) says that there exists a stage such that for each later stage, no further empirical effects are encountered. That involves only two quantifiers, so the restriction is easily satisfied.

Restriction 2 (well-foundedness of forcibility) *If pattern b is forcible at e, then there exists pattern b' of which b is a sub-pattern such that b' is in $\triangle_e$.*

One would expect that if (A, B, C) is in $\triangle_e$, then there should be further experience e' such that (B, C) is in $\triangle_e$; but that is not necessarily the case.[13] It simplifies the following theory to ignore those cases as well. Let $*$ denote concatenation.

Restriction 3 (graceful decrementation) *If $A * B * c$ is in $\triangle_e$, then there exists proper extension e' of e compatible with K such that $B * c$ is in $\triangle_{e'}$ and exactly one anomaly occurs along e' properly after the end of e.*

If g is an answer pattern, let $g * \triangle_e$ denote the set of all $g * g'$ such that g' is an element of $\triangle_e$. Say that an *anomaly* occurs at finite, non-empty input sequence e compatible with K if and only if there exists a non-empty, finite answer pattern $A * g$ such that:

1. $A * g * \triangle_e \subseteq \triangle_{e\_}$;
2. no g' in $\triangle_e$ begins with answer A.

Suppose that two marbles are seen simultaneously at stage e in the counting problem. This anomaly is represented in figure (Figure 4.6).The fact that answer pattern $A * g$ is non-empty ensures that nature moves down some path in $\triangle_e$. Thus, seeing a black marble is not an anomaly after the noise that announces it.

If w is a world in K, then let $c(w, e)$ denote the number of anomalies that occur along w properly after e. If A is an answer, let $c(A, e)$ denote the least $c(w, e)$ such that w is in $K_e \cap A$. Call $c(w, e)$ the *conditional anomaly complexity* of w (or of A) given e, and similarly for $c(A, e)$.

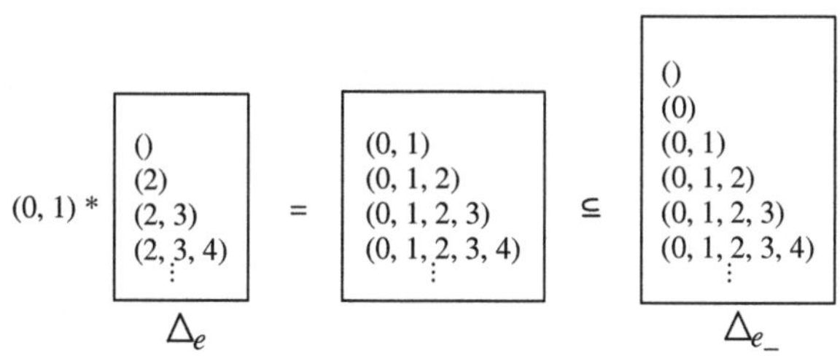

Figure 4.6. Simultaneous observation of two marbles

[13] Suppose that you have to determine the total number of marbles and the time of the first marble if the total count is 2. If no marbles appear in e yet, then we have that $(0, 1, 3)$ is in $\triangle_e$. But upon seeing the first marble at stage k in e', $(1, (2, k), 3)$ is in $\triangle_{e'}$, so $(1, 3)$ is not in $\triangle_{e'}$.

Then let *unconditional* anomaly complexity be given by $c(w) = c(w, ())$ and $c(A) = c(A, ())$, where () is the empty input sequence.

Marbles are still anomalies in the marble-counting problem, but the preceding definitions don't see the marbles; they see only the structural "shadow" each marble occurrence casts against the branching topology of the marble counting problem. The noise announcing a marble is anomalous, but seeing a marble after the noise is not. Seeing two marbles after the noise is anomalous, however. If several marbles are visible and some of them might disappear permanently at any time, then disappearances of marbles count as anomalies and simple worlds have more marbles than complex ones. Refutations of lower polynomial degrees and the discovery that a linear function depends upon an independent variable also count as anomalies in the corresponding problems (assuming that the data consist of ever-tighter open intervals around the dependent variable).

8 OCKHAM'S RAZOR, SYMMETRY, AND STALWARTNESS

Answer A is *simplest* at e if and only if

$$c(A, e) = \min_{B \in \Pi_e} c(B, e).$$[14]

A method satisfies *Ockham's razor* at e just in case the answer output by the method at e is '?' or is simplest at e. *Symmetry* at e requires that the method output at e either "?" or the unique answer that minimizes $c(A, e)$. *Stalwartness* at e requires that if the scientist's output A at e_- is uniquely simplest at e, then the scientist produces A also at e.

Ockham's razor may be defined in terms of simplicity rather than complexity, using a standard rescaling trick familiar from information theory. Define *conditional simplicity* as:

$$s(A, e) = \exp(-c(A, e)).$$

This definition reveals an interesting connection between Ockham's razor and Bayesian updating, for it follows immediately from the definition of $c(A, e)$ that:

$$c(A, e) = c(A \cap K_e) - c(K_e).$$

[14] In light of lemma 14.7 in the appendix, this condition is equivalent to $c(A, e) = 0$.

Applying the definition of $s(A, e)$ to both sides of the preceding equation yields:

$$s(A, e) = \frac{s(A \cap K_e)}{s(K_e)},$$

which is the usual definition of Bayesian updating. Then Ockham's razor requires that one choose the uniquely simplest hypothesis, where simplicity degree is updated by conditionalization. Nothing about coherence or probability has been presupposed, however, so Bayesians who seek Ockham's razor in prior probabilities updated by conditioning put the arbitrary cart before the essential horse.

9 SYMMETRICAL SOLVABILITY

Not every problem has a symmetrical solution. For example, suppose that the problem is to say not only how many marbles appear, but when each of them appears. In this problem, every answer compatible with e is simplest at e, since only patterns of unit length are forcible. That may seem counter-intuitive, since particle counts are analogous to free parameters and times of appearance are analogous to settings of those parameters, so it would seem that answers involving more free parameters are more complex. But it must be kept in mind that the same possibilities could be parameterized in different ways, and simplicity depends upon which parametrization the question asks about. If the problem is to count marbles, then worlds with more marbles are more complex, whenever the marbles arrive. If it is to count n-bles, then worlds with more n-bles are more complex, regardless of when the marbles arrive. If the problem is to identify particular worlds, the parametric structure of the problem disappears and complexity is flattened. Examples of the latter sort are excluded from consideration by the following restriction.

Restriction 4 (symmetrical solvability) *Only problems with symmetrical solutions are considered in the results that follow.*

In typical applications, restriction 4 can be sidestepped by coarsening or refining the question in a manner that disambiguates the intended parametrization. It is also worth mentioning that restrictions 2 and 4 are logically independent given restriction 1.[15]

[15] The problem of identifying individual worlds in which at most finitely many marbles occur satisfies the determinacy assumption (restriction 1) and the well-foundedness assumption (restriction 2) but not the symmetrical-solvability assumption (restriction 4), whereas

10 EFFICIENCY DEFINED

Let $C_e(n)$ denote the set of all worlds in K_e such that $c(w, e) = n$. Refer to $C_e(n)$ as the nth *anomaly complexity class* at e.[16] Complexity classes depend only on the structure of the problem to be solved, so they are not mere matters of description.

Let σ be a solution to (K, Π) and let e be compatible with K. Let the worst-case timed retractions over $C_e(i)$ be the supremum of the timed retraction costs incurred by σ over worlds in $C_e(i)$. As mentioned above, the idea is to examine worst-case bounds over anomaly complexity classes rather than over answers. Accordingly, define:

1. solution σ is *efficient* at e with respect to a given cost if and only if for each solution σ' that agrees with σ along e_- and for each n, the worst case cost bound of σ over $C_e(n)$ is less than or equal to that of σ';
2. solution σ is *strongly beaten* at e with respect to a given cost if and only if there exists solution σ' that agrees with σ along e_- such that for each n such that $C_e(n)$ is non-empty, the worst case cost bound of σ over $C_e(n)$ is greater than that of σ';
3. solution σ is *weakly beaten* at e with respect to a given cost if and only if there exists solution σ' that agrees with σ along e_- such that for each n, the worst case cost bound of σ' over $C_e(n)$ is less than or equal to that of σ and there exists n such that the worst-case cost bound of σ' over $C_e(n)$ is less than that of σ.

Notice that there is no imposed bias or weighting, probabilistic or otherwise, in favor of lower complexity classes or simple worlds in the preceding definitions. There are just dominance relations over worst-case bounds on structurally motivated complexity classes. That is as it must be if the efficiency argument for Ockham's razor is to avoid the narrow circularity of standard, Bayesian explanations.

11 NESTED PROBLEMS

The marble counting problem and the problem of finding the true polynomial degree of a curve both have the attractive feature that there exists

the disappearing marble example described earlier satisfies restrictions 1 and 4 but not restriction 2, for a symmetrical solution could simply wait until tomorrow to see how many marbles there are and could then guess the current number of marbles at each stage.

16 The complexity classes are sets (subsets of K).

a uniquely simplest answer for each possible evidential circumstance e. But there may be more than one maximally simple answer, as in the black and white marble counting problem when the noise is heard. Accordingly, say that a problem is *nested* if there exists a uniquely simplest answer at each e compatible with K. Nested problems allow for branching paths, but have the property that there is a uniquely simplest answer at each stage of inquiry, as in the two-color counting problem when no noise is heard prior to seeing the marble. In that case, nature can choose which color to present at each stage, but the current count is always the uniquely simplest answer. Other familiar scientific questions with this structure include finding the set of all independent variables a linear equation depends upon and the inference of conservation laws in particle physics [18].

12 THE MAIN RESULTS

For brevity, these assumptions govern all the results that follow. All proofs are presented in the appendix.

1. (K, Π) is a problem satisfying restrictions 1–4;
2. the cost under consideration is timed retractions;
3. e is a finite input sequence compatible with K.

The main result is that, in general, every deviation from Ockham's razor incurs a strong beating. Hence, the argument for Ockham's razor is stable, in the sense that you always have a motive to return to Ockham's fold no matter how prodigal you have been in the past.

Proposition 12.1 (efficiency stably implies Ockham's razor) *If solution σ violates Ockham's razor at e, then σ is strongly beaten in terms of timed retractions at e.*

The same is true of stalwartness.

Proposition 12.2 (efficiency stably implies stalwartness) *If solution σ violates stalwartness at e, then σ is strongly beaten in terms of timed retractions at e.*

Symmetry is a stronger principle than Ockham's razor and its general vindication is correspondingly weaker: violating symmetry results in a weak beating at the first violation rather than a strong beating at each violation.[17]

[17] For example, suppose at e that a curtain will be opened tomorrow that reveals either a marble emitter or nothing at all. The question is whether there is an emitter behind the

Proposition 12.3 (efficiency implies symmetry) *If solution σ violates symmetry at e, then σ is weakly beaten at the first moment e' along e at which symmetry is violated.*

Again, being beaten is no sin if every solution is beaten. To clinch the argument, stalwart, symmetrical solutions are efficient. That amounts to an existence proof, given that the problem is symmetrically-solvable, since every symmetrically solvable problem is solvable by a stalwart, symmetrical method.[18] The efficiency is also stable if the problem under consideration is nested.

Proposition 12.4 (symmetry and stalwartness imply efficiency)

1. *If the problem is nested and σ is a stalwart, Ockham solution from e onward, then σ is efficient at e.*
2. *If σ is a stalwart, symmetrical (and, hence, Ockham) solution at every stage, then σ is efficient at every stage.*

In nested problems, all solutions are partitioned into the strongly beaten ones and the stalwart Ockham ones. This duplicates the situation in the counting problem.

Corollary 12.5 *If the problem is nested and σ is a solution, then the following statements are equivalent:*

1. *σ is efficient at each e;*
2. *σ is weakly beaten at no e;*
3. *σ is strongly beaten at no e;*
4. *σ is stalwart and Ockham at each e.*

More generally, the possibility of weakly beaten, non-symmetrical methods must be allowed.

curtain and if so, how many marbles it will emit. The no-emitter world and the marble-free emitter world are both simplest in this example, so symmetry requires that one suspend judgment between the corresponding answers until the curtain is opened. Suppose that you flout symmetry and guess that you are in the marble-free emitter world. Had you refrained from choosing, you would have had no retractions in complexity class $C_e(0)$, but you have incurred at least one retraction in class $C_e(0)$, so you are weakly beaten (every solution, including you, retracts at least k times after e in the worst case in class $C_e(k)$). You are not strongly beaten, however, because you do as well as possible in each class $C_e(k)$ such that k exceeds zero.

18 For a symmetrical solution converges to the uniquely simplest answer in each world and is not prevented from doing so by hanging onto a uniquely simplest answer until it is no longer uniquely simplest.

Corollary 12.6 *If σ is a solution, then the following statements are equivalent.*

1. *σ is efficient at each e;*
2. *σ is weakly beaten at no e;*
3. *σ is stalwart and symmetrical (and, hence, Ockham) at each e.*

13 CONCLUSION AND PROSPECTS

A very general, structural theory of simplicity and of Ockham's razor has been presented, according to which Ockham's razor does not point at the truth, but keeps one on the most direct route thereto. Indeed, choosing only the uniquely simplest hypothesis compatible with experience and hanging onto it until its uniquely simple status is undermined is demonstrably equivalent to minimizing timed retractions prior to convergence to the truth. This result provides a relevant, non-circular connection between simplicity and finding the true theory. No standard, alternative account of simplicity does so.

The results suggest that the scientific realism debate is not a genuine debate. The anti-realist is correct that simplicity cannot function as a magical divining rod for truth. The realist is correct that simplicity, nonetheless, provides the best possible advice for finding the truth, because it keeps one on the straightest possible path thereto. The results also provide some solace for scientists who employ off-the-shelf data-mining procedures that employ a wired-in prior bias toward simplicity. Such methods really are more efficient at finding the truth, even though they cannot be said to divine or point at the truth.[19] Finally, the results reverse the common impression that convergence considerations impose no constraints on the course of inquiry in the short run. It has been demonstrated that timed retraction efficiency leaves just one choice open to a convergent scientist: how long to wait for evidence to accumulate before leaping to the uniquely simplest hypothesis in light of the data. Which answer to choose and when to drop it are both uniquely determined.

Like all new ideas, the proposed account of Ockham's razor suggests a range of potential improvements and generalizations. (1) Efficiency with respect to total number of erroneous answers produced prior to convergence

[19] Simulation studies suggesting the contrary notwithstanding. When an Ockham procedure seems to have a higher chance of producing the true answer in a randomly chosen example than non-Ockham procedures, the underlying sampling distribution over worlds is biased toward simple worlds. That is just a motorized version of the circular Bayesian argument.

is equivalent to the symmetry principle and, hence, entails Ockham's razor. The same is true if efficiency is defined in terms of weak Pareto-dominance with respect to timed retractions and errors jointly. Other combinations of costs can be considered. (2) Penalizing total retracted content rather than just retractions yields the intuitive result that one should only retract to "one black or one white" when the noise announcing a new marble is heard. (3) It remains to apply the preceding ideas with equal rigor and generality to statistical and causal inference (see [12] for some preliminary ideas). (4) It also remains to explore realistic recommendations when finding the Ockham hypothesis is computationally infeasible (see [11] for more preliminary ideas). (6) Finally, the symmetrical solvability and well-foundedness restrictions can and should be weakened.

14 APPENDIX

In the following results, (K, Π) is assumed to be an empirical problem satisfying restrictions 1–4, and e, e' range over finite input sequences. Also, let $\omega[k]$ denote the sequence $(\omega, \ldots, \omega)$ in which ordinal ω is repeated exactly k times.

Proof of proposition 7.1. Let p be a play sequence in the g-forcing game in problem (K, Π) at e. Let p_S be the sub-sequence consisting of the scientist's plays, and let p_N be the corresponding sub-sequence for nature. Let W be the winning condition for nature. In light of Martin's (1975) theorem, it suffices to show that W is a Borel set. Then $p \in W$ if and only if:

1. $p_N \in K_e$ and
2. (a) $\neg((\exists n)(\forall m \geq n)\ p_S(m) \neq$ '?' and $p_N \in p_S(m))$ or
   (b) g is a sub-sequence of p_S.

Condition $p_N \in K_e$ is Borel because K is assumed to be Borel and the condition of extending e is clopen. Condition $p_S(m) \neq$'?' is clopen. Since (K, Π) is solvable, each cell in Π is Σ^0_2, since w is in answer A if and only if there exists a time such that for each later time the solution converges to A. Hence, the condition that $p_N \in p_S(m)$ is Σ^0_2 Borel. Finally, the condition that g is a sub-sequence of p_S is open. Borel conditions are preserved under first-order quantification and Boolean connectives, so W is Borel. ■

Proof of proposition 12.1. Let σ be a solution that violates Ockham's razor at e (which need not be the first violation). So $\sigma(e) = A$, where A is not a simplest answer compatible with e. Let σ' agree with σ along e

and then produce the simplest answer compatible with e' if it exists and '?' otherwise, for each e' properly extending e. Since (K, Π) is symmetrically solvable (restriction 4), σ' solves (K, Π), because σ' converges, in each world, to whatever the assumed symmetrical solution converges to in that world. Let r be the timed retraction cost common to both methods σ and σ' along e_- (recall that r is a finite, ascending sequence of natural numbers).

Suppose that $C_e(k)$ is non-empty. There exists a pattern $B * b$ of length at least $k+1$ in $\triangle_e$ (by lemma 14.3). Since A is not a simplest answer, $B \neq A$ (by lemma 14.5). There exists w in $B \cap K_e$ along which $B * b$ remains forcible after e (by lemma 14.4). Since σ is a solution, σ retracts A after e along w, say at e' of length j. Now $B * b$ is still forcible given e', so there exists w' in $C_{e'}(k)$ along which σ can be made to repeat each successive entry in $B * b$ an arbitrary number of times (by lemma 14.9). Since $B * b$ is forcible at e' and $B * b$ is in $\triangle_e$, no anomaly occurs along e' after e (by lemma 14.1). Hence, w' is in $C_e(k)$. So the worst-case timed retraction bound for σ over $C_e(k)$ is at least $r * j * \omega[k]$, where it will be recalled that $\omega[k]$ denotes the sequence $(\omega, \ldots, \omega)$, with ω repeated k times and $*$ indicates concatenation. But since σ' retracts after e only at anomalies (by lemma 14.8), the worst-case timed retraction bound for σ' over $C_e(k)$ is at most $r * i * \omega[k]$, where $i < j$ is the length of e. Since $r * i * \omega[k] < r * j * \omega[k]$ and C_k is an arbitrary, non-empty complexity class, σ' strongly beats σ at e in terms of timed retractions. ■

Proof of proposition 12.2. Let σ be a solution that violates stalwartness at e (which need not be the first violation). So for some answer A that is uniquely simplest at e, $\sigma(e_-) = A$ but $\sigma(e) \neq A$. Let σ' be a solution constructed as in the proof of proposition 12.1, and let r be the timed retraction cost incurred along e_- by both σ and σ'. Let i be the length of e. Then σ incurs timed retraction cost $r * i$ along e, but σ' incurs only r. Let $C_e(k)$ be non-empty. So there exists a pattern b in $\triangle_e$ of length at least $k+1$ (by lemma 14.3). There exists w in $C_e(k)$ along which σ can be made to repeat each successive entry in b an arbitrary number of times (by lemma 14.9). So the worst-case timed retraction bound for σ over $C_e(k)$ is at least $r * i * \omega[k]$. Since $\sigma'(e_-) = A$ and σ' is stalwart at e and A is simplest at e, $\sigma'(e) = A$, so the timed retraction cost of σ' along e is just r. Since σ' retracts after e only at anomalies (by lemma 14.8), the worst-case timed retraction bound for σ' at e is at most $r * \omega[k]$. Since $r * \omega[k] < r * i * \omega[k]$ and C_k is an arbitrary, non-empty complexity class, σ' strongly beats σ at e in terms of timed retractions. ■

Proof of proposition 12.3. Suppose that σ is a solution that violates the symmetry principle (somewhere). Then there exists finite input sequence e compatible with K such that σ violates symmetry at e, but not at any proper sub-sequence of e. So $\sigma(e) = A$, where A is not the uniquely simplest answer compatible with e. Let σ', r, and $\omega[k]$ be as in the proof of proposition 12.1.

Since A is not uniquely simplest at e, there exists world w in $C_0(e)$ such that w satisfies some answer $B \neq A$ (by lemma 14.7). Since σ is a solution, σ converges to B in w, so there exists some e' properly extending e and extended by w such that $\sigma(e') \neq A$. So the timed retractions of σ along e' are at least $r * j$, where j is the length of e'. So the worst case timed retractions of σ over $C_e(0)$ are at least $r * j$. Let w' be an arbitrary element of $C_e(0)$. Then σ' never retracts in w after e (by lemma 14.8). It is possible that σ' retracts at e. So the worst case timed retractions of σ' over $C_e(0)$ are less than or equal to $r * i$, where $i < j$ is the length of e. Observe that $r * i < r * j$.

Now consider non-empty complexity class $C_e(k)$, for arbitrary $k \geq 0$ and let w be in $C_e(k)$. Then there exists pattern b in $\triangle_e$ of length at least $k + 1$ (by lemma 14.3).

Case A: σ retracts at e if σ' does. Then the worst case timed retractions of both methods along e are exactly the same, say r', and the worst-case timed retraction bound for σ' over $C_e(k)$ is no worse than $r' * \omega[k]$. Also, there exists w' in $C_e(k)$ along which σ produces the successive entries along b after e with arbitrarily many repetitions (by lemma 14.9). Hence, the worst-case timed retractions of σ after e are at least as bad as $\omega[k]$, so the worst-case timed retraction bound for σ over $C_e(k)$ is at least $r' * \omega[k]$. But since σ' retracts after e only at anomalies (by lemma 14.8), the worst-case timed retraction bound for σ' over $C_e(k)$ is at most $r' * \omega[k]$.

Case B: σ' retracts at e and σ does not. Since e is the first symmetry violation by σ and $\sigma(e_-) = \sigma(e)$, answer $A = \sigma(e)$ is uniquely simplest at e_- but not at e. So there exists w in $C_e(0) - A$ such that w is not in $C_{e_-}(0)$ (by lemma 14.7). So $c(w, e) = 0$ but $c(w, e_-) > 0$. Hence, e is an anomaly. So there exists pattern $B * d$ such that no pattern in $\triangle_e$ begins with B and $B * d * \triangle_e \subseteq \triangle_{e_-}$. Since the uniquely simplest hypothesis A at e_- begins each forcible sequence in $\triangle_{e_-}$ (by lemma 14.6), $B = A$, so no pattern in $\triangle_e$ begins with A. So pattern b begins with some answer $D \neq A$. So there exists world $w' \in D \cap K_e$ such that for each e' extending e and extended by w', b is forcible at e' (by lemma 14.2). Since σ is a solution, σ converges to D in w' and, hence, retracts A at some e' properly extending e and extended by w'. Let j be the length of e', so $j > i$, where i is the length of e. Then b is still forcible at e', so there exists w'' in $D \cap C_{e'}(k)$ along which the successive entries in b are produced with arbitrary repetitions (by lemma 14.9). Since b is still forcible at e' and b is in $\triangle_e$, no anomalies occur after

e along e' (by lemma 14.1), so w'' is also in $C_e(k)$. Hence, the worst-case timed retraction bound for σ over $C_e(k)$ is at least $r * j * \omega[k]$. But since σ' retracts after e only at anomalies (by lemma 14.8), the worst-case timed retraction bound for σ' over $C_e(k)$ is at most $r * i * \omega[k] < r * j * \omega[k]$. ■

Proof of proposition 12.4.1. Let σ' be a solution to a nested problem that is Ockham and stalwart from e onward. Since the problem is nested, σ' is also symmetrical from e onward. Let σ agree with σ' along e_{-}. Suppose that $C_e(k)$ is non-empty. There exists a pattern b of length at least $k+1$ in $\triangle_e$ (by lemma 14.3).

Case A: σ retracts at e if σ does. Then let r denote the identical costs of σ and σ' along e. Since σ' is symmetrical and stalwart from e onward, the worst-case timed retraction bound for σ' over $C_e(0)$ is less than or equal to $r * \omega[k]$ (by lemma 14.8). There exists w' in $C_e(k)$ along which σ can be made to repeat each successive entry in b an arbitrary number of times (by lemma 14.9), so the worst-case timed retraction bound for σ over $C_e(0)$ is at least $r * \omega[k]$.

Case B: σ' retracts at e and σ does not. Since (K, Π) is nested, there exists a uniquely simplest answer B at e. So every pattern in $\triangle_e$ begins with B (by lemma 14.6), so b begins with B. Let i be the length of e. Then since σ' retracts only at anomalies after e (by lemma 14.8), the worst-case timed retraction bound for σ' over $C_e(k)$ is less than or equal to $r * i * \omega[k]$. Since σ' is stalwart at e and retracts at e, answer $A = \sigma'(e_{-}) = \sigma(e_{-})$ is not uniquely simplest at e, so $A \neq B$. There exists w in $B \cap K_e$ along which b remains forcible after e (by lemma 14.4). Since σ is a solution, σ must retract A in w after e, say by e'. Now b is still forcible given e', so there exists w' in $C_{e'}(k)$ along which σ can be made to repeat each successive entry in b an arbitrary number of times (by lemma 14.9). Since b is forcible at e', no anomaly occurs along e' after e (by lemma 14.1). Hence, w' is in $C_e(k)$. So letting $j > i$ be the length of e', the worst-case timed retraction bound for σ over $C_e(k)$ is at least $r * j * \omega[k] > r * i * \omega[k]$. ■

Proof of proposition 12.4.2. Let σ' be a stalwart, symmetrical solution at every e. Let σ agree with σ along e_{-}. Now consider non-empty complexity class $C_e(k)$, for arbitrary $k > 0$ and let w be in $C_e(k)$. Then there exists pattern b in $\triangle_e$ of length at least $k+1$ (by lemma 14.3).

Case A: σ retracts at e if σ' does. Follow the argument for case A in the proof of proposition 12.3, observing that a stalwart, symmetrical solution retracts only at anomalies (by lemma 14.8).

Case B: σ' retracts at e and σ does not. Since σ' is always symmetrical and stalwart and σ' retracts at e, answer $A = \sigma'(e_{-}) = \sigma(e_{-})$ is uniquely simplest at e_- but not at e. Pick up from here in case B of the proof of proposition 12.3, again observing that a stalwart, symmetrical solution retracts only at anomalies (by lemma 14.8). ■

Proof of corollary 12.5. (1) implies (2) implies (3) by definition. (3) implies (4) by propositions 12.1 and 12.2. (4) implies symmetry and stalwartness since the problem is nested. Symmetry and stalwartness imply (1) by proposition 12.4.1. ■

Proof of corollary 12.6. (1) implies (2) by definition. (2) implies (3) by propositions 12.3 and 12.2. (3) implies (1) by proposition 12.4.2. ■

Lemma 14.1 (anomaly freedom) *Let b be in $\triangle_e$ and let b be forcible at e' properly extending e. Then for all e'' properly extending e and extended by e':*

1. *b is in $\triangle_{e''}$ and*
2. *e'' is not an anomaly.*

Proof. Suppose that b is in $\triangle_e$ and b is forcible at e' properly extending e. Let e'' properly extend e and be extended by e'. Then b is forcible at e'' since b is still forcible at e'. Suppose for contradiction that b is not in $\triangle_{e''}$. Then since b is forcible at e'', there exists b' forcible at e'' such that b is a subsequence of b' but b is not an initial segment of b'. But then b' is forcible at e, so b is not in $\triangle_e$. Contradiction. So b is in $\triangle_{e'}$. Again, let e'' be an arbitrary input sequence properly extending e and extended by e'. Then it has just been shown that b is in both $\triangle_{e''}$ and $\triangle_{e''_-}$. Suppose that e'' is an anomaly. Then there exists $A * g$ such that $A * g * \triangle_{e''} \subseteq \triangle_{e''_-}$ and no element of $\triangle_{e''}$ begins with A. So $A * g * b$ is in $\triangle_e$. But b does not begin with A, so b is not an initial segment of $A * g * b$. Hence, b is not in $\triangle_e$. Contradiction. ■

Lemma 14.2 (forcibility is asymptotic) *Let $A * a$ be forcible given e. Then there exists a world w in $K_e \cap A$ extending e such that for each finite initial segment e' of w, $A * a$ is forcible given e'.*

Proof. Suppose $A * b$ is forcible given e. Suppose for contradiction that the consequent of the lemma is false. Then for each w in $A \cap K_e$ there exists e' extending e and extended by w such that $A * b$ is not forcible given e'. For each w in $A \cap K_e$, let e_w be the shortest such e'. For each e_w, $A * b$ is not forcible at e_w, so since the forcing games in (K, Π) are all determined (by restriction 1), there exists a solution σ_w for (K_{e_w}, Π_{e_w}) that never produces $A * b$ after e_w. Let σ solve (K, Π) and let σ^* be just like σ except that control is shifted permanently to σ_w when e_w is encountered. So σ^* is a solution that never produces $A * b$ after seeing some e_w. Let $\sigma^\dagger$ be like σ^* except that $\sigma^\dagger$ produces "?" along each e_w and at each e not extended by some e_w such that σ returns A at e. Then $\sigma^\dagger$ is still a solution, since σ^* converges to the truth over $K_e \cap A$ (the question marks eventually end in each w in $K_e \cap A$) and over $K_e - A$ (σ does not converge to A in any such world, so again, the question marks end eventually in each w in $K_e - A$). But $\sigma^\dagger$ doesn't produce $A * b$ after e along any e' extending e. So $A * b$ is not forcible given e. Contradiction. ■

Lemma 14.3 (forcible pattern existence) *Suppose that $C_e(n)$ is non-empty. Then there exists a finite pattern in $\triangle_e$ of length at least $n + 1$.*

Proof. Let w be in $C_e(0)$. In the base case, nature can force the answer A true in w from an arbitrary solution. For induction, suppose that w is in $C_e(n + 1)$. Let e' be the first anomaly along w after e. So there are n anomalies occurring in w after e'. By the induction hypothesis, there exists pattern a in $\triangle_{e'}$ of length at least $n + 1$. Since e' is an anomaly, there exists pattern $A * b$ such that $A * b * a$ is a pattern in $\triangle_{e'_-}$. Hence, $A * b * a$ has length at least $n + 2$. Since $A * b * a$ is forcible at e'_-, $A * b * a$ is forcible at e as well. So there exists some pattern d in $\triangle_e$ of which $A * b * a$ is a sub-pattern (by restriction 2), so d has length at least $n + 2$. ■

Lemma 14.4 (nature's starting point) *Let $A * a$ be in $\triangle_e$. Then there exists a world w in $C_e(0) \cap A$ such that for each finite initial segment e' of w that extends e, $A * a$ is in $\triangle_{e'}$.*

Proof. Let $A * a$ be in $\triangle_e$. So $A * a$ is forcible given e. By lemma 14.2, there exists w in $K_e \cap A$ such that $A * a$ is forcible along each initial segment of w extending e. Let e' properly extend e and be extended by w. Then $A * a$ is in $\triangle_{e'}$ and e' is not an anomaly (by lemma 14.1). Hence, w is in $C_e(0)$. ■

Lemma 14.5 (simplest answer forcible first) *Let answer A be the first entry in some pattern in $\triangle_e$. Then A is a simplest answer.*

Proof. Suppose that $A * b \in \triangle_e$. Then there exists w in $A \cap C_e(0)$ (by lemma 14.4). So $c(A, e) = 0$. ■

Lemma 14.6 (uniquely simplest answer and forcibility) *Let answer A be uniquely simplest at e. Then each pattern in $\triangle_e$ begins with A.*

Proof. Suppose that for some answer $B \neq A$, pattern $B * a$ is in $\triangle_e$. Then by lemma 14.5, B is simplest at e. So A is not uniquely simplest. ■

Lemma 14.7 (simple world existence) *Let K_e be non-empty. Then there exists a world w in $C_e(0)$.*

Proof. Suppose there exists w in K_e. If $c(w, e) = 0$, we are done. So suppose $c(w, e) = k > 0$. Then (by lemma 14.3) there exists $A * a$ in $\triangle_e$ of length $k + 1$. So there exists w' in $A \cap C_e(0)$ (by lemma 14.4). ■

Lemma 14.8 (simplest answer defeated only by anomalies) *Let K_e be non-empty, let e be non-empty, and let A be an answer in Π such that A is uniquely simplest at e_- and A is not uniquely simplest at e. Then e is an anomaly.*

Proof. Let K_e, e be non-empty. Then K_{e_-} is non-empty, so by lemma 14.7, $C_{e_-}(0)$, $C_e(0)$ are non-empty. So since A is uniquely simplest at e_- but not at e, we have $C_{e_-}(0) \subseteq A$ but $C_e(0) \nsubseteq A$. So there exists w in $C_e(0) - C_{e_-}(0)$. Hence, $c(w, e_-) > 0$ and $c(w, e) = 0$, so e is an anomaly. ■

Lemma 14.9 (forcing lemma) *Let σ be a solution and let pattern a of length at least $k + 1$ be in $\triangle_e$ and let m be a natural number. Then there exists w in $C_e(k)$ such that after e, σ produces a_0 successively for m times and then a_1 successively for m, times, ... and finally a_k successively for m times.*

Proof. Let natural number m be given. In the base case, let pattern (A) be in $\triangle_e$. Then there exists world $w \in A \cap C_e(0)$ such that (A) remains in w from e onward (by lemma 14.4). Since A is true in w and σ is a solution, σ converges to A in w, so σ produces A at least m times in succession after e in w.

For induction, let $A * a$, be forcible at e, where a is a finite answer pattern of length $k + 1$. There exists a world w in $A \cap K_e$ such that $A * a$ is in $\triangle_{e'}$, for each finite, initial segment e' of w extending e (by lemma 14.4). Since σ is a solution, σ converges to A in w. Nature can wait m steps after the onset of convergence until σ produces A at least m times after e in w. Let e' extend e such that a is in $\triangle_{e'}$ and exactly one anomaly occurs along e' after e (by restriction 3). So by the induction hypothesis, there exists w' in $C_e(k)$ such that, after e, σ produces a_0 successively for m times and then a_1 successively for m, times, ... and finally a_k successively for m times. Hence, σ produces A successively for m times followed by a_0 for m times, etc. Since exactly one anomaly occurs along e' after the end of e and k anomalies occur along w after e', w' is in $C_e(k + 1)$. ■

REFERENCES

[1] Chart, D. (2000). "Discussion: Schulte and Goodman's Riddle", *The British Journal for the Philosophy of Science* 51, 147–149.

[2] Forster, M. and Sober, E. (1994). "How to Tell When Simpler, More Unified, or Less

ad Hoc Theories Will Provide More Accurate Predictions", *The British Journal for the Philosophy of Science* 45, 1–35.

[3] Freivalds, R. and Smith, C. (1993). "On the Role of Procrastination in Machine Learning", *Information and Computation* 107, 237–271.

[4] Garey, M. and Johnson, D. (1979). *Computers and Intractability*, New York: Freeman.

[5] Glymour, C. (1980). *Theory and Evidence*, Princeton: Princeton University Press.

[6] Goodman, N. (1983). *Fact, Fiction, and Forecast*, 4th ed., Cambridge (Mass.): Harvard University Press.

[7] Harman, G. (1965). "The Inference to the Best Explanation", *Philosophical Review* 74, 88–95.

[8] Hitchcock, C. (ed.) (2004). *Contemporary Debates in the Philosophy of Science*, Oxford: Blackwell.

[9] Jain, S., Osherson, D., Royer, J.S. and Sharma A. (1999). *Systems That Learn: An Introduction to Learning Theory*, 2nd ed., Cambridge (Mass.): MIT Press.

[10] Kelly, K. (2002). "Efficient Convergence Implies Ockham's Razor", in *Proceedings of the 2002 International Workshop on Computational Models of Scientific Reasoning and Applications* (CMSRA 2002), Las Vegas, USA, June 24–27.

[11] Kelly, K. (2004). "Uncomputability: The Problem of Induction Internalized", *Theoretical Computer Science* 317, 227–249.

[12] Kelly, K. and Glymour, C. (2004). "Why Probability Does Not Capture the Logic of Scientific Justification", in Hitchcock, C. [8], 94–114.

[13] Kuhn, T.S. (1970). *The Structure of Scientific Revolutions*, Chicago: University of Chicago Press.

[14] Laudan, L. (1981). "A Confutation of Convergent Realism", *Philosophy of Science* 48, 19–48.

[15] Mitchell, T. (1997). *Machine Learning*, New York: McGraw-Hill.

[16] Popper, K. (1968). *The Logic of Scientific Discovery*, New York: Harper and Row.

[17] Schulte, O. (1999). "Means-Ends Epistemology", *The British Journal for the Philosophy of Science* 50, 1–31.

[18] Schulte, O. (2000a). "Inferring Conservation Principles in Particle Physics: A Case Study in the Problem of Induction", *The British Journal for the Philosophy of Science* 51, 771–806.

[19] Schulte, O. (2000b). "What to Believe and What to Take Seriously: A Reply to David Chart Concerning the Riddle of Induction", *The British Journal for the Philosophy of Science* 51, 151–153.

[20] Sklar, L. (1977). *Space, Time, and Spacetime*, Berkeley: University of California Press.

[21] Van Fraassen, B. (1981). *The Scientific Image*, Oxford: Clarendon Press.

5

INDUCTION OVER THE CONTINUUM

IRAJ KALANTARI

Department of Mathematics, Western Illinois University,
Macomb, IL 61455, U.S.A., i-kalantari@wiu.edu

Abstract: We exploit the analogy between the well ordering principle for nonempty subsets of ℕ (the set of natural numbers) and the existence of a greatest lower bound for non-empty subsets of $[a, b)$ [1] to formulate a principle of induction over the continuum for $[a, b)$ analogous to induction over ℕ. While the gist of the idea for this principle has been alluded to, our formulation seems novel. To demonstrate the efficiency of the approach, we use the new induction form to give a proof of the compactness of $[a, b]$. (Compactness, which plays a key role in topology, will be briefly discussed.) Although the proof is not fundamentally different from many familiar ones, it is direct and transparent. We also give other applications of the new principle.

1 INTRODUCTION

When teaching a first course in analysis recently, I formulated a proof of the Heine-Borel Theorem. Upon a search of archives, I learned that W. L. Duren Jr. [4] had a close and similar idea, except that his analogy (and formulation) is to Zorn's Lemma as applied to a chain of intervals to get a maximal element, while the present analogy (and formulation) is to ordinary induction (and so perhaps more easily accessible). When several other theorems of analysis readily found proofs through this "inductive approach", I thought to share the idea.

[1] For real numbers a and b, $[a, b)$ is the set of real numbers between a and b, including a and excluding b; $[a, b]$ is the set of real numbers between a and b, including a and b.

M. Friend, N.B. Goethe and V.S. Harizanov (eds.), Induction, Algorithmic Learning Theory, and Philosophy, 145–154.
© 2007 *Springer*.

While ordinary induction on $\mathbb{N}$ and *transfinite induction* on ordinals both hinge upon the underlying *well ordering* structures present, and while Zorn's Lemma, the Axiom of Choice, or the *well ordering principle for every set* can play interrelated roles with those induction forms, we confine the focus of this paper to *ordinary induction* and *induction on the continuum.*

In Section 2, we state and show equivalence between some principles involving induction over the set of natural numbers. The proofs are presented to help carry the analogy for the similar principles involving induction over the continuum presented in Section 3. After briefly discussing *compactness* in Section 4, we prove the Heine-Borel theorem in Section 5, and give further applications of the new principle in Section 6. Some brief historical remarks and miscellanea are presented in Sections 7 and 8.

2 ORDINARY INDUCTION

When we wish to establish the truth of an assertion $\forall n P(n)$, where n ranges over $\mathbb{N} = \{0, 1, 2, \cdots\}$, and $P(n)$ is a predicate about n, we may define $S = \{n \in \mathbb{N} : P(n)\}$ and seek to demonstrate that $S = \mathbb{N}$ by exploiting the principle of:

Induction Over $\mathbb{N}$ (**ION**). For any $S \subseteq \mathbb{N}$, if

(1) $0 \in S$, and
(2) $\forall k[\, k \in S \Rightarrow k + 1 \in S\,]$,

then $S = \mathbb{N}$.

In using **ION**, we find efficiency and satisfaction on a few counts. For example, demonstrating (1) is often a simple verification, and establishing (2) is a boot-strapping process propelled by the assumption of $k \in S$ in pursuit of concluding $k + 1 \in S$. Furthermore, when the predicate $P(n)$ is free of quantifiers, we seem to avoid the 'magic' of a proof by contradiction which often masks and mystifies some mathematically meaningful underlying processes.

The dual principle for $\mathbb{N}$, ordered under "$\leq$", sometimes used to explain the truth of **ION**, as well as used for an indirect proof of $\forall n P(n)$, is the following:

Well Ordering Principle (**WOP**). If $T \subseteq \mathbb{N}$ and $T \neq \emptyset$, then T has a least element.

Because we wish to find conceptual parallels, we restate **ION** for $S \subseteq \mathbb{N}$ in the following equivalent form (known as *strong induction*):

ION. If
(1) $\exists k[\ (k \geq 0) \wedge ([0,k) \subseteq S)\ ]$, and
(2) $\forall k[\ [0,k) \subseteq S \Rightarrow (\exists l > k)[0,l) \subseteq S\ ]$,
then $S = \mathbb{N}$.

Here we have $[0,k) =_{\text{def}} \{j \in \mathbb{N} : 0 = j \vee j < k\}$. (So $[0,0) = \{0\}$.) At this stage, we make the following familiar observation.

Theorem 2.1 WOP ***iff*** **ION.**

Proof. ($\Rightarrow$) Assume **WOP**. Further suppose that S, a subset of $\mathbb{N}$, satisfies the properties (1) and (2) of **ION**. Presume $S \neq \mathbb{N}$. Then $S' = \mathbb{N} - S$, the complement of S, is a nonempty subset of $\mathbb{N}$. Applying **WOP** to S', let z be the least element of S'; note $0 < z$ by (1) of **ION**. Clearly, $[0,z) \cap S' = \emptyset$. So $[0,z) \subseteq S$, and by (2) of **ION**, there is $y > z$ such that $[0,y) \subseteq S$. Thus z is not the least element of S'; a contradiction. Hence our presumption is false and $S = \mathbb{N}$.

($\Leftarrow$) Assume **ION**. Further let $T \subseteq \mathbb{N}$ and presume T does not have a least element. We will show $T = \emptyset$. Consider $T' = \mathbb{N} - T$. Clearly $0 \in T'$ because otherwise 0 would be in T and its least element, contradicting the hypothesis. Thus T' satisfies condition (1) of **ION**. Next, for arbitrary k, assume $[0,k) \subseteq T'$. But then k cannot belong to T as otherwise it would be T's least element. So $k \in T'$. So, for some $l > k$, $[0,l) \subseteq T'$. (Here l is possibly just $k+1$.) So T' satisfies condition (2) of **ION**. Thus, **ION** applies to T' and $T' = \mathbb{N}$. Thus $T = \emptyset$. ∎

The proofs above are trivial but they are included to help carry the intended analogy that will follow next.

3 INDUCTION OVER THE CONTINUUM

We also recall a familiar property for nonempty, bounded subsets of $\mathbb{R}$ (the set of real numbers) under the ordering "$\leq$", for any reals $a, b \in \mathbb{R}$ with $a < b$:

The Greatest Lower Bound Principle (**GLBP**). If $T \subseteq [a,b)$ and $T \neq \emptyset$, then T has a greatest lower bound (in $[a,b)$).

Consider the interval $[a,b)$ with $a, b \in \mathbb{R}$, $a < b$, and $S \subseteq [a,b)$. We formulate an "induction scheme" over $[a,b)$ as follows.

Induction Over the Continuum (**IOC**). If
(1) $\exists x(x \geq a) \wedge ([a,x) \subseteq S)$, and

(2) $\forall x[\,[a, x) \subseteq S \Rightarrow (\exists y > x)[a, y) \subseteq S\,]$,
then $S = [a, b)$.

Here, we have $[a, x) =_{\text{def}} \{t \in [a, b) : a = t \vee t < x\}$. So $[a, a) = \{a\}$. Similarly to the case for **ION**, we establish the next two results.

Theorem 3.1 *If* **GLBP**, *then* **IOC**.

Proof. Assume **GLBP**. Further suppose that S, a subset of $[a, b)$, satisfies the properties (1) and (2) of **IOC**. Presume $S \subsetneq [a, b)$. Then $S' = [a, b) - S$ is a bounded, non-empty subset of $[a, b)$. Applying **GLBP**, let z be the greatest lower bound of S'; note $a < z < b$ by (1) of **IOC**. Clearly, $[a, z) \cap S' = \emptyset$. So $[a, z) \subseteq S$, and by (2) of **IOC**, there is $y > z$ such that $[a, y) \subseteq S$. Thus z is not the greatest lower bound of S'; a contradiction. Hence our presumption is false and $S = [a, b)$.[2] ■

Theorem 3.2 *If* **IOC**, *then* **GLBP**.

Proof. Assume **IOC**. Further let $T \subseteq [a, b)$ and presume T does not have a greatest lower bound. We will show $T = \emptyset$. Consider $T' = [a, b) - T$. Clearly $a \in T'$ because otherwise a would be in T and its greatest lower bound, contradicting the hypothesis. Thus T' satisfies condition (1) of **IOC**. Next, for arbitrary x, assume $[a, x) \subseteq T'$. But then x is a lower bound for T and is not, by hypothesis, its the greatest lower bound. So there exists y with $y > x$ such that y is a lower bound for T. Consequently, $[a, y) \subseteq T'$. So T' satisfies condition (2) of **IOC**. Thus, **IOC** applies to T' and $T' = [a, b)$. Thus $T = \emptyset$. ■

Theorems 3.1 & 3.2 combined and compared to Theorem 2.1 demonstrate the analogy: **WOP** is to **ION** as **GLBP** is to **IOC**. As **GLBP** is logically equivalent to the *completeness* of $[a, b]$, **IOC** could be assumed as an axiom in place of the completeness axiom.

[2] As **ION** often is metaphorically described through "domino theory", it seems that the motion of a "curling stone" can serve as a metaphorical description for **IOC**. Indeed it was surprising to excogitate the following quatrain, which seems to capture the idea of **IOC** closely:

The Curling Stone slides; and, having slid,
Passes me toward thee on this Icy Grid,
If what's reached is passed for'll Crystals amid,
Th'Stone Reaches thee in its Eternal Skid.

By Harak A'Myomy (12th century), translated by Walt Friz De Gradde (1897).

4 COMPACTNESS

The theorem of our central interest in this paper, which is about the *compactness* of $[a, b]$, helped streamline the development of analysis and topology. It could be said that *compact* is to pursuits in topology as *finite* is to pursuits in set theory. In the theory of sets, finite sets behave more tamely than infinite sets; in topology, compact sets behave more tamely than noncompact sets. For that reason, the concept was pursued mathematically in several different planes (including mathematical logic) simultaneously or independently during the last two centuries.

Definition 4.1 *A subset of $\mathbb{R}$, K, is **compact** if whenever a (possibly infinite) family $\mathcal{O}$ of open subsets of $\mathbb{R}$ covers K (that is the union of members of $\mathcal{O}$ contains K), there is a finite subfamily of $\mathcal{O}$ that covers K.*

To motivate this concept as well as the focus of this paper, the Heine-Borel Theorem (in the next section), we recall the following result:

A function which is continuous[3] *on $[a, b]$ is uniformly continuous*[4] *there.*

Clearly, continuity of a function at $x \in [a, b]$, is an ε, δ process where δ depends both on x and ε. However, uniform continuity guarantees a δ which depends on ε but applies to *any* point x in $[a, b]$.

While the proof of the above result using **IOC** is directly possible, we delay that until the subsequent section. In the following section, we prove the Heine-Borel Theorem, which is traditionally used to derive the above result and many other fundamental theorems of analysis.

5 THE HEINE-BOREL THEOREM

The fact that $[a, b]$ is a compact subset of $\mathbb{R}$ is established in what appears to be three distinct ways: through "there is a finite subcover for every infinite open cover"; through "an indirect proof to get a nested sequence of intervals leading to an application of Cantor's Nested Interval Theorem"; or through "every bounded sequence has a convergent subsequence". In the first mentioned (and the more frequently presented) type of proof, there seems to be a few "twists" necessitated by the indirect approach (see, for example,

[3] $f : \mathbb{R} \to \mathbb{R}$ is*continuous on* $L \subseteq \mathbb{R}$ if

$$(\forall x \in L)(\forall \varepsilon > 0)(\exists \delta_{(x,\varepsilon)} > 0)(\forall y \in L)[\ |x - y| < \delta_{(x,\varepsilon)} \Rightarrow |f(x) - f(y)| < \varepsilon\ ].$$

[4] $f : \mathbb{R} \to \mathbb{R}$ is *uniformly continuous on* $L \subseteq \mathbb{R}$ if

$$(\forall \varepsilon > 0)(\exists \delta_\varepsilon > 0)(\forall x \in L)(\forall y \in L)[\ |x - y| < \delta_\varepsilon \Rightarrow |f(x) - f(y)| < \varepsilon\ ].$$

Royden's *Real Analysis*, [11]); although the proof below is quite similar, twists are absent.

Theorem 5.1 (Heine-Borel) *For any $a, b \in \mathbb{R}$ with $a < b$, the interval $[a, b]$ is compact.*

Proof. Let $\mathcal{O}$ be an open cover for $[a, b]$. Set

$$S = \{t : t \in [a, b) \text{ and } [a, t] \text{ is contained in a finite cover from } \mathcal{O}\}.$$

Firstly, there exists $x \geq a$ such that $[a, x) \subseteq S$, as $a \in V_a \in \mathcal{O}$ for some open V_a. Secondly, assume $[a, x) \subseteq S$. Since, $x \in V_x$ for some open $V_x \in \mathcal{O}$, there exist $y > x$ and $x' < x$ with $x \in (x', y) \subseteq V_x \cap [a, b]$. As $x' \in S$, the finite cover for $[a, x']$ together with V_x confirm that $[a, y) \subseteq S$.

Thus the two statements in **IOC**'s hypothesis are satisfied and accordingly we have $S = [a, b)$. Finally, as $b \in V_b \in \mathcal{O}$ for some open V_b, there exists a $b' < b$ with $b' \in V_b \cap S$; so $[a, b]$ is contained in the finite cover comprised from the finite cover for $[a, b']$ together with V_b. So $[a, b]$ is compact. ■

6 OTHER APPLICATIONS

In this section, we use **IOC** to prove a sample of familiar theorems of elementary analysis.

Theorem 6.1 (Heine) *If f, a function from $[a, b]$ to $\mathbb{R}$, is continuous, then it is uniformly continuous.*

Proof. Assume f is continuous on $[a, b]$. To show f is uniformly continuous there, let $\varepsilon > 0$ be given. Also, let

$$S = \{t : t \in [a, b) \wedge (\exists \delta_\varepsilon > 0)(\forall u, v \in [a, t])[\ |u - v| < \delta_\varepsilon \Rightarrow |f(u) - f(v)| < \varepsilon\ ]\}.$$

We first use **IOC** to show $S = [a, b)$. Since f is continuous at a, $f(a)$ is defined and we have, trivially,

$$(\exists \delta_\varepsilon > 0)(\forall u, v \in [a, a))[\ |u - v| < \delta_\varepsilon \Rightarrow |f(u) - f(v)| < \varepsilon\ ].$$

So $[a, a) = \{a\} \subseteq S$ and S satisfies condition (1) of **IOC**.

Next, for an arbitrary $x \in [a, b)$, assume $[a, x) \subseteq S$. We wish to show, for some $y > x$, we have $[a, y) \subseteq S$. Consider x and $\frac{\varepsilon}{2}$. Since f is continuous at x, there exists $\delta_{(x, \frac{\varepsilon}{2})} > 0$, such that, for any x', if $|x - x'| < \delta_{(x, \frac{\varepsilon}{2})}$, then $|f(x) - f(x')| < \frac{\varepsilon}{2}$. (Assume $\delta_{(x, \frac{\varepsilon}{2})}$ is smaller than $x - a$ and $b - x$.)

Consider $t = x - \frac{1}{2}\delta_{(x,\frac{\varepsilon}{2})}$. Since $t \in S$, we have, there exists $\delta_\varepsilon > 0$ such that, for any $u, v \in [a, t]$, if $|u - v| < \delta_\varepsilon$, then $|f(u) - f(v)| < \varepsilon$.

Further let $y = x + \frac{1}{2}\delta_{(x,\frac{\varepsilon}{2})}$. We claim that $[a, y) \subseteq S$. We will actually show $[a, y] \subseteq S$. To establish this fact, we note that since $t \in S$ and $a \leq t' < t$ imply $t' \in S$, we need only show $y \in S$. To see this claim, for the given $\varepsilon > 0$, we offer δ^*_ε to be the minimum of δ_ε and $\frac{1}{2}\delta_{(x,\frac{\varepsilon}{2})}$. Next, consider arbitrary $u, v \in [a, y]$ with $|u - v| < \delta^*_\varepsilon$. If u, v are both in $[a, t] \subseteq S$, since $\delta^*_\varepsilon \leq \delta_\varepsilon$, by hypothesis we have:

$$|f(u) - f(v)| < \varepsilon.$$

If either u or v is in $(t, y]$, then both u and v are closer than $\delta^*_\varepsilon \leq \frac{1}{2}\delta_{(x,\frac{\varepsilon}{2})} < \delta_{(x,\frac{\varepsilon}{2})}$ to x, and so the continuity of f at x applies and (invoking the triangle inequality) we have:

$$|f(u) - f(v)| = |f(u) - f(x) + f(x) - f(v)| < |f(u) - f(x)| + |f(x) - f(v)| < \varepsilon.$$

This completes the proof of the claim, and so condition (2) of **IOC** is satisfied also. Thus, by **IOC**, $S = [a, b)$. Hence for any t with $a \leq t < b$ we have

$$(\exists \delta_\varepsilon > 0)(\forall u, v \in [a, t])[\ |u - v| < \delta_\varepsilon \Rightarrow |f(u) - f(v)| < \varepsilon\ ].$$

A similar argument applied to continuity of f at b shows that the same is true for $t = b$. Since this is true for any given $\varepsilon > 0$, f is uniformly continuous over $[a, b]$. ■

Theorem 6.2 (Cousin) *Let $\mathcal{C}$ be a collection of closed subintervals of $[a, b]$ such that for every $x \in [a, b]$ there is a corresponding 'fineness' $r_x > 0$, such that $\mathcal{C}$ contains every subintervals of $[a, b]$ with length smaller than or equal to r_x and containing x. Then there exist $x_0 = a < x_1 < x_2 < \cdots < x_n = b$ such that $[x_i, x_{i+1}]$ belongs to $\mathcal{C}$ for every $0 \leq i < n$; that is, $\mathcal{C}$ contains a partition of $[a, b]$.*

Proof. Let $S = \{t : t \in [a, b)$ and $\mathcal{C}$ contains a partition of $[a, t]\}$. Since r_a exists, for every nonnegative $t' < r_a$, $[a, a + t']$, which is in $\mathcal{C}$, is a partition of itself, putting each $a + t'$ in S. Thus $[a, r_a) \subseteq S$. Next, assume $[a, x) \subseteq S$ for $x \in [a, b)$. Working with r_x, and applying the induction hypothesis to $[a, x - \frac{r_x}{2}]$, find a partition of $[a, x - \frac{r_x}{2}]$ in $\mathcal{C}$, add to that partition the interval $[x - \frac{r_x}{2}, x + \varepsilon]$, for any $0 < \varepsilon \leq \frac{r_x}{2}$, and end up with a partition for $[a, x + \varepsilon]$ in $\mathcal{C}$. Thus $[a, x + \frac{r_x}{2}) \subseteq S$.

It is now clear that S satisfies the hypothesis for **IOC**, and therefore $S \supseteq [a, b) \supseteq [a, b - r_b]$. Since $(b - r_b) \in S$ and $[b - r_b, b] \in \mathcal{C}$, we have the desired partition for $[a, b]$. ■

Theorem 6.3 *(A form of the Intermediate Value Theorem) Let f be a continuous function over $[a, b]$ with no roots, and $f(a) > 0$. Then $f(x) > 0$ for all $x \in [a, b]$.*

Proof. Let $S = \{t : t \in [a, b)$ and f is positive over $[a, t]\}$, and apply **IOC.**■

The reader can apply **IOC** to similar theorems to find similar proofs.

7 THE RELEVANT (TELEGRAPHIC) HISTORY

It was E. Heine [7] who first (1872) implicitly proved what is now called the 'Heine-Borel' theorem while showing *if f is continuous on $[a, b]$, then f is uniformly continuous there.* Later (1895), Cousin [3] proved similar findings and he too implicitly used the Heine-Borel result. It was Borel [1] who made this result explicit in his covering theorem that any countably infinite open cover for a bounded and closed interval of $\mathbb{R}$ can be replaced with a finite subcover. Finally, Lebesgue [8] and Lindelöf [9] independently showed Borel's result is also true in case the original cover is uncountably infinite.

O. Veblen [14] proved $[a, b]$ is compact iff $[a, b]$ is closed (he did not use the term 'compact'). Later, in [15] he defined a *linear continuum*, without the use of a metric, and observed that the same method of [14] can apply to linear continua to draw similar conclusions.

Weierstrass's theorem, *a continuous function over $[a, b]$ attains its maximum at some point of* $[a, b]$, is one of the important results of analysis and related to compactness. The term *compact* was first used by Frèchet in his thesis [6] in which, motivated to generalize Weierstrass's theorem (above) for abstract topological spaces, he described a certain phenomenon which is closely related to the modern usage of the word "compact".

Most classic textbooks that prove the Heine-Borel Theorem as a 'covering' theorem use a proof similar to the one in Royden's; I have not seen an elementary textbook that has adopted Duren's method [4]. The advanced (and comprehensive) textbook, *Real Analysis* by Bruckner-Bruckner-Thomson [2], starts with Cantor's Nested Interval Theorem & Cousin's Theorem, and considers the concepts of "full" and "additive" for a collection of closed intervals to pave the way for establishing the basic results of elementary analysis.

It should be noted that Duren attributes the origin of his approach to L.R. Ford [5] who examines proofs for "statements" that are "interval-additive"; that is, those properties that hold in the union of two overlapping intervals whenever they hold in each of the two intervals. Shanahan [12, 13] rediscovers the same "additive" approach.

Moss and Roberts [10] also isolate a theme common among elementary analysis theorems akin to the approaches by Ford, Duren, Shanahan, or through **IOC**. Namely, they establish results when they find a common theme of a transitive relation on $[a, b]$ which links the points from a to b through neighborhoods whenever the relation is such that every $x \in [a, b]$ has a neighborhood whose every point to the left of x is related to its every point to the right of x.

8 MISCELLANEA

It should be clear that **IOC**, seen as a form of *increasing* induction, can be modified to yield a form of *decreasing* induction over $(a, b]$ with equivalent results. To formulate a similar form of **IOC** for $[a, b]$, the second clause has to be altered to permit $x = b$, and all mentioned intervals of the form $[a, y)$ for $y > x$ would have to be intersected with $[a, b]$ before being required to be included in S. Furthermore the proof for the principle will have to carry the burden of two cases: when $x < b$ and when $x = b$. In that event, the proofs for some of the applications will be shorter as the last 'capping' step will be already in the principle and not needed as an additional step. However, the analogy to ordinary induction would be lost.

Our heuristic observation above has been that in the context of linear orderings on $\mathbb{N}$ and $\mathbb{R}$: **WOP** is to **ION** as **GLBP** is to **IOC**. Perhaps the reader can find other interesting analogies or other applications for the formulation of **IOC** in this paper.

9 EPILOGUE

Finally, purely for efficiency and concision, we note that we could define $[0, k) =_{\text{def}} \{j : 0 \leq j < k)\}$ (so $[0, 0) = \emptyset$), and similarly define $[a, x) =_{\text{def}} \{t : a \leq t < x)\}$ (so $[a, a) = \emptyset$), and state **ION** and **IOC** in the following condensed form: if S, a subset of the universe, has some 'inductive expansion' property, then as $\emptyset \subseteq S$, S is all of the universe. Formally:

ION. If $\forall k[\ [0, k) \subseteq S \Rightarrow (\exists l > k)[0, l) \subseteq S\ ]$, then $S = \mathbb{N}$.
IOC. If $\forall x[\ [a, x) \subseteq S \Rightarrow (\exists y > x)[a, y) \subseteq S\ ]$, then $S = [a, b)$.

ACKNOWLEDGEMENTS

I am grateful to Sara Kalantari whose curiosity prompted the approach taken here. I thank Dan Velleman, Tamara Lakins, Bahman Kalantari, Ali Enayat, and John Chisholm for encouragement, correspondence and thoughtful advice. I also thank the referee for helpful remarks.

REFERENCES

[1] Borel, E. (1895). "Sur Quelques Points de la Théorie des Fonctions", *Annales Scientifiques de l'École Normale Supérieure, Paris*, 3, 1–55.

[2] Bruckner, A.M., Bruckner, J.B. and Thomson, B.S. (1997). *Real Analysis*, Upper Saddle River (New Jersey): Prentice-Hall.

[3] Cousin, P. (1895). "Sur les Fonctions de n-Variables Complexes", *Acta Mathematica* 19, 1–61.

[4] Duren, Jr., W.L. (1957). "Mathematical Induction in Sets", *American Mathematical Monthly* 64, 19–22.

[5] Ford, L.R. (1957). "Interval-Additive Propositions", *American Mathematical Monthly* 64, 106–108.

[6] Frèchet, M. (1906). "Sur Quelques Point du Calcul Fonctionnel", *Rendiconti di Palermo* 22, 1–74.

[7] Heine E. (1872). "Die Elemente der Funktionenlehre", *Crelles Journal* 74, 172–188.

[8] Lebesgue, H. (1904). *Leçons sur l'Intégration*, Paris: Gauthier-Villars.

[9] Lindelöf, E.L. (1903). "Sur Quelques Points de la Théorie des Ensembles", *Comptes Rendus Hebdomadaire des Seances de l'Academie des Sciences, Paris*, 137, 697–700.

[10] Moss, R.M.F. and Roberts, G.T. (1968). "A Creeping Lemma", *American M athematical Monthly* 75, 649–652.

[11] Royden, H.L. (1968). *Real Analysis*, New York: MacMillan.

[12] Shanahan, P. (1972). "A Unified Proof of Several Basic Theorems of Real Analysis", *American Mathematical Monthly* 79, 895–898.

[13] Shanahan, P. (1974). "Addendum to 'A Unified Proof of Several Basic Theorems of Real Analysis'", *American Mathematical Monthly* 81, 890–891.

[14] Veblen, O. (1904). "The Heine-Borel Theorem", *Bulletin of the American Mathematical Society* 10, 436–439.

[15] Veblen, O. (1905). "Definitions in Terms of Order Alone in the Linear Continuum and in Well-Ordered Sets", *Transactions of the American Mathematical Society* 6, 165–171.

II

PHILOSOPHY PAPERS

6

LOGICALLY RELIABLE INDUCTIVE INFERENCE

OLIVER SCHULTE
Department of Philosophy and School of Computing Science, Simon Fraser University, Burnaby, B.C. V5A 1S6, Canada, oschulte@sfu.ca

Abstract: This paper is an overview of formal learning theory that emphasizes the philosophical motivation for the mathematical definitions. I introduce key concepts at a slow pace, comparing and contrasting with other approaches to inductive inference such as confirmation theory. A number of examples are discussed, some in detail, such as Goodman's Riddle of Induction. I outline some important results of formal learning theory that are of philosophical interest. Finally, I discuss recent developments in this approach to inductive inference.

1 INTRODUCTION: CONVERGENCE TO THE TRUTH

The purpose of this article is to provide a brief, philosophically engaging discussion of some of the key mathematical concepts of formal learning theory. Understanding these concepts is essential for following the philosophical and mathematical development of the theory. The reader may find further discussion and defence of the basic philosophical ideas in this volume, as well as in sources such as [22, 33, 12, 14, 39, 41].

Learning theory addresses the question of how we should draw conclusions based on evidence. Philosophers have noted since antiquity that if we are interested in questions of a general nature, the evidence typically does not logically entail the answer. To start with a well-worn example, any finite number of black ravens is logically consistent with some future raven not

M. Friend, N.B. Goethe and V.S. Harizanov (eds.), Induction, Algorithmic Learning Theory, and Philosophy, 157–178.
© 2007 *Springer.*

being black. In such cases, logical deduction based on the evidence alone does not tell us what general conclusions to draw. The question is what else should govern our inferences. One prominent idea is that we should continue to seek something like a logical argument from evidence as premises to theory as conclusion. Such an argument is not guaranteed to deliver a true conclusion, but something other than truth. For example, we may seek a type of argument to the effect that, given the evidence, the conclusion is probable, confirmed, justified, warranted, rationally acceptable etc.[1]

Formal learning theory begins with an alternative response to the underdetermination of general conclusions by evidence. Empirical methods should reliably deliver the truth just as logical methods do. But unlike deduction, *inductive inquiry need not terminate with certainty.* In the learning-theoretic conception of inductive success, a method is guaranteed to eventually arrive at the truth, but this does not mean that after some finite time, the method yields certainty about what the right generalization is: An inquirer can be in the possession of the truth without being certain that she is. A philosophical forerunner of this idea is Peirce's notion that science would find the truth "in the limit of inquiry", but need never yield certainty [31]. As his fellow pragmatist William James put it, "no bell tolls" when science has found the right answer [20]. Reichenbach's pragmatic vindication of induction applied this conception of empirical success to the problem of estimating probabilities (interpreted as limits of relative frequencies) [34]. Reichenbach's student Hilary Putnam showed how the idea could be developed into a general framework for inductive inference [32, 33].[2] The notion of success in the limit of inquiry is subtle and requires some getting used to. I will illustrate it by working through two simple examples.

2 FIRST EXAMPLE: BLACK RAVENS

Consider the problem of investigating whether all ravens are black. Imagine an ornithologist who tackles this problem by examining one raven after another. There is exactly one observation sequence in which only black ravens are found; all others feature at least one nonblack raven. Figure 6.1 illustrates the possible observation sequences.

[1] For a fairly detailed but brief comparison of formal learning theory with various ways of cashing out this idea, see [25].

[2] The cognitive scientist Mark Gold independently developed the same conception of inductive inference as Putnam to analyze language acquisition [15].

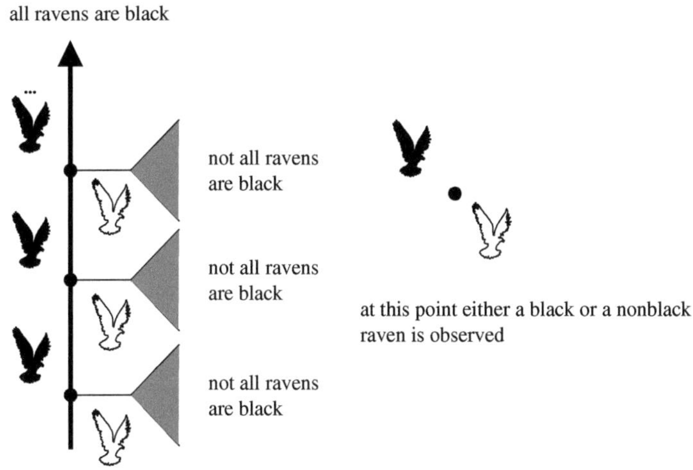

Figure 6.1. Data sequences and alternative hypotheses for the raven color problem

If the world is such that only black ravens are found, we would like the ornithologist to settle on this generalization. (It may be possible that some nonblack ravens remain forever hidden from sight, but even then the generalization "all ravens are black" at least gets the observations right.) If the world is such that eventually a nonblack raven is found, then we would like the ornithologist to arrive at the conclusion that not all ravens are black. This specifies a set of goals of inquiry. For any given inductive method that might represent the ornithologist's disposition to adopt conjectures in the light of the evidence, we can ask whether that method measures up to these goals or not. There are infinitely many possible methods to consider; we will look at just two, a sceptical one and one that boldly generalizes. The bold method conjectures that all ravens are black after seeing that the first raven is black. It hangs on to this conjecture unless some nonblack raven appears. The skeptical method does not go beyond what is entailed by the evidence. So if a nonblack raven is found, the skeptical method concludes that not all ravens are black, but otherwise the method does not make a conjecture one way or another. Figure 6.2 illustrates both the generalizing and the skeptical method.

Do these methods attain the goals we set out? Consider the bold method. There are two possibilities: either all observed ravens are black, or some nonblack raven is found. In the first case, the method conjectures that all ravens are black and never abandons this conjecture. In the second case, the method concludes that not all ravens are black as soon as the first nonblack raven is found. Hence no matter how the evidence comes in, eventually the

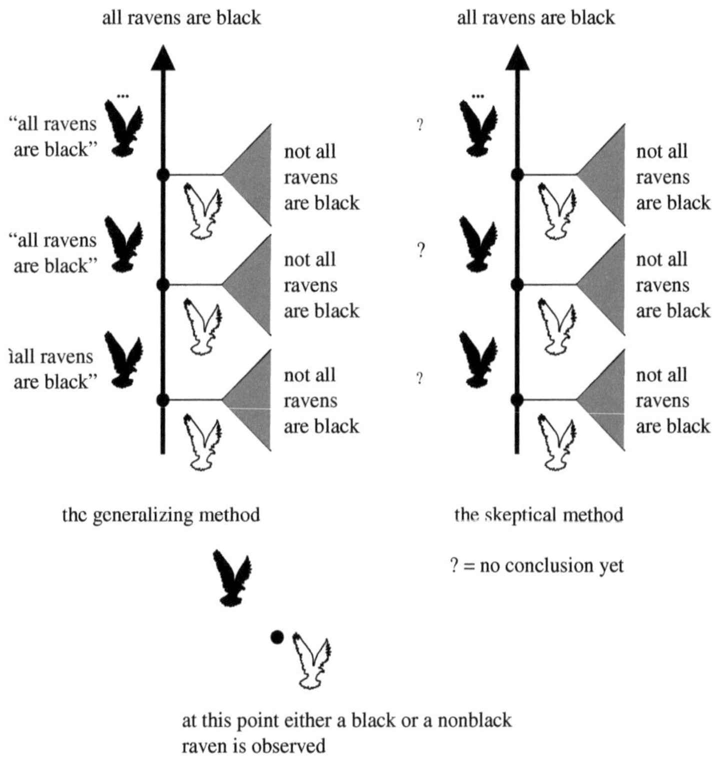

Figure 6.2. The generalizer and the skeptic in the raven color problem

method gives the right answer as to whether all ravens are black and sticks with this answer.

The skeptical method does not measure up so well. If a nonblack raven appears, then the method does arrive at the correct conclusion that not all ravens are black. But if all ravens are black, the skeptic never takes an "inductive leap" to adopt this generalization. So in that case, the skeptic fails to provide the right answer to the question of whether all ravens are black.

This illustrates how means-ends analysis can evaluate methods: the bold method meets the goal of reliably arriving at the right answer, whereas the skeptical method does not. Note the character of this argument against the skeptic: The problem, in this view, is not that the skeptic violates some canon of rationality, or fails to appreciate the "uniformity of nature". The learning-theoretic analysis concedes to the skeptic that no matter how many black ravens have been observed in the past, the next one could be white. The issue is that if all observed ravens are indeed black, then the skeptic never answers the question "are all ravens black?". Getting the right answer to that question requires generalizing from the evidence even though the generalization could be wrong.

3 SECOND EXAMPLE: THE NEW RIDDLE OF INDUCTION

Let us go through a second example to reinforce the notion of reliable convergence to the right answer.

Nelson Goodman posed a famous puzzle about inductive inference known as the (New) Riddle of Induction [16]. Our next example is inspired by his puzzle. Goodman considered generalizations about emeralds, involving the familiar colors of green and blue, as well as certain unusual ones:

> Suppose that all emeralds examined before a certain time *t* are green... Our evidence statements assert that emerald *a* is green, that emerald *b* is green, and so on... Now let me introduce another predicate less familiar than "green". It is the predicate "grue" and it applies to all things examined before *t* just in case they are green but to other things just in case they are blue. Then at time *t* we have, for each evidence statement asserting that a given emerald is green, a parallel evidence statement asserting that emerald is grue.

The question is whether we should conjecture that all emeralds are green rather than that all emeralds are grue when we obtain a sample of green emeralds examined before time *t*, and if so, why.

Clearly we have a family of grue predicates in this problem, corresponding to different "critical times" *t*; let's write $grue(t)$ to denote these. Following Goodman, I refer to "projection rules" in discussing this example. A projection rule succeeds in a world just in case it settles on a generalization that is correct in that world. Thus in a world in which all examined emeralds are found to be green, we want our projection rule to converge to the proposition that all emeralds are green. If all examined emeralds are $grue(t)$, we want our projection rule to converge to the proposition that all emeralds are $grue(t)$. Note that this stipulation treats green and grue predicates completely on a par, with no bias towards either. As before, let us consider two rules: the "natural" and the "gruesome" projection rules. The natural projection rule conjectures that all emeralds are green as long as only green emeralds are found; if a blue emerald is found, say at stage *n* for the first time, the rule conjectures that all emeralds are $grue(n)$. The "gruesome" rule keeps projecting the next grue predicate consistent with the available evidence. Expressed in the green-blue vocabulary, the gruesome projection rule conjectures that after observing some number of n green emeralds, all future ones will be blue. Figures 6.3 and and 6.4 below illustrate the possible observation sequences and the two methods mentioned in this model of the New Riddle of Induction.

How do these rules measure up to the goal of arriving at a true generalization? Suppose for the sake of the example that the only serious possibilities under consideration are that either all emeralds are green or that all emeralds

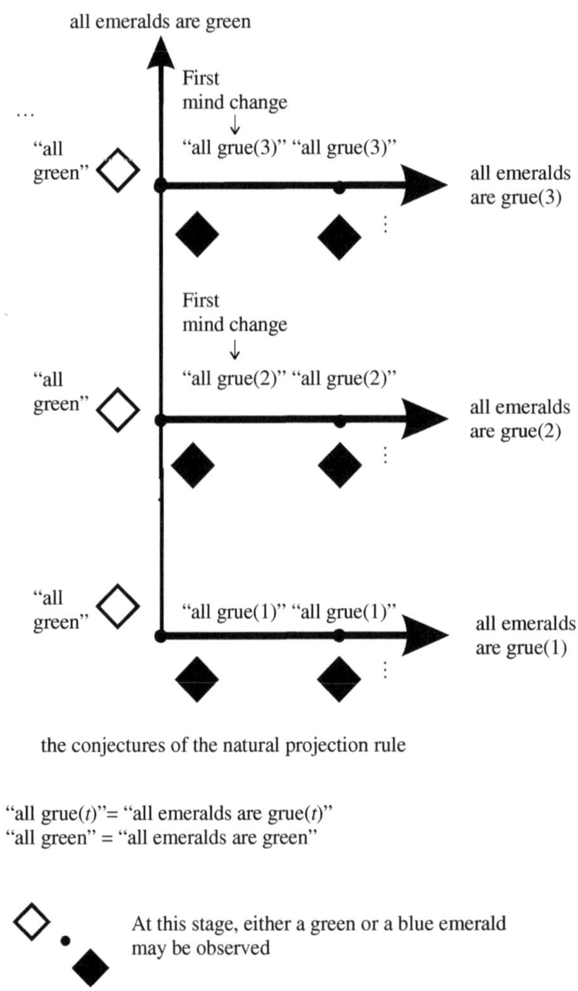

Figure 6.3. The natural projection rule in the new riddle of induction

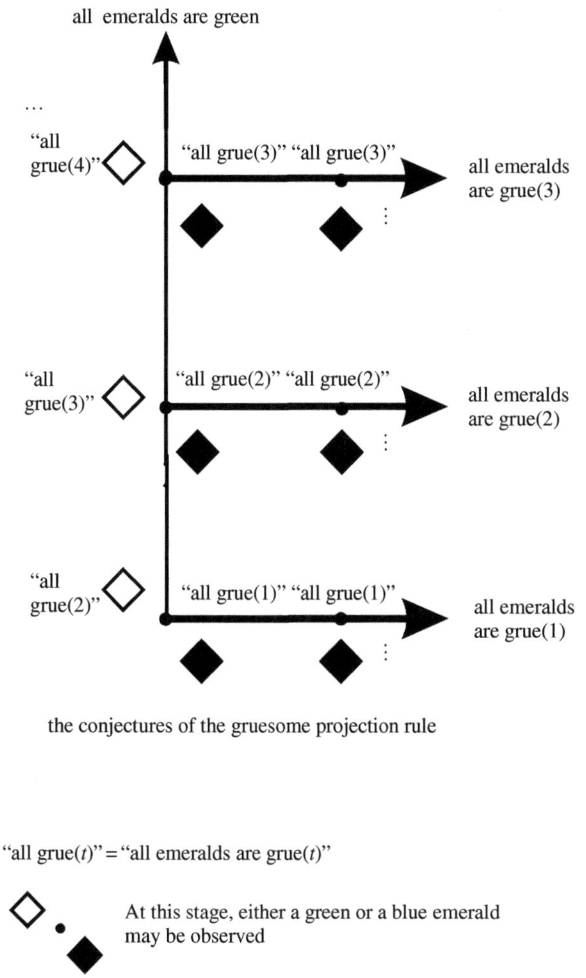

Figure 6.4. The gruesome projection rule in the new riddle of induction

are *grue*(*t*) for some critical time *t*. Then the natural projection rule settles on the correct generalization no matter what the correct generalization is. For if all emeralds are green, the natural projection rule asserts this fact from the beginning. And suppose that all emeralds are *grue*(*t*) for some critical time *t*. Then at time *t*, a blue emerald will be observed. At this point the natural projection rule settles on the conjecture that all emeralds are *grue*(*t*), which must be correct given our assumption about the possible observation sequences. Thus no matter what evidence is obtained in the course of inquiry—consistent with our background assumptions—the natural projection rule eventually settles on a correct generalization about the color of emeralds.

The gruesome rule does not do as well. For if all emeralds are green, the rule will never conjecture this fact because it keeps projecting grue predicates. Hence there is a possible observation sequence—namely those on which all emeralds are green—on which the gruesome rule fails to converge to the right generalization. So means-ends analysis would recommend the natural projection rule over the gruesome rule. Some comments are in order.

(1) As in the previous example, nothing in this argument hinges on arguments to the effect that certain possibilities are not to be taken seriously *a priori*. In particular, nothing in the argument says that generalizations with grue predicates are ill-formed, unlawlike, or in some other way *a priori* inferior to "all emeralds are green".
(2) The analysis does not depend on the vocabulary in which the evidence and generalizations are framed. For ease of exposition, I have mostly used the green-blue reference frame. However, grue-bleen speakers would agree that the aim of reliably settling on a correct generalization requires the natural projection rule rather than the gruesome one, even if they would want to express the conjectures of the natural rule in their grue-bleen language rather than the blue-green language that I have used. (For more on the language-invariance of means-ends analysis see [37, 38].)
(3) Though the analysis does not depend on language, it does depend on assumptions about what the possible observation sequences are. The example as I have described it seems to comprise the possibilities that correspond to the color predicates Goodman himself discussed. But means-ends analysis applies just as much to other sets of possible predicates. Schulte [38] and Chart [7] discuss a number of other versions of the Riddle of Induction, in some of which means-ends analysis favors projecting that all emeralds are grue on a sample of all green emeralds.

4 RELIABLE CONVERGENCE TO THE TRUTH: GENERAL CONCEPTS AND DEFINITIONS

Now that we have seen two examples of the basic idea, let us encapsulate it more generally in a mathematical definition. I begin with the description of an inductive or learning problem, which involves a specification of *possible observations, alternative hypotheses* and which hypotheses count as *correct* given a total body of evidence. Then I define the concept of an inductive method, and finally specify Putnam's and Gold's notion of empirical success for inductive methods.

4.1 Inductive Problems: Observations, Data Streams, Hypotheses, Background Knowledge and Correctness

In both examples, we have a *set of possible hypotheses* that an inquirer could adopt in the course of inquiry. In the ravens example, the set is {"all ravens are black", "not all ravens are black"}. In the Riddle of Induction, the (infinite) set of hypotheses is {"all emeralds are green", "all emeralds are $grue(1)$", "all emeralds are $grue(2)$", ..., "all emeralds are $grue(n)$", ...}. In realistic examples, the hypotheses may be considerably more complex. For instance, in language learning models the set of alternatives is the set of all grammars that may govern the language spoken in the learner's (child's) native environment [30]. In models of scientific inquiry, the alternative theories could be sets of conservation principles [40], or models of cognitive functioning [13, 4].

Another part of the specification of a learning problem is a set of *evidence items*. In the raven example, there are two kinds of evidence items "a black raven is observed", or "a nonblack raven is observed". In the Riddle of Induction, the set of evidence items is {green emerald, blue emerald}. In more realistic applications, we have many more, even infinitely many, evidence items. For example, an evidence item may be a measurement of a quantity, or set of quantities, in a physical experiment. In studying particle dynamics, the set of evidence items comprises all interactions among elementary particles that we may observe in particle accelerators [40]. In cognitive psychology, an evidence item could be the behavior profile of a subject in an experiment [13].

A *data stream* is an infinite sequence of evidence items. We write ε for a typical data stream, ε_i for the ith datum observed in the data stream ε, and $\varepsilon|n$ for the first n data observed along ε. For example, if ε is the data stream along which only green emeralds are observed, then ε_i = "green" for all i, and $\varepsilon|n$ is <"green", "green",...,"green"> for n repetitions of "green". If ε is the data stream on which all emeralds are $grue(1)$, then ε_1 = "green", ε_i = "blue" for all $i > 1$, and $\varepsilon|n$ = <"green", "blue", ..., "blue"> with $n - 1$ repetitions of "blue".

An inquirer may have background knowledge relevant to the question under investigation. For example, a particle physicist may assume that all particle reactions satisfy relativity theory. In a language learning problem, we may restrict attention only to languages with computable (total recursive) grammars. In such cases, the inquirer may be willing to rule out certain observations *a priori*. We can model the inquirer's background assumptions as a set K of data streams that represents the set of all infinite observation sequences that may arise for all the inquirer knows.

Definition 4.1 (Evidence Items and Empirical Background Knowledge) *Let E be a set of* ***evidence items****.*

1. *A data stream* ε *is an infinite sequence of evidence items. That is,* ε_n *is a member of E for each n.*
2. *The initial sequence comprising the first n observed data along* ε *is denoted by* $\varepsilon|n$*.*
3. *The inquirer's background knowledge is represented by a set of data streams K that may occur for all the inquirer knows.*

In applications of learning theory, we assume that for every data stream there is a hypothesis that is *correct* for the data stream. For example, the hypothesis "all emeralds are green" is correct for the data stream on which only green emeralds are observed. The hypothesis "not all ravens are black" is correct on any data stream on which some nonwhite raven is observed. The correctness relation between data streams and hypotheses is part of the specification of the inductive problem. Learning theory is agnostic about what correctness is. In the examples we have considered, correctness amounts to empirical adequacy: the goal is to find a generalization that makes the right predictions about what will be observed when. Correct hypotheses may be the true ones, or the simplest true ones, or simply the empirically adequate hypotheses. Another way to put it is that the correctness relation C expresses the inquirer's goals: if the total (infinite) observational data were such and such, as found in a data stream ε, then the inquirer wants to adopt a hypothesis H such that $C(H, \varepsilon)$ holds. Thus learning theory per se does not recommend to an inquirer what hypotheses she should view as correct for a total body of evidence. Rather, the theory helps the inquirer find a correct hypothesis from the partial body of evidence actually available at a given stage of inquiry.

Without going into details, it may be useful to indicate how the model of inquiry I have outlined so far corresponds to the language learning models much studied in formal learning theory. In language learning models [19], the evidence items are called "strings" and the counterpart of a data stream is a "text". The alternative hypotheses are (indices for) "languages"; a language is a set of evidence items, which models the view of a language as a set of strings.

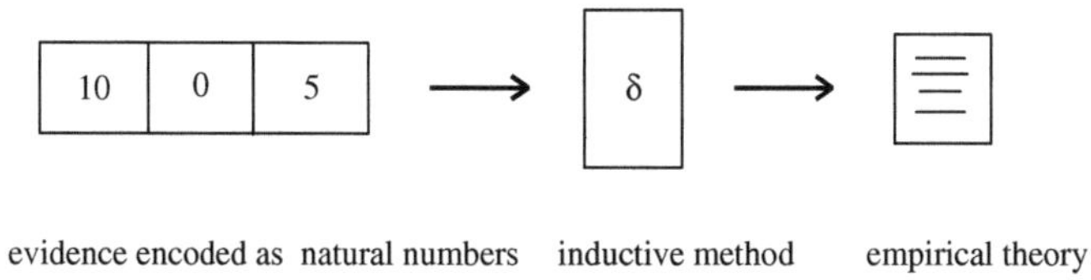

Figure 6.5. An inductive method takes as input an evidence sequence and outputs a hypothesis. Discrete evidence items can be generically represented by natural numbers (e.g., 0 for "black raven", 1 for "white raven"

4.2 Inductive Methods and Inductive Success

After observing a finite sequence of evidence items, an inquirer produces a hypothesis—her guess as to the right answer. Mathematically, this corresponds to a function that assigns a hypothesis to a finite data sequence. We also allow an inquirer to refrain from adopting an answer, which is indicated by a "?" for "no guess yet". Such a function is a mathematical representation of an inquirer's disposition to output guesses in response to evidence. Following some philosophical tradition, we refer to such a function as an inductive method, or method for short. Figure 6.5 illustrates the notion of a method.

Definition 4.2 (Inductive Methods) *Let E^* denote the set of finite evidence sequences, and let $\mathcal{H}$ be a collection of alternative hypotheses. An inductive method is a function $\delta : E^* \to \mathcal{H} \cup \{?\}$ such that for each finite data sequence E, the output $\delta(E)$ is either a hypothesis H or the vacuous output ?.*

Some comments will clarify the concept of a method in relation to other concepts and terminology.

(1) Philosophers often discuss functions from evidence to belief without calling them methods. A fairly common alternative term is "rule". For example, Goodman discusses "projection rules" for generalizing from observed emeralds. In his analysis of knowledge, Nozick does use the term "method" for a doxastic disposition [29]. Learning-theoretic analysis applies to any disposition that gives rise to belief given evidence, whether such a disposition is called "method" or not. An alternative term for method in learning theory is simply "learner", and recently the term "scientist" has come into use [19, 28].

(2) The notion of method as given in Definition 4.2 is neutral about the interpretation of adopting a hypothesis: "outputting" a hypothesis can model various epistemic attitudes that an inquirer may take towards her theory, such as belief, full belief, posit, acceptance, entertaining, etc. In

fact, learning theory is even more agnostic about the concept of belief than Definition 4.2 suggests because the framework can accommodate just about any concept of belief, including degrees of belief as in a probabilistic theory, or degrees of confirmation as in confirmation theory. For example, Putnam investigated whether Carnap's inductive methods (his "c-functions") arrive at the right answer, in the sense that whatever the true generalization is, eventually the true generalization always receives degree of confirmation greater than $1/2$ [32]. Or we can ask whether the degree of belief of a Bayesian agent in the true generalization will come arbitrarily close to 1 ([21]; [9] Ch. 9.6; [28], Sec. 3.6.9).

In general, to apply learning theory it suffices to have a notion of an (epistemic) state s and a correctness relation $Correct(\varepsilon, s)$ that specifies the correct states for the agent to be in, given that the total observational facts are as described by the data stream ε. The point is that learning theory does not presuppose, and hence does not depend on, a particular analysis of belief or epistemic attitudes. Rather, the theory addresses the question of how best to change one's belief, however understood.

(3) The notion of method as given in Definition 4.2 is agnostic about internal facts concerning how the agent arrives at her hypothesis. In effect, the definition views a method as a black box, as suggested in Figure 6.5. Learning theory focuses on the behavior and performance of epistemic dispositions, not on their internal structure. As a consequence, learning theoretic analysis applies to any recommendation for how we should reason from evidence to theory: whether the proposal is to follow a certain style of argument (e.g., probabilistic), seek a certain kind of conformation (e.g., Carnap's c-functions [5] or Glymour's bootstrap confirmation [16]), or adopt some set of normative criteria for rational belief formation: we can always ask whether those ways of producing belief would lead an inquirer to the correct hypothesis (cf. [39]).

With Definitions 4.1 and 4.2 in hand, we are ready to define Putnam's and Gold's conception of empirical success.

Definition 4.3 (Reliable Convergence to the Correct Hypothesis) *Let E be a set of evidence items, $\mathcal{H}$ a set of alternative hypotheses, C a correctness relation that specifies which hypotheses are correct for each data stream ε comprising observations drawn from E.*

1. *A method δ converges to, or identifies, a correct hypothesis H on a data stream $\varepsilon \iff H$ is correct for ε and there is a stage n such that $\delta(\varepsilon|n') = H$ for all stages $n' \geq n$.*

2. *A method δ is reliable for, or identifies, $\mathcal{H}$ given background knowledge $K \iff$ for all data streams ε consistent with K (i.e., ε in K), the method δ converges to a correct hypothesis on ε.*

To illustrate this definition, we verified in Section 6.3 that the natural projection rule reliably identifies a true generalization about emerald colors given the set of alternatives {"all green", "all $grue(1)$", ...}. The gruesome method that keeps predicting that the next emerald is blue fails to converge to "all emeralds are green" on the data stream featuring only green emeralds. Definition 4.3 envisions a method converging to a single hypothesis; in algorithmic learning theory, this corresponds to "EX-learning"—see the introductory paper in this volume.

Part of the traditional concept of a method, for example in Mill and arguably in Aristotle, is that a method should be a step-by-step reasoning procedure. The definition above does not require that a method should be easy to follow. In modern terms, a step-by-step procedure of the sort sought by traditional philosophers corresponds to an *algorithm* which by Church's thesis can be implemented on a Turing machine. It is therefore natural to require that methods should be algorithmic or computable. Such an algorithm provides a step-by-step procedure for following the method. Much of formal learning theory studies algorithmic methods, so much so that the subject is often referred to as algorithmic learning theory (as in the title of this volume) or computational learning theory.

Some striking results in algorithmic learning theory examine what norms of inductive reasoning help agents with bounded cognitive powers and which hinder them in attaining the aims of inquiry. The point is not the trivial one that the deductive abilities of agents limited to the reasoning powers of a Turing machine fall short of ideal logical omniscience. Rather, it turns out that computable inquiry sometimes requires fundamentally different strategies than inquiry by idealized agents. In such cases, trying to approximate or "get as close as possible" to the ideal norm can be a bad strategy for computable agents seeking to identify a correct hypothesis.

For example, consider the seemingly banal *consistency principle*: do not accept a hypothesis that is inconsistent with the data (see for example Hempel's "conditions of adequacy" for a definition of scientific confirmation [17]). Kelly and Schulte describe an inductive problem with an empirical hypothesis H such that a step-by-step reasoning procedure can reliably identify in the limit whether or not H is correct, but inductive methods even with infinitely uncomputable reasoning powers cannot do so—if they are required to satisfy the consistency principle. (For another restrictiveness result along these lines, see ([28], Prop. 60)). Intuitively, the main reason why the consistency principle restricts the potential of computable inquiry is that an agent with bounded logical powers cannot immediately recognize when a hypothesis is inconsistent with the data, but must first gather *more data*. The consistency principle rules out this inductive strategy because it mandates that an agent should reject a hypothesis as soon as it is refuted. For further discussion of the differences between methodology for logically omniscient agents and those with bounded deductive abilities, see [24]; [22], Ch. 6, 7, 10; [28].

5 ADDITIONAL EPISTEMIC GOALS: FAST AND STABLE CONVERGENCE TO THE TRUTH

The seminal work of Putnam and Gold focused on reliable convergence to a correct hypothesis. A major extension of their approach is to consider cognitive desiderata *in addition to finding a correct hypothesis* (such desiderata are called "identification criteria" in the computer science literature [6]). In this section, I consider two epistemic aims that have received considerable attention from learning theorists: stable and fast convergence to a correct theory.

The motivation for examining convergence speed is that other things being equal, we would like our methods to arrive at a correct theory sooner rather than later. A venerable philosophical tradition supports the idea that stable belief is a significant epistemic good. Since Plato's Meno, philosophers are familiar with the idea that stable true belief is better than unstable true belief, and epistemologists such as Sklar [43] have advocated similar principles of "epistemic conservatism". Church's thesis tells us that a major reason for conservatism in paradigm debates is the cost of changing scientific beliefs [26]. In this spirit, learning theorists have examined methods that minimize the number of times that they change their theories before settling on their final conjecture.

As it turns out, the idea of adding cognitive goals in addition to finding a correct hypothesis addresses a long-standing objection to identification in the limit. Reichenbach's student Salmon criticized his teacher's pragmatic vindication of induction on the grounds that the vindication, even if successful, leaves belief underdetermined in the short run [35]. The reason is that while Reichenbach's straight rule is guaranteed to approach the true probability of an event, so are infinitely many other rules. For example, consider a rule δ

that estimates the probability of a coin coming up heads to be 1 for 1,000 tosses no matter what the outcome of the tosses is. After 1,000 tosses, δ switches to following the straight rule. Thus in the limit of inquiry, the rule δ converges to the same answer as the straight rule does.

From this example it is easy to see the general pattern: Suppose that δ is a reliable method; let e be any evidence sequence, and H be any hypothesis. Then there is a method δ' that outputs H on e and follows the reliable method δ on any other evidence. So δ' converges to the same hypothesis as δ and thus δ' is reliable. This shows that any conjecture H on any evidence e is consistent with long-run reliability.

The situation changes drastically if we take into account other aspects of empirical success. Several general recent results show that maximizing stable belief, or minimizing mind changes, strongly constrains the conjectures of optimal inductive methods in the short run. I will illustrate the power of additional epistemic goals in the two simple traditional examples already considered.

First, we need to define what it is for an inductive method to succeed with respect to an epistemic goal. For a given epistemic desideratum, a method may perform well in some circumstances but not in others. To compare the performance of methods with regard to a range of possible ways the world might be—more precisely, with regard to all the data streams consistent with background knowledge—we may apply two familiar principles from decision theory: **admissibility** and **minimax**. A method is *admissible* iff it is not *dominated*. In general, an act A dominates another act A' if A necessarily yields results at least as good as those of A', and possibly better ones, where a given collection of "possible states of the world" determines the relevant sense of necessity and possibility. An act A *minimaxes* if the worst possible outcome from A is as good as the worst possible outcome from any other act.

For the two epistemic desiderata of minimizing time-to-truth and reversals of opinion, applying the two decision-theoretic criteria of admissibility and minimax yields $2 \times 2 = 4$ identification criteria. It turns out that two of these, admissibility for mind changes and minimaxing convergence time, are feasible only for empirical questions that pose no genuine problem of induction; more precisely, they are feasible only if the data are eventually guaranteed to entail which hypothesis is correct. (For the details see [37].) Thus learning theorists have focused on minimaxing theory changes and admissibility with respect to convergence time. I will discuss minimizing reversals of opinion in the remainder of this section and the next and then return to admissibility with respect to time-to-truth.

5.1 Stable Convergence to a Correct Hypothesis

We say that a method δ **changes its mind** on a data sequence $e_1, ..., e_n, e_{n+1}$ if the method's output on the previous data $e_1, ..., e_n$ is not? (i.e., $\delta(e_1, ..., e_n) \neq ?$) and differs from its output at stage $n+1$ (i.e., $\delta(e_1, ..., e_n) \neq \delta(e_1, ..., e_n, e_{n+1})$). No mind changes occur on the empty data sequence.

Definition 5.1 (Stable Belief: Minimizing Mind Changes) *Suppose that δ is a reliable discovery method for alternative hypotheses $\mathcal{H}$ given background knowledge K.*

1. *The number of mind changes of δ on data stream ε is given by $MC(\delta, \varepsilon) \equiv |\{n : \delta$ changes its mind on $\varepsilon|n\}|$.*
2. *The method δ succeeds with at most n mind changes given $K \iff MC(\delta, \varepsilon) \leq n$ for all data streams ε consistent with K.*
3. *The method δ **minimaxes mind changes** given hypotheses $\mathcal{H}$, background knowledge $K \iff$ there is no other reliable method δ' for $\mathcal{H}$ such that the maximum number of times that δ might change its mind, given background knowledge K, is greater than the same maximum for δ'.*

The New Riddle of Induction turns out to be a nice illustration of minimizing mind changes. Consider the natural projection rule (conjecture that all emeralds are green on a sample of green emeralds). If all emeralds are green, this rule never changes its conjecture. And if all emeralds are $grue(t)$ for some critical time t, then the natural projection rule abandons its conjecture "all emeralds are green" at time t—one mind change—and thereafter correctly projects "all emeralds are $grue(t)$". Hence the natural projection rule changes its mind at most once in the New Riddle of Induction (see Figure 6.3). Remarkably, rules that project grue rather than green do not do as well. For example, consider a rule that conjectures that all emeralds are $grue(3)$ after observing one green emerald. If two more green emeralds are observed, the rule's conjecture is falsified and it must eventually change its mind, say to conjecture that all emeralds are green (suppose that green emeralds continue to be found). But then at that point, a blue emerald may appear, forcing a second mind change. This argument can be generalized to show that the aim of minimizing mind changes allows only the green predicate to be projected on a sample of all green emeralds ([37], Prop. 11). Figure 6.6 illustrates in a typical case how an unnatural projection rule may have to change its mind twice or more. From the insight illustrated in Figure 6.6, we can establish the optimality of the natural projection rule.

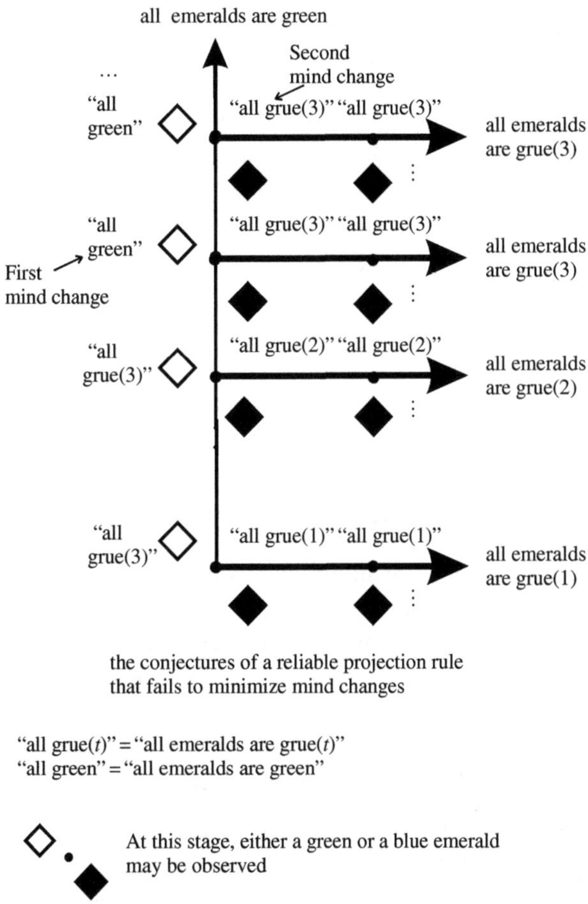

Figure 6.6. A reliable projection rule that projects a grue predicate on an all green sample of emeralds can be forced to change its mind twice

Proposition 5.2 *Let δ be any projection rule (inductive method) that reliably identifies a true generalization about emerald colors in the Riddle of Induction and changes its conjecture at most once. Let e be any finite sequence featuring only green emeralds (i.e., e is of the form <green emerald, green emerald, ...>. Then either δ(e) = ?—the method makes no guess—or δ(e) = "all emeralds are green".*

Less formally, the proposition says that after observing a sequence of emeralds consistent with "all emeralds are green", an optimal method must conjecture "all emeralds are green" or else withhold opinion. The criteria of reliable convergence to the truth and stable belief do not determine how many instances exactly are required for inference to "build up enough confidence" and "take an inductive leap". These goals do determine that (1) a reliable method must eventually take an inductive leap, and (2) when the method does adopt a universal generalization in the Riddle of Induction, on a sample of all green emeralds that generalization must be "all emeralds are green".

In the ravens example, the results of the analysis are similar. A reliable method that minimaxes retractions may withhold opinion on a sample of all black ravens, but if it does generalize beyond the data, it must conjecture that all ravens are black rather than that some nonblack raven will appear in the future. Our two examples illustrate the typical pattern for methods that achieve as much stable belief as possible: minimizing mind changes determines the what of inductive generalizations, but not the when. (For more precise statements and proofs of this principle, see [27, 45].)

5.2 Fast Convergence to a Correct Hypothesis

Let us return to the idea of minimizing time-to-truth. Formally, we may develop this success criterion as follows. Define the **convergence point** of a method δ on a data stream ε to be the time at which the method starts to converge to an answer. That is, $CP(\delta, \varepsilon) \equiv$ the least n such that $\delta(\varepsilon|n) = \delta(\varepsilon|n')$ for all $n' \geq n$. For a set of alternative hypotheses $\mathcal{H}$ and given background knowledge K, an inductive method δ **dominates** another inductive method δ' with respect to convergence time $\iff$

1. background knowledge K entails that δ converges no later than δ' does (i.e., $CP(\delta, \varepsilon) \leq CP(\delta', \varepsilon)$ for all $\varepsilon \in K$), and
2. there is some data stream, consistent with background knowledge K, on which δ converges before δ' does (i.e., there is $\varepsilon \in K$ such that $CP(\delta, \varepsilon) < CP(\delta', \varepsilon)$).

A method δ is data-minimal given K if no other reliable method for $\mathcal{H}$ dominates δ with respect to convergence time ([22], Ch. 4.8; see also [28], Def. 28).

There is a theorem that characterizes the properties of data-minimal methods [38], Th. 8; [28], Ex. 39. A consequence of the theorem is that data-minimal methods always adopt a definite belief—that is, they never output "?". Intuitively, suspending belief loses time, because the method could have begun converging to a true belief instead. For our examples, it follows that the natural projection rule in the Riddle of Induction and the bold generalizer in the ravens problem are the only reliable data-minimal methods that minimax retractions.

6 FURTHER EXTENSIONS AND APPLICATIONS

This section indicates some further extensions and developments of the theory of reliable inquiry with additional epistemic values.

(1) Many problems do not allow a finite bound on mind changes, although there is still an intuitive sense that some methods achieve stable belief more than others. Freivalds showed how the notion of a finite mind change bound can be extended to an ordinal or transfinite bound [11]. This well-studied criterion considerably enhances the range of inductive problem in which the goal of minimizing mind changes is feasible [19].

(2) Although problems such as the Riddle of Induction and generalizing about black ravens may appear very different on the surface, there is a common structure to problems that can be solved with at most 1 mind change, as Figures 6.1 and 6.3 suggest. This holds true for any finite and even transfinite mind change bounds. The common deep structure of problems solvable with a given mind change bound can be explicated in terms of point-set topology (cf. [38, 22], Ch. 4; [27], Sec. 3). For language and function learning problems, which are commonly studied in Computational Learning Theory, the mind change complexity of an inductive problem is characterized by Cantor's classic concept of accumulation order ([2]; [27], Th. 1).

The fact that the goals of true and stable belief place such strong constraints on inductive inference allows us to evaluate specific inference methods with respect to how well they serve these goals. Pursuing this question almost always leads to insights into the inductive problem under investigation, increases our understanding of known learning methods, and can lead to the development of new methods. I conclude this introduction with some brief illustrations of applying this kind of learning-theoretic analysis in some fairly realistic inference problems.

(1) An inductive problem that arises in particle physics is to find a set of conservation laws that correctly predict which reactions among elementary particles are possible [40]. A prominent type of conservation law consists of additive conservation laws , also known as selection rules. It can be shown that there is a unique optimal method for inferring selection rules [40]. It turns out that the standard set of laws that particle physicists have actually adopted makes exactly the same predictions as the output of the learning-theoretically optimal method [42].

(2) Angluin introduced the well-known concept of a "pattern" for describing a set of strings [1]. For example, the pattern 0xx1 describes such strings

as 0001, 0111, 000001, 011111. A one-variable pattern is a pattern that contains at most one distinct variable, such as 0xx1. Angluin provided an inference algorithm for identifying a one-variable pattern in the limit that does not, however, minimize mind changes ([27], Sec. 5). Luo and Schulte describe a different algorithm that is mind change optimal (moreover, their algorithm requires time only linear in the length of a data sequence *e* to produce a conjecture for *e*).

(3) Kelly has generalized the idea of reliable inference with bounded mind changes to settings of statistical inference concerned with statistical theories that determine the distribution of observed variables [23], Sec. 11. In that setting Kelly argues that the standard practice of testing statistical point hypothesis testing is mind change optimal. Another application of Kelly's analysis are problems of causal inference. In causal inference, a basic problem is to find which variables are directly causally linked to each other (e.g., there is a direct connection between "tar content in lung" and "lung cancer" which mediates the indirect connection between "smoking" and "lung cancer"). Standard methods for causal inference conjecture that there is no direct link between two variables unless and until a direct connection is conclusively verified (by statistical tests). Kelly argues that this inference method is mind change optimal ([23], Sec. 11).

In conclusion, formal learning theory provides a rich set of concepts for analyzing the complexity of inductive problems and the performance of inductive methods. In applications, these analytical tools have yielded insights into the learning problem, validated existing learning methods and led to the development of new ones. One goal of this article was to lay out some of the basic concepts and techniques that underlie learning-theoretic analysis to invite the development of further applications.

REFERENCES

[1] Angluin, D. (1980). "Finding Patterns Common to a Set of Strings", *Journal of Computer and System Sciences* 21, 46–62.

[2] Apsitis, K. (1994). "Derived Sets and Inductive Inference", in Arikawa, S. and Jantke, K.P. [3], 26–39.

[3] Arikawa, S. and Jantke, K.P. (eds.) (1994). *Proceedings of ALT 1994*, Berlin: Springer-Verlag.

[4] Bub, J. (1994). "Testing Models of Cognition Through the Analysis of Brain-Damaged Performance", *The British Journal for the Philosophy of Science* 45, 837–855.

[5] Carnap, R. (1952). *The Continuum of Inductive Methods*, Chicago: University of Chicago Press.

[6] Case, J. and Smith, C. (1983). "Comparison of Identification Criteria for Machine Inductive Inference", *Theoretical Computer Science* 25, 193–220.

[7] Chart, D. (2000). "Discussion: Schulte and Goodman's Riddle", *The British Journal for the Philosophy of Science* 51, 147–149.

[8] Drew, M.S. and Schulte, O. (2006). "An Algorithmic Proof That the Family Conservation Laws are Optimal for the Current Reaction Data", *High Energy Physics Preprint Archive* (preprint: http://arxiv.org/abs/hep-ph/0602011).

[9] Earman, J. (1992a). *Bayes or Bust?*, Cambridge (Mass.): MIT Press.

[10] Earman, J. (ed.) (1992b). *Inference, Explanation and Other Frustrations*, Berkeley: University of California Press.

[11] Freivalds, R. and Smith, C. (1993). "On the Role of Procrastination in Machine Learning", *Information and Computation* 107, 237–271.

[12] Glymour, C. (1991). "The Hierarchies of Knowledge and the Mathematics of Discovery", *Minds and Machines* 1, 75–95.

[13] Glymour, C. (1994). "On the Methods of Cognitive Neuropsychology", *The British Journal for the Philosophy of Science* 45, 815–835.

[14] Glymour, C. and Kelly, K. (1992). "Thoroughly Modern Meno", in Earman, J. [10].

[15] Gold, E.M. (1967). "Language Identification in the Limit", *Information and Control* 10, 447–474.

[16] Goodman, N. (1983). *Fast, Fiction and Forecast*, 4th ed., Cambridge (Mass.): Harvard University Press.

[17] Hempel, C.G. (1965). *Aspects of Scientific Explanation*, New York: Free Press.

[18] Houser, N. and Kloesel, C. (eds.) (1992). *The Essential Peirce* 1, Bloomington: Indiana University Press.

[19] Jain, S., Osherson, D., Royer, J.S. and Sharma, A. (1999). *Systems That Learn: An Introduction to Learning Theory*, 2nd ed., Cambridge (Mass.): MIT Press.

[20] James, W. (1982). "The Will to Believe", in Thayer, H.S. [44].

[21] Juhl, C. (1997). "Objectively Reliable Subjective Probabilities", *Synthese* 109, 293–309.

[22] Kelly, K. (1996). *The Logic of Reliable Inquiry*, Oxford: Oxford University Press.

[23] Kelly, K. (2004). "Justification as Truth-Finding Efficiency: How Ockham's Razor Works", *Minds and Machines* 14, 485–505.

[24] Kelly K. and Schulte, O. (1995). "Church's Thesis and Hume's Problem", *Proceedings of the IX International Joint Congress for Logic, Methodology and the Philosophy of Science*, Dordrecht: Kluwer.

[25] Kelly, K., Schulte, O. and Juhl, C. (1997). "Learning Theory and the Philosophy of Science", *Philosophy of Science* 64, 245–267.

[26] Kuhn, T.S. (1970). *The Structure of Scientific Revolutions*, Chicago: University of Chicago Press.

[27] Luo, W. and Schulte, O. (2005). "Mind Change Efficient Learning", in *Learning Theory: 18th Annual Conference on Learning Theory (COLT 2005), Lecture Notes in Artificial Intelligence* 3559, Auer, P. and Meir, R. (eds.), Bertinoro (Italy): Springer-Verlag, 398–412.

[28] Martin. E. and Osherson, D. (1998). *Elements of Scientific Inquiry*, Cambridge (Mass.): MIT Press.

[29] Nozick, R. (1981). *Philosophical Explanations*, Cambridge (Mass.): Harvard University Press.

[30] Osherson, D., Stob, M. and Weinstein, S. (1986). *Systems That Learn: An Introduction to*

Learning Theory for Cognitive and Computer Scientists, Cambridge (Mass.): MIT Press.

[31] Peirce, C.S. (1878). "How to Make Our Ideas Clear", in Houser, N. and Kloesel, C. [18], 124–141.

[32] Putnam, H. (1963). "'Degree of Confirmation' and Inductive Logic", in Schilpp, P.A. [36].

[33] Putnam, H. (1975). "Probability and Confirmation", in *Mathematics, Matter, and Method*, Cambridge: Cambridge University Press.

[34] Reichenbach, H. (1949). *The Theory of Probability*, London: Cambridge University Press.

[35] Salmon, W.C. (1991). "Hans Reichenbach's Vindication of Induction", *Erkenntnis* 35, 99–122.

[36] Schilpp, P.A. (ed.) (1963). *The Philosophy of Rudolf Carnap*, La Salle (Ill.): Open Court.

[37] Schulte, O. (1996). "Means-Ends Epistemology", *The British Journal for the Philosophy of Science* 79-1, 141–147.

[38] Schulte, O. (1999). "The Logic of Reliable and Efficient Inquiry", *The Journal of Philosophical Logic* 28, 399–438.

[39] Schulte, O. (2000a). "Review of Martin and Osherson's 'Elements of Scientific Inquiry'", *The British Journal for the Philosophy of Science* 51, 347–352.

[40] Schulte, O. (2000b). "Inferring Conservation Laws in Particle Physics: A Case Study in the Problem of Induction", *The British Journal for the Philosophy of Science* 51, 771–806.

[41] Schulte, O. (2005). "Formal Learning Theory", in Zalta, E. [46].

[42] Schulte, O. and Drew, M.S. (2006). "Algorithmic Derivation of Additive Selection Rules and Partice Families from Reaction Data" (preprint: http://arxiv.org/abs/hep-ph/0602011).

[43] Sklar, L. (1975). "Methodological Conservatism", *Philosophical Review* LXXXIV, 374–400.

[44] Thayer, H.S. (ed.) (1982). *Pragmatism*, Indianapolis: Hackett.

[45] Wei, L. and Oliver, S. (2006). *Logic and Computation,* 204:989–1011.

[46] Zalta, E. (ed.) (2005). *The Stanford Encyclopedia of Philosophy*, Summer 2005 ed., (The Metaphysics Research Lab, Stanford University) (www-version: http://plato.stanford.edu/archives/sum2005/entries/learning-formal/).

7

SOME PHILOSOPHICAL CONCERNS ABOUT THE CONFIDENCE IN 'CONFIDENT LEARNING'

MICHÈLE FRIEND
Department of Philosophy, The George Washington University, Washington, D.C. 20052, U.S.A., michele@gwu.edu

Abstract: A learner is engaged in "**confident learning**" when we are guaranteed that the hypotheses put out by the learner will converge in the limit. Whence the word "confidence"? We are confident that the learner will eventually learn. The question raised in the paper is: what does the learner really learn? Friend applies the **Putnam permutation argument** to the scenario of confident learning to undermine our confidence.

It is true that the learner learns an **equivalent** algorithm to that put out by the informant. However, it only has to be equivalent in some respects, in order to converge. That is, the algorithm does not have to be **identical** to the algorithm put out by the informant. As a result, the learning is somewhat superficial. Our confidence can only be placed at a superficial level of learning, which might be alright for some purposes, but not for others.

1 INTRODUCTION

There are six sections in this paper: this introductory section, a discussion about what confident learning is which will mainly consist in definitions, a general philosophical presentation of the problem, a discussion about a result from model theory which underpins the problem with "confident learning", and finally, a section on some questions which arise in the philosophy of language from the above discussion concerning the metaphor of "learning".

Much of the work in the early sections is exploration of the concepts, and the vocabulary used to discuss those concepts. This is necessary since there

M. Friend, N.B. Goethe and V.S. Harizanov (eds.), Induction, Algorithmic Learning Theory, and Philosophy, 179–197.
© 2007 *Springer*.

is a lot of slip and ambiguity in the language used in learning theory and its surrounding disciplines.

We shall narrow the discussion by considering a scenario wherein a learner is presented with a class of (formal) first-order languages, and some data (example well-formed formulas) from one of the languages. The learner then has to guess which language, or class of languages, the data comes from. The scenario could be re-expressed in terms of a learner learning a set or learning a function. Using this terminology, we are thinking of particular first-order languages as functions, and of the class of first-order language as a class of functions. A learner has the task of learning a function from a set of data in the form of n-tuple arguments and single values. The apparent limitations imposed, by the scenario discussed here, will be expanded in the fifth section because it is worth exploring the parameters of the philosophical problem.

2 CONFIDENT LEARNING

This section is mainly composed of definitions, with a few interjections for giving the intuitive, or less formal, idea behind the definitions. Through the examination of the definitions we can distil a clear picture of the notion of confident learning. Apologies are made in advance for tedium of style. Terms in bold are the ones being defined. Terms which are italicized, will receive a definition subsequently.

A *learner* is engaged in "**confident learning**" when we are guaranteed that *the hypotheses put out by the learner* will *converge in the limit*. What does all this mean?

A "**learner**" is thought of as a Turing machine with some classes of algorithm for generating *grammars*. "**Grammars**" are names for *languages* ([3], p. 449).

This is non-standard terminology in some circles, so it is worth discussing the point of using the vocabulary this way. Grammars can be thought of as elaborate functions which allow the learner to discriminate the "good" from the 'bad' data, which is given to the learner by the *informant*. "Good data" will be well-formed formulas. "Bad data" will be ill-formed formulas.

A "**language**" is always a formal language, or rather, can be expressed as a formal language. We shall be interested in *classes of first-order formal languages*. A "**class of first-order languages**" is a class whose members each contain a list of *vocabulary* and a function for generating well-formed formulas. What counts as "**Vocabulary for a first-order language**"? The vocabulary will be taken from the following: brackets,[1] *logical connectives* and *operators*. **Operators** modify the meaning of whole expressions or sentences in some way, as opposed to just being components in the sentence. Familiar examples of operators are the universal quantifier over object-level variables, the existential quantifier over object-level variables and modal operators over well-formed formulas. The vocabulary will also include first-order variables, proposition variables, first-order predicate variables, first-order function variables and first-order relation variables. The vocabulary also possibly includes constants: first-order constants, predicate constants, function constants and relation constants.

An example of a first-order constant is a number, such as 0. An example of a first-order predicate constant is "is a number". An example of a function constant is addition. An example of a relation constant is identity. Intuitively, first-order languages spread out from *first-order monadic logic*, with no constants, to its extensions to make a class of languages. We can extend first-order monadic logic by adding more first-order operators, such as "provability" or "logical necessity". We can also extend first-order monadic logic by adding polyadic relation or function variables or constants. We might extend first-order monadic logic to, say, first-order group theory or to first-order arithmetic. "**Logical connectives**" minimally include negation plus one binary connective chosen from: conjunction, disjunction, implication and the biconditional. Equivalently, we can have negation plus several binary connectives. A minimal version of "**first-order monadic logic**" contains only: brackets,[2] logical connectives, the universal first-order quantifier, an infinite supply of first-order variables and an infinite supply of predicate letters. We now have a precise idea of the class of first-order languages. It is a subclass of this which our learner learns confidently. The learner learns on the basis of data given out by an *informant*.

An "**informant**" is usually thought of as a Turing machine. An informant might not be a Turing machine. The intuitive idea is that the informant is a little like a teacher in a class room, or like an adult teaching a child how to speak correctly. An informant will give out both *positive* and *negative data*. The informant tells the learner whether the data is positive or negative. A "**positive datum**" is an ordered n-tuple which is a member of the class

[1] Strictly speaking these are "structural", and should not really be included in the "vocabulary", however, the distinction between structural, and meaningful, vocabulary is not important here.

[2] Strictly speaking, we can dispense with brackets or parentheses by adopting a convention about ordering of the vocabulary, as was developed by the Poles. The system is known as 'Polish notation'. It is provably equivalent to the more usual use of brackets or parentheses.

to be learned. The n-tuple consists in a string of symbols taken from the vocabulary. If the data is given by an informant, then there will also be a label informing the learner whether the datum is a member of the class of languages to be learned or not. If the datum is not from the class to be learned, then it is **negative datum**. The data are not necessarily presented in any order. However, it is understood that no data is missed out *in the limit*.

The learner begins by "listening" that is, by just sending out *question marks*. Each **question mark** solicits a new datum from the informant. Based on the information received, at some point the learner will start to generate *hypotheses*. The **hypotheses** concern the generation of data. The hypotheses can take the form of a definition of a function, a definition of a class of functions, or simply a series of data. Usually, all three types of hypothesis can be compared to the information which the teacher has.[3]

The intuitive idea comes from thinking of a child who learns to speak.[4] At first the child is simply exposed to examples of good speech, and then later initiates speech. In the mathematical modeling of this sort of scenario, we can make a simplifying assumption about the learner not initiating speech, but simply learning to discriminate positive from negative examples. Here, we shall adopt this simplifying assumption, since the added complication of a learner initiating data is unnecessary for the future philosophical considerations.

We shall be using some vocabulary in a different way from how it is used in learning theory. In our scenario, the learner is trying to guess a class of grammars of first-order theories. We need some definitions here too, in order to make our example precise. A "**well-formed formula**" is a string of vocabulary symbols which can be attributed a truth value, given an *interpretation*. To be a candidate for truth-value attribution, a well-formed formula has to be generated according to the following rules.

3 This might not be the case if the teacher, say, is a native speaker of a language, and the learner hypothesises a rule of grammar with which the teacher is unfamiliar. The rule might be correct, but this can only be checked by an outside observer, or if the teacher learns some grammar rules (a way of articulating the algorithm the teacher is using to generate the data).

4 The "intuitive idea" comes from modelling the psychological data. "Learning theory" in computer science started as an attempt to computationally model the psychological data. There is now a healthy dialogue between psychological learning theory and computational learning theory. This exchange of ideas is already aluded to in Gold ([3], pp. 447–448).

1. A proposition variable or constant is a well-formed formula.
2. A predicate letter followed by an object constant (name) is a well-formed formula.
3. An n-place relation or function letter followed by n object constants is a well-formed formula.
4. A predicate letter immediately followed by an object variable which falls within the scope of a first-order quantifier is a well-formed formula.
5. A negated predicate letter immediately followed by an object variable which falls within the scope of a first-order quantifier is a well-formed formula.
6. A well-formed formula falling within the scope of an operator is a well-formed formula.

A string of symbols of the vocabulary of the first-order theory with free variables is not well-formed, because we cannot attribute a truth value to it. So we shall call this a *formula*. A "**formula**" is understood to be a string of symbols from the vocabulary of the theory which would be well-formed iff the first-order variables were either all replaced by object constants (names) or were bound by quantifiers, or other operators.

7. The negation of a formula, where all the object-level variables are bound, is a well-formed formula.
8. The conjunction, disjunction, implication or biconditional of two well-formed formulas, is a well-formed formula.
9. An n-place function or relation letter followed by n variables or constants, such that the variables all fall within the scope of a first-order quantifier, is a well-formed formula.
10. The negation of a well-formed formula is a well-formed formula. Note: the difference between 6 and 10 is simply to allow negation within the scope of a quantifier and to allow a formula whose main operator is a quantifier to be negated.
11. A binary connective between two well-formed formulas is a well-formed formula.
12. Brackets are arranged in the following way: between any pair of brackets there is to be only one main connective. Not more than one connective in any well-formed formula may appear without a pair of brackets around it.
13. Nothing else is a well-formed formula.

A "**specified class of formal languages**" is a class of formal languages. It is specified in the sense that the teacher has a definition of this class. The learner's task is to pick out one member, or a subclass, of the class, namely a particular language or class of languages.

Allow me to introduce a new character: a *judge*. A "**judge**" is the one who pronounces that confident learning is taking place, or that the learner has converged on the data being fed to it by the informant.[5] There are several techniques for determining whether or not convergence has been reached. The techniques are applied independent of the activity of the learner or the teacher. We can imagine the techniques being applied by the computer scientist as he, or she, observes the learning process. Less dramatically, the techniques are incorporated into the mathematical description of the learning scenario.

Before we discuss the notion of convergence any further, we should be aware that there is a difference between how the philosopher uses the terms "*syntax*" and "*semantics*" and how the learning theorist uses these terms. The learning theorist associates **semantics** with data and **syntax** with rules or algorithms. The learning theorist distinguishes semantics from syntax by means of presentation of the material. The form, or grammar, of a language might be presented axiomatically, but need not, one could simply give a list of examples. If one simply gave a list of examples then this would count as semantics for the learning theorist. Thus, specifying a class of languages can be done syntactically or semantically.

In contrast, the philosopher associates semantics with meaning and syntax with form or grammar. Exactly where the difference is carved out, even in a presentation of a formal language such as first-order logic, is a source of controversy amongst philosophers. The philosopher distinguishes semantics from syntax by means of aspects of a language. For the philosopher, the distinction between syntax and semantics is *intensional*. For the learning theorist, the distinction is extensional.

The **intension** of a defined term is the definition by which we pick out the **extension**. For example, I can give a definite description which allows me to pick out an extension, such as "the people presently in this room". The extension of 'the people presently in this room' is the singleton set {Michèle Friend} at the time and place of writing.

In our learning scenario, the learner is guessing the intension behind the extension being fed to it. That is, the learner is given data in the form of correct or incorrect examples. From this list, the learner tries to piece together a function or description which will generate all and only positive data. Matching vocabulary: the extension of "the grammar of first-order

[5] We are not interested in how it is that the judge makes his judgment. This is a mathematical problem. Moreover, it is the stuff of learning theory to determine under what conditions this judgment can or cannot be made.

language λ" is the set of all well-formed formulae of the first-order language λ. The intensional description of the grammar is set of rules for forming well-formed formulae together with the vocabulary of the language. The informant gives the learner the extension of the "grammar of the first-order language λ", i.e., example formulae. The learner is trying to guess the grammar. This will allow the learner to discern formulae belonging to the first-order language λ from formulae which do not belong to that language. For example, a formula with a constant which does not belong to the first-order language λ will be one which does not belong to the language λ.

In our scenario, we are only interested, at this stage, with a learner learning to discriminate between well, and ill, formed formulae for a particular language λ, or between those of a class of first-order languages. When the learner has learned to do this, we say that "the learner has converged on the data". In learning theory, **convergence** is reached when the hypotheses put out by the learner are the same as those put out by the teacher. Note that there is no order to the data being put out. Therefore, we can only really know that the data matches in the limit. Matching of data occurs under BC-learning. "BC" stands for "behaviourally correct". Intuitively, this phrase is very appealing. The learner can perfectly imitate the teacher, in respect of the data set it is producing. In the constrained context of this paper, we can take this to mean that the learner works out what are, and what are not, well-formed formulae of the language. That is, the learner which has converged (on the data) can recognise all and only the positive (correct) data being fed to it as complying with its hypothesised grammar (algorithm for generating well-formed formulae of the first-order language λ), or class of languages.

That is, the learner is generating an algorithm for the grammar of the first-order language λ. Also, in the language adopted here, the learner is developing an intensional definition of the set of well-formed formulae of the language. The hypothesised grammar is a means of testing whether or not a formula supplied by the informant is part of the language. When the learner is given a new datum, it checks whether or not it could generate the supplied sentence using its hypothesised grammar. The judge judges that "**convergence**" is reached when a datum and all future positive data can be discriminated by the algorithm hypothesized by the learner.

If a learner is engaged in confident learning, then it is guaranteed to converge in the limit. A learner converges "**in the limit**" if it converges after a finite amount of time. The finite amount of time could be very long, longer than the age of the physical universe (assuming this is finite).

3 THE PROBLEM

We can now utter our first important philosophical point: we might be tempted to think that when convergence is reached, the learner has guessed **the true grammar**. For, this would give us an explanation as to how it is that the learner can recognise all and only the right sentences (the positive data supplied by the informant).

However, we should question our temptation, and ask whether or not extensional convergence implies intensional identity. For, this is what we suppose when we say that "the learner has guessed **the true grammar**".

Here are some problems. The obvious one, which has an apparently easy solution, has to do with identifying grammars. Two presentations of a grammar might be equivalent, but not identical, because the presentations are in a different order. So, really what we want is for the learner to guess a member of the equivalence class of presentations of grammars for a first-order language λ. In natural language we can think of different texts explaining Finnish grammar. They are not identical in their presentation, but insofar as they are correct and thorough, they all belong to the equivalence class of grammars of Finnish, assuming that there are no open problems in Finnish grammar. So, it seems reasonable to require that the learner's algorithm for discriminating well-formed from ill-formed formulae to be equivalent to that of the informant. We all understand what we mean by equivalent in this context, so we have solved the problem.

However, things are not so simple. The learner may learn a member of the equivalence class of grammars, at one level, but not at another level. The learner might still have an incorrect grammar (in an intensional sense). The slide between what is "correct" and what is "incorrect" exploits the gap between equivalence and identity, and that between "extensional" and "intensional". This problem, which is much harder to dismiss, concerns a result from model theory which, famously, was used by Putnam against the philosophical position called "metaphysical realism" [8, 9]. We shall not be interested in Putnam's use of the arguments against metaphysical realism. Instead, we are interested in what the arguments have to say about "confident learning", and learning a language. To be precise we need more definitions.

Two grammars are **identical** if they are the same in all intensional and extensional properties. This is not to say that they share all possible, or all conceivable, properties. However, the properties not captured by intensionality and extensionality are going to be mathematically incidental, if not to say bizarre.

For example, consider two copies of the "same" text book. That is, the two copies are by the same author, printed by the same publisher, in the same year, are of the same edition. The only distinguishing feature is spatio-temporal remove. The properties which distinguish them will have to do with their separate interactions with people, book shelves, desk tops, bags and so on. To say that 'the' first word cited in the first book is different from the first word in the second book because of their different relations to humans and their different locations in space and time is just bizarre. More importantly, it is not a mathematical difference. It is a socio-historical spatio-temporal difference. These bizarre differences are not of concern to us here.

Two grammars are "**equivalent**" if they share a (hopefully) specified set of properties. Often which properties are salient is not specified because this is not important, since the context makes it clear which properties are important and which are not. To be precise we should say that two grammars are equivalent in such and such respects, rather than simply "equivalent".

There are two arguments which make precise in what respects two grammars might be equivalent but not identical in a sense which matters. One argument comes from the Downward Löwenheim-Skolem theorem. The other has been called, by Putnam and others, "the permutation argument". I shall rehearse the permutation argument, and then discuss what philosophical conclusions we should draw from the argument. More importantly, I shall discuss who should be worried by the argument. For, it is not clear that computer scientists should feel in the least bit threatened by the argument; in fact, there is a sense in which the computer scientist is aware of it anyhow. He, or she, does not dwell on the argument precisely because the argument does not affect his, or her, immediate concerns.

I shall only briefly discuss the Löweheim-Skolem argument. In this, we use the Downward Löweheim-Skolem property, which applies to first-order theories (of first-order arithmetic or first-order set theory). A formal theory has the Downward Löweheim-Skolem property iff every set of well-formed formulae of the formal theory which have a model, has a countable model. Using the Downward Löweheim-Skolem property, we can generate non-standard models of the theory. That is, we can give a semantics which satisfies all the axioms of the theory. The model is non-standard in the sense that, viewed from another model, it has a different size than it reports to have from within the model. For example, a set might be called "uncountable" from within the non-standard model, but within the standard model, "the same set" is countable ([5], pp. 275–282). It turns out that in a theory which is first-order, the cardinal notions of 'uncountable' or 'finite' are relative to the model of the universe which is given. We only notice the relativity from outside the model: from within another model, or from a second-order perspective. A second-order theory is categorical. That is, there is no relativity of size. Cardinality is invariant across models satisfying the second-order axioms (of arithmetic or set theory). Another way of putting this is that, in a second-order theory, sets are picked out uniquely up to isomorphism, where the isomorphism function includes a measure of cardinality of set. We have to be philosophically careful here, since isomorphism is an equivalence relation, and equivalence relations fall short of identity, so there will still be some features which are missed out. These might, or might not, be important for our purposes. To show this discrepancy between the equivalence relation and identity, we should turn to the permutation argument which has wider application than the argument which uses the Downward Löwenheim-Skolem property, since the Downward Löwenheim-Skolem property is a property of first-order theories, not of full second-order theories.

4 RESULTS FROM MODEL THEORY

The permutation argument applied to confident learning runs: consider the grammar guessed by the learner. Imagine that the learner has reached convergence. That is, the learner is recognising all and only well-formed formulae of the first-order language the learner was supposed to learn. In this case, the judge (of what is happening between the teacher and the leaner) judges that the learner has learned a member of an equivalence class of grammars of the first-order language. That is, the learner has learned, to all intents and purposes (uniquely up to isomorphism which defines an equivalence class), the same grammar as the intended grammar the teacher conveys by giving extensional examples. The learner has achieved BC-learning. The problem is that we cannot tell, from the fact that the learner has converged on the extension of the grammar, that the learner has learned the intended grammar (the right intension),[6] or even a member of the intended class of grammars. So, the learner might not have learned **the true grammar**.

We'll begin with the argument, and then discuss an example. Take the grammar used by the informant to generate the positive data. Call this grammar the **intended grammar**. Introduce a function π which switches two constants (so for us, these have to be a particular object, predicate, relation or function constant, or a logical connective such as "$\vee$") in the language. Call the constant to be switched "a". π takes "a" as an argument and gives the value $\pi(a)$. π is a one-to-one function from the vocabulary of

[6] Note that the extensions of the grammars of the informant and the learner will be in the same equivalence class. Where the difference lies is in the grammars (at the intensional level).

the language onto the same vocabulary. So if 'a' belongs to the vocabulary so does $\pi(a)$. In our case, π has to switch two symbols of the same type around. For example, π might switch two binary connectives, or two object-level constants. Consider an n-place function f in the intended grammar which gives a rule for forming well-formed formulae. In the case of f this will be given in the intended grammar or derived from the intended grammar. In the case of $f\pi$, this will not be derived, but will be a perverse permutation (on the rules for forming well-formed formulae). Define a new function $f\pi$ by:

$$f\pi(\pi(a_1), \ldots \pi(a_n)) \text{ iff } (a_1, \ldots a_n).$$

Be careful about the biconditional. This is not the biconditional of truth. Rather it says that if one side is a well-formed formula, then so is the other side. Remember that in this learning scenario, we are simply interested in recognising well-formed formulae, not truth-functionally equivalent formulae.

We can guarantee that $\pi(a)$ is different from 'a' in at least one of the places of f. Consider an $f\pi$ where "a" can occupy the kth place, but where $\pi(a)$ is different from "a". This is enough to guarantee the difference between f and $f\pi$. The names "a" and $\pi(a)$ denote different vocabulary symbols. The intended grammar can be permuted by replacing f in the intended grammar by $f\pi$. Call this the **permuted grammar**. All and only well-formed formulae will be recognised by the learner and the teacher. The judge judges that the intended and the permuted grammars are the same. They have converged. So the extension of the intention is equivalent (by the biconditional statement above). However, the two grammars: the intended grammar and the permuted grammar, are different in the conditions under which the formulae are true. The learner might learn a permuted grammar rather than the intended grammar, and there will be perfect convergence in extension.

You might think that finding such a permuting function is impossible. However consider, for example, "$f\pi$" to be a disjunction of implications such as: if the third binary connective in a well-formed formula is $\vee$, read it as an $\wedge$, or if the third binary connective in a well-formed formula is $\wedge$, read it as an $\vee$. If we add this rule, or function, to the rules for making well-formed formulae, the learner will recognise all the well-formed formulae of the first-order language. Indeed, the learner has (as its hypothesised permuted grammar) an algorithm for recognising, and even for generating, all and only the well-formed formulae of the first-order language, but the learned grammar is aberrant. The class of well-formed formulae recognised by the learner will be isomorphic to the intended grammar, but the *meaning* of a fraction of the well-formed formulae will be different. This seems puzzling.

The judgment that the meaning is different is important. The learner will exhibit BC-learning. It will have converged with the data being fed to it by the teacher. The judge tells us that if the learner has converged, then we can be confident that it will continue to recognise, or generate, all and only well-formed formulae of the intended first-order language. So what is the problem?

The problem will come later with theorem recognition and the recognition of tautologies. It is the satisfaction conditions which will be different for the class of formulae which have been recognised using the function which produces an unintended interpretation. So on one level, the learner has learned the intended first-order language. On another level the learner has not really learned the language since language learning also involves satisfaction of formulae; and it is the satisfaction of formulae which the philosopher identifies with meaning, i.e., semantics.

At this stage, one might think that the problem will go away if we ask the learner to learn not just formulae but tautologies or theorems. However, a similar perversion of the intended class of learned material can be created at those levels too.

Say the learner was meant to discriminate all the tautologies of the language instead. Then the particular function constant we chose would not do. There would be a discrepancy between the data of the informant and that of the learner. This would probably eventually show up, but until it did, the judge would not allow us to say with confidence that the learner had confidently learned. So we need to find another permuting function.

Keep the biconditional above for f and $f\pi$. This time we have tautologies on either side of the biconditional so the biconditional reads: "f is a tautology if and only if $f\pi$ is a tautology". Now consider: if a well-formed formula is syntactically proved or semantically proved, (so is a theorem or tautology), and contains one negated bracketed conjunction to the left of the main binary connective, then switch this to a disjunction, each disjunct of which is negated, and vice versa. This is just an application of part of DeMorgan's laws to a sub-formula of a theorem or tautology.

After careful consideration, we find that the tautology, or theorem, will remain so, so the extension of the equivalence class of theorems the teacher generates, will be the equivalence class of theorems recognised by the learner. But the two classes are still intentionally different. The shape of the proof (whether semantic or syntactic) will be different; and this is a difference in intension, but not in extension.

What can we learn from this?

5 PHILOSOPHICAL REMARKS

The immediate problem of learning theorems or tautologies will affect those philosophers who think that we understand a connective by means of its introduction and elimination rules in natural deduction. Many constructivists argue for this.[7] Following Wittgenstein, the associated slogan is: "meaning is use", and the constructivists interpret this to mean that using the connectives is what gives them meaning. Use of the connectives is entirely constrained by the introduction and elimination rules, that is, by the proof theory. To borrow from Wittgenstein again: it is enough, for full understanding, to "play the language game". To put the immediate problem in Wittgenstinian: the learner will have learned to play the language game; but there is still a lingering sense in which it will not have really learned the language. So, perfect skills in discriminating well, from ill, -formed formulae is not enough to show that we are using the right rules. Similarly, perfect skills in recognising theorems or tautologies is not enough to guarantee that a learner has the right (intended) introduction and elimination rules.

It gets worse. The realist does not fare any better. Even if someone is thought to understand the connectives by means of their truth-table definitions; we see that the full truth tables will be quite elaborate, and perverse, in the case of the permuted "f", i.e., $f\pi$. Thus, the learner will not have learned the correct meaning of $\vee$, $\wedge$ and negation. So, again, it is outside the confines of the learning task that we discover the discrepancy between the intended description behind what is being learned, and the extension of the intended function, algorithm or language.[8]

Alarmingly, for the philosopher, the problem generalises quite a lot. The permutation argument can be applied to any circumstances which can be described as BC-learning which is a convergence measure on Confident learning. That is, provided that the way of testing that something has been learned is through some sort of recognition, or generation, of a class of objects, such as sentences, then the permutation problem appears. Thus the permutation argument can be applied to less formal circumstances, and to any language: formal or natural. The lesson being learned here is not, as Putnam argues, that metaphysical realism is wrong, although the problems are related, but rather that the gap between intended function, algorithm,

[7] This is the basis of the sequent calculus developed by Dag Prawitz in [6, 7].

[8] It should be obvious that this is very close to the problems raised by Kripke in his interpretation of Wittgenstein's rule following considerations. This is also related to what Quine, Dummett and Putnam have called "the inscrutability of reference". See the bibliography at the end for references.

grammar, or whatever, and its extension is not necessarily visible from within a context,[9] but might only be detected from outside (at a meta-level (relative to the context)), by articulating the function, algorithm or grammar. The philosophical reason why the learning theorist is not affected by these results is that he, or she, stops at the judgment made by the judge. Convergence in the limit is enough, as judged within the chosen context. The judge does not step out of the context to pronounce judgment.

The learning theorist stays within the context, as it were. Or, insofar as he, or she, departs from it, all he, or she, has to do is register the fact that the function, algorithm or grammar being learned by the learner might be wrong in some way other than behaviourally! The learning theorist recognises when it is that confident learning is taking place. He, or she, has set parameters for this. There will be applications where confident learning is sufficient for the task at hand, and furthermore, it will be possible, in the right circumstances to design systems (sets of algorithms) which confidently learn. This is the philosophical reason why the learning theorist is not affected by this problem.

Nevertheless, it is instructive to explore explicitly what are the limitations of confident learning. That is, it is worth exploring exactly how the learner might still get something wrong even when it is exhibiting BC-learning.

Let us return to the alarmed philosopher. There is the problem articulated above about some different philosophical positions. What is common to all of them is pinning down what it is that captures meaning. If we want to discuss **meaning**, we should keep with philosophical tradition and re-introduce the notions of semantics and syntax, for the extension of these terms is ambiguous. Semantics (for the philosopher) has to do with meaning. Syntax (for the philosopher) has to do with form.

Above, we had two theses: that the meaning of the connectives is completely contained in the manipulation rules. This is the constructivist position. This is contrasted to the realist position which claims that the meaning of the connectives is contained in their truth-table definitions. For the realist, semantic proofs in first-order logic are either proofs constructed by means of truth tables or semantic tableaux and syntactic proofs are natural deduction proofs. For the constructivist, the form and the meaning merge. Manipulation rules for the symbols give them their meaning. The understanding is through the form, and nothing else. Natural deduction proofs give the meaning of the connectives. There is no difference in appearance between syntactic proofs and semantic proofs.

9 This point about "inside" and "outside" a context is discussed very nicely in Hallett: "Putnam and the Skolem Paradox" in [1], pp. 66–97.

Sliced either the realist way or the constructivist way, the philosopher's distinction between syntax and semantics will not help with meaning or understanding. For, neither explanation is enough to block aberrant or perverse (permuted) understanding. For, permuted understanding is not really understanding. Meaning has not been conveyed, if there is a perversion somewhere *en route*. So both positions are missing something in their account of meaning. That is, we can no longer simply state that the semantic part of a first-order theory is the interpretation (a domain for the variables to range over, together with a designation for the constants), and the syntactic part of the theory is the grammar. We can no longer blithely say such things because the learner can behave as though it understands, and yet at another level, it clearly does not. That is, there is a shortfall in our account of meaning, or understanding, or learning, of a formal system. Call this the 'inadequacy of our account of learning' problem.

The problem, for the philosopher, is that the extension of the function or language being learned is not uniquely determined by an equivalence class of intensional descriptions of that function or language. Thus, the notions of intensional description and extension of the description are not absolute. They are relative to a context. Quine calls this the "indeterminacy of translation" ([10], p. 27). However, as we saw above the problem regenerates from task to task: learning to recognise well-formed formulae can be made precise if we also ask the learner to learn about the truth conditions of formulae. But it might still have an aberrant way of understanding the truth conditions and how they are generated. So, this problem goes beyond the inscrutability of reference, since the problem regenerates whenever there is a gap between equivalence and identity. Equivalence is not enough for understanding; only identity is. Call this the "equivalence falls short" problem.

More broadly: consider what the learning theorist calls "general learners". These are learners whose Turing machine capacities set up for generating hypothesised algorithms are unknown. Examples of general learners are people, who, for example, learn languages. People learn languages without necessarily having a thorough background in grammar. This will certainly be the case in English. In French one learns a good deal of grammar, whether one is a native or not.[10] We know from the above result that there are probably cases of something quite close to confident learning when ordinary people learn a language. We have to leave a little leeway since language changes over time, but let us ignore this. Is semantic (extensional) correctness sufficient

[10] Traditionally, in France, until they go to Lycée, children spend just under half their classroom time explicitly learning grammar, spelling and general language skills.

for communication? Probably, it is in most circumstances. However, rather grave misunderstandings can be had due to a small aberration in a rule, or in an algorithm.

So maybe we should turn the question around. When will confident learning of a language be shown to be inadequate? Answer: when we have disagreements **about** a language. That is, when we adopt a more general (meta-) perspective on a language. In other words, the meta-perspective is essential to (directly proportional to) depth of understanding. We cannot be certain to communicate, or learn, if we are only prepared to stick to one level—the object level.

Lesson: even if a group of people has learned to speak a language perfectly: in terms of vocabulary and structure of sentences (so no one utters ungrammatical or categorical nonsense) then they might still have a job agreeing on, for example, the class of grammars which formalise the language. Thus, there is no unique standard for fully capturing the grammatical structure of a natural language which we can lift from use of the language (under the plausible hypothesis that we engage in confident learning). The same applies to sophisticated computers trying to come up with a grammar for a language. In fact, there are an infinite number of perverted grammars for any one extension in the form of an infinite stream of data. For this reason it would be a miracle (have a probability zero) that we should get **the true grammar**. Call this the "many perversions" problem.

6 CONCLUSION

For the conclusion, let us be explicit about how our three problems above are related to each other, and then discuss solutions. The solutions are reached by playing the Modus Ponens/Modus Tollens[11] game where we turn a modus tollens argument into a modus ponens argument.

The "many perversions" problem is just an expansion on the "equivalence falls short" problem. The former tells us that equivalence does not only just fall sort. It completely misses the mark. It would be a miracle (has a probability of 0) for a learner engaged in confident learning (whose possible algorithms is unknown or an open set) to converge and get the intended

[11] There is a saying in philosophy that "one person's Modus Ponens is another person's Modus Tollens". This is just a factor of solutions which do not sit easily with everyone. By discussing arguments under their Modus Ponens versions and their Modus Tollens versions, we increase our understanding.

function, algorithm or grammar; where **the intended function**, algorithm or grammar is that which underpins (or is used to generate) the data being fed by the teacher to the learner.

So who does this affect? Answer: anyone who is concerned about the "inadequacy of our account of learning" problem. The problem will trouble anyone who thinks that meaning, or understanding, are absolute notions; even if he, or she, thinks that these notions are only absolute in ideal cases, such as logic. This group of people includes anyone who thinks, along with Frege and others, that logic is pure and untarnished. For example, we are told, by Frege and others, that logic is free from ambiguity and vague expressions. However, the permutation argument shows that meaning or understanding of a formal language (in an absolute sense of "meaning") cannot be located either in the proof theory or in the model theory of the language exclusive of anything else. We need, and intuitively have, a further component in the form of a meta-perspective. So, in particular, meaning is not located in what the philosophers call the "semantics of a language". There is a residual ambiguity guaranteed by the gap between the equivalence class of some extension and the identity of the intentional description which generates the extension.

To see some solutions, let us play the Modus Ponens/Modus Tollens game. The Modus Ponens version of the solution to the "inadequacy in our account of learning" problem will be to say something like: "we need to consider not only the proof theory and the model theory of a language to find the meaning, but also the relationship between the two". In other words, to find meaning, we also have to step outside the system itself, to move to a meta-level. Of course, the problem will occur again at the next level, but maybe this will be enough for now. Meaning is relatively stable. But to say that "this is enough for now" is to give up on an absolute or unique notion of meaning. For many philosophers this is intolerable.

The Modus Tollens version of the solution to the "inadequacy in our account of learning" problem will be to say right away that meaning is simply not an absolute notion. We can now join up with the other two problems. What counts, in the end, is communicating well enough. That is, it is usually enough to get the right equivalence class.[12] Furthermore, if, later, we find that there is a problem, then we can seek to clarify by moving to a meta-level, or by looking at what we have learned from without. In other words, the Modus

[12] It amuses me to think here of the many empty conversations one has, where one simply goes through the motions of prompting agreeing, qualifying etc. without really taking an interest about the truth or content of what is being said. Socially, it is often enough to communicate at a primitive social level, instead of a more scientific quest for "truth and understanding".

Tollens thinker will end up with a relativistic position saying something about learning being a process, not an object, and being quite grateful that we happen to move so deftly to meta-levels when communicating.

The Modus Ponens thinker, who here is being associated with a more realist stance, or absolutist stance might acknowledge that there is a problem whenever there is a gap between equivalence and identity. Then there are two possible routes to take. One is to bite the bullet and say that indeed there are an infinite number of ways in which we might still go wrong even if we have the extension of a function, algorithm or grammar right, and it is just good luck that we get things right at the level of the function too. In other words a miracle takes place. Zero probability of an event occurring does not imply that the event does not take place. This is intolerable because positing miracles as explanations is not "good form" in philosophy. The other route is to say that at some point up the meta-levels we will reach identity. Thus, meaning is absolute, but not at ground level. Furthermore, there is a direction to pursue to increase our understanding. Furthermore, it might even be (ideally?) possible for us to get there. Temporarily, this will be a more comfortable solution than the relativistic one of the Modus Tollens thinker. However, the comfort will be short lived because we have to make explicit what we mean by "ideally possible", and this is not obvious.

In short, confident learning is only judged in terms of BC-learning, or extensionally correct learning. A learner engaged in confident learning, which has converged, might be far from intensionally correct in its learning. Thus, there is no way that we can be confident that in confident learning the learner has learned the unique underlying function or language.

REFERENCES

[1] Clark, P. and Hale, B. (eds.) (1994). *Reading Putnam*, Oxford: Basil Blackwell.

[2] Glymour, C., Kelly, K. and Juhl, C. (1994). "Reliability, Realism, and Relativism", in Clark, P. and Hale, B. [1], 98–160.

[3] Gold, E.M. (1967). "Language Identification in the Limit", *Information and Control* 10, 447–474.

[4] Hallett, M. (1994). "Putnam and the Skolem Paradox", in Clark, P. and Hale, B. [1], 66–97.

[5] Machover, M. (1996). *Set Theory, Logic and their Limitations*, Cambridge: Cambridge University Press.

[6] Prawitz, D. (1965). *Natural Deduction: A Proof-Theoretical Study*, Stockholm: Almqvist and Wiksell.

[7] Prawitz, D. (1974). "On the Idea of a General Proof Theory", *Synthese* 27, 63–77.

[8] Putnam, H. (1975). "'Degree of Confirmation' and Inductive Logic", *Mathematics, Matter, and Method*, Cambridge: Cambridge University Press, 270–292.

[9] Putnam, H. (1981). "A Problem About Reference" and "Appendix", in *Reason, Truth and History*, Cambridge: Cambridge University Press, 22–48, 217–218.

[10] Quine, W.V.O. (1960). *Word and Object*, Cambridge (Mass.): MIT Press.

8

HOW TO DO THINGS WITH AN INFINITE REGRESS

KEVIN T. KELLY
Department of Philosophy, Carnegie Mellon University, Pittsburg, PA 15213, U.S.A., kk3n@andrew.cmu.edu

Abstract: Scientific methods may be viewed as procedures for converging to the true answer to a given empirical question. Typically, such methods converge to the truth only if certain empirical presuppositions are satisfied, which raises the question whether the presuppositions are satisfied. Another scientific method can be applied to this empirical question, and so forth, occasioning an empirical regress. So there is an obvious question about the point of such a regress. This paper explains how to assess the methodological worth of a methodological regress by solving for the strongest sense of single-method performance that can be achieved given that such a regress exists. Several types of regresses are "collapsed" into corresponding concepts of single method performance in this sense. The idea bears on some other issues in the philosophy of science, including Popper's falsificationism and its relationship to Duhem's problem.

1 CONFIRMATION AND NATURALISM

Here is a familiar but unsatisfying approach to the philosophy of science. Science seeks to "justify" empirical hypotheses. Usually, evidence does not and never will entail them, so they must be "justified" in some weaker way. So there must be a relation of "partial support" or "*confirmation*" or "empirical rationality" falling short of full (deductive) support that justifies them. The principal task of the philosophy of science is to "explicate" the relation of empirical justification from practice and from historical examples. Any feature of scientific method or procedure that is not derived from this relation is extraneous to the philosophy of science *per se*, although it may be

M. Friend, N.B. Goethe and V.S. Harizanov (eds.), Induction, Algorithmic Learning Theory, and Philosophy, 199–217.
© 2007 *Springer.*

of tangential psychological, sociological, or purely computational interest. Thus, virtues such as confirmation, explanation, simplicity, and testing are normatively and philosophically relevant, but the logic of discovery (procedures for inventing new hypotheses) and procedural efficiency are beside the point (e.g., [13]).

The trouble with this approach is that explicating the justification relation (supposing it to be possible at all) does not begin to explain why justification *should* be as it is rather than some other way. Convincing, *a priori* answers are not forthcoming and attempts to provide them are no longer in fashion. One responds, instead, with the *naturalistic* view that if scientific standards are to be justified, that justification must *itself* be scientific (i.e., empirical). The next question is how scientific reasoning can justify itself. Circular justifications are more popular than infinite regresses of justification in the philosophical literature (somehow an infinite regress of justifications never "fires" or "gets started"), but it is hard to explain what the point of circles or regresses of justification could possibly be without first knowing what the point of justification, itself, is. And the familiar, confirmation-theoretic philosophy of science under consideration provides no such explanation.

2 A PROCEDURAL PARADIGM

Consider an alternative paradigm for the philosophy of science, according to which scientific methods are *procedures* aimed at *converging to correct answers* rather than relations between hypotheses and finite bodies of evidence. Procedures are justified not by embodying some abstract relation of empirical justification between theory and evidence at every stage, but because they find correct answers both reliably and efficiently. Computational efficiency is relevant, since it brings one to the truth faster. The logic of discovery is also relevant, because the concept of convergence to a correct answer applies as much to methods producing hypotheses as to methods assessing given hypotheses. This is the perspective of *computational learning theory*, an approach to inductive inference proposed by Hilary Putnam ([19, 20]) and E.M. Gold ([2, 3]) and subsequently developed largely within computer science.[1]

Here is a more precise formulation of the idea. Empirical methods are procedures or dispositions that receive successive inputs from nature and output successive guesses in response. Like computational procedures,

[1] For book-length reviews of the technical literature, cf. [17, 6]. For attempts to relate the ideas to the philosophy of science, cf. [14], [7], [9], and [11].

inductive methods may be judged as solutions to problems. An *empirical problem* is not a particular situation but a *range of serious possibilities* over which the method is required to succeed. Success in a possibility means converging to a correct answer on the stream of inputs that would be received if that possibility were actual. Correctness may be truth or something weaker, such as empirical adequacy. It may also involve pragmatic components, such as being a potential answer to a given question; or, following Thomas Kuhn, it might be something like ongoing puzzle-resolving effectiveness in the unbounded future. The precise choice of the correctness relation is not the crucial issue. What does matter is that there is a potentially endless stream of potential inputs and that the standard notions of correctness transcend any finite amount of data. So achieving correctness reliably—in each of a broad range of cases—occasions the problem of induction: that no answer unverified by a finite sequence of inputs is guaranteed to be correct.

There are several different senses of convergent success, some of which are more stringent than others (cf. [7]). Let a hypothesis be given. It would be fine if a scientific procedure could eventually halt with acceptance or rejection just in case the hypothesis is respectively correct or incorrect. Call this notion of success *decision with certainty*. But some hypotheses are only *verifiable with certainty* (halt with acceptance if and only if the hypothesis is correct) or *refutable with certainty* (halt with rejection if and only if the hypothesis is false). Other hypotheses are only *decidable in the limit*, meaning that some method eventually stabilizes to acceptance if the hypothesis is correct and to rejection otherwise. There are also hypotheses for which it is only possible to stabilize to acceptance just in case the hypothesis is correct (*verification in the limit*) or to stabilize to rejection just in case the hypothesis is incorrect (*refutation in the limit*). Between decision in the limit and verification and refutation with certainty, one may refine the notion of success by asking how many *retractions* are necessary prior to convergence.[2] Kuhn and others have emphasized the tremendous social cost of retracting fundamental theories and, furthermore, the number of retractions required prior to convergence may be viewed as a notion of convergent success in its own right that bridges the concepts of certainty and limiting convergence with an infinite sequence of refined complexity concepts.

A given, empirical problem may be solvable in one of the above senses but not in another. The best sense in which it is solvable may be said to be its *empirical complexity*. This is parallel to the theory of computability and computational complexity. In fact, the complexity classes so defined are already

[2] Cf. [11] for an explanation of Ockham's razor in terms of retraction minimization.

familiar objects in analysis and computability theory [4]. In the philosophy of science, one speaks vaguely of *underdetermination* of theory by evidence. Elsewhere, I have proposed that degrees of underdetermination correspond to degrees of empirical complexity ([7], [9]). That yields a comprehensive framework for comparing and understanding different inference problems drawn from different contexts, as well as a unified perspective on formal and empirical inquiry ([7], Chapters 6, 7, 8, and 10; [10]), something that has bedeviled the confirmation-theoretic approach from the beginning.

Many methodological ideas familiar to philosophers of science emerge naturally from the procedural framework just described (cf. [11]). One such idea is *Duhem's problem*, which turns on the observation that individual hypotheses in a scientific theory are refutable only in the context of other "auxiliary hypotheses". The problem is how to assign credit or blame to a hypothesis when falsifying instances may be due to a false auxiliary hypothesis with which the hypothesis has been forced to keep company. Here is how the problem looks from the perspective of learning theory. A hypothesis that is not refutable with certainty may be refutable with certainty *given* some other auxiliary hypotheses, which is the same as saying that the conjunction of the hypothesis with the other hypotheses is refutable with certainty. Indeed, there may be many potential sets of auxiliary hypotheses that make a given hypothesis refutable with certainty.

One can enumerate the possible systems of auxiliaries thought of so far and accept H as long as the first system of $H+$ auxiliaries consistent with receipt of the current inputs is not refuted. If the first such system is refuted, then H is rejected and one selects the first such system consistent with the data and with H. If new systems of auxiliaries are thought of, they can be added to the end of the queue of auxiliaries thought of already. This procedure verifies H in the limit so long as "creative intuition" eventually produces systems of auxiliaries covering all relevant possibilities admitted by H. So verifiability in the limit corresponds to the intuitive epistemic perplexity occasioned by Duhem's problem. That is important, because many issues in the philosophy of science (realism, conventionalism, observability, theory-ladenness, and paradigms) cluster around Duhem's problem.

One can (and, I suggest, should) think of the Kuhnian [12] distinction between "normal" and "revolutionary" science along similar, procedural lines. A "paradigm" is a hypothesis that is not refutable in isolation but that becomes refutable when "articulated" with auxiliary hypotheses. Normal science involves the selection of auxiliary hypotheses compatible with the paradigm and with experience that make the paradigm refutable. Revolutionary science involves choice among paradigms. The crisp, stepwise solvability of "normal" problems reflects the constraints imposed by the presumed paradigm. Revolutionary science is far less crisp, since each paradigm can be articulated in an infinite variety of ways.[3]

The preceding points are illustrated naturally and concretely when the hypothesis in question concerns a *trend*. Questions about trends often generate controversy, whether in markets or in nature, because any finite set of evidence for the trend *might* be a local fluctuation around an unknown equilibrium. This sort of ambiguity permeated the debate between uniformitarian and progressionist geologists in the nineteenth century (cf. Ruse [21]). Progressionists[4] held that geological history exhibits progress due to the cooling of the Earth from its primordial, hot state, whereas Lyell reinterpreted all apparent trends as local fluctuations on an immensely expanded time scale. If progressionism is articulated with a particular schedule of appearance for finitely many fossils, it can be decided with two retractions starting with "no". Just say "no" until all the fossil types are seen to appear in the fossil record as early as anticipated (remember that it may take arbitrarily long to find a fossil that appears as early as anticipated); then say "yes" until some fossil type is observed to appear earlier than expected, after which say "no" again.[5] In historical fact, Lyell claimed to have refuted progressionism when the Stonesfield mammals were found in Jurassic strata, prior to progressionist expectations, but it was open to the progressionists simply to "re-articulate" their paradigm with an accelerated schedule accommodating the new find, so progressionism is not really refutable with certainty. In fact, the progressionists were free to accelerate their schedule repeatedly, and no finite set of fossils could possibly refute all possible schedules, so without further reframing, the debate allows for a potentially endless give-and-take. To verify progressionism in the limit, do the following: enumerate the possible schedules of progress (assuming

3 Much more can be said about this [9]. For example, one can also provide a naturalistic account of theory-laden data in learning theoretic terms.

4 The progressionists were called "catastrophists" because the early cooling of the Earth was supposedly accompanied by catastrophic changes unobserved nowadays.

5 Attentive readers may have noticed that if the progressionists were to always posit exactly the currently observed earliest appearances for each fossil type then at most one retraction is required per re-articulation. That would be true if progressionism were flexible enough to articulate with arbitrary schedules. But the progressionist paradigm also included prior ideas about which fossil forms were more "advanced" than others, so if a "rudimentary" form were to appear earlier than expected, still more rudimentary forms would have to appear even earlier than that, and it is possible that no such examples had yet been found. In that case, progressionists would have to set a new schedule in which some rudimentary forms are expected earlier than the earliest known examples. And then the right rule is to say "no" until such examples are found, "yes" after they are found, and "no" after still earlier examples are found.

them to be presented as discretely presented rules). Apply the preceding two-retraction method to each schedule. If the first schedule for which the method says "yes" does not change when a new observation is made, say that progressionism is true. Otherwise, say that progressionism is false. If progressionism is true, it is true in virtue of some schedule. Eventually, all the fossil types appear as soon as predicted and no fossil type ever appears earlier, so from that point onward the two-retraction method stabilizes to "yes" on the true schedule. For each schedule prior to the true one, the method eventually stabilizes to "no" (either because no fossil of a given type ever appeared early enough or because some fossil type is seen too early). So the enumeration method converges to "yes" when the first true schedule's method has stabilized to "yes" and all prior schedules' methods have stabilized to "no". If progressinism is false, then every schedule's method eventually converges to "no", so the enumeration method outputs "no" infinitely often.

Uniformitarianism, on the other hand, is *refutable* in the limit: it looks bad as long as the two-retraction method says "yes" for a fixed schedule and looks good when the current schedule is refuted.

Global warming [1] provides a more recent example of an awkward trend question. Is the current warming trend a chaotic spike no greater than historical spikes unaccompanied by corresponding carbon dioxide doses or is it larger than any historical spike unaccompanied by current carbon dioxide levels? Newly discovered high spikes in the glacial record make us doubt global warming and increasing temperatures higher than discovered spikes in the glacial record make us more confident that carbon dioxide levels are the culprit.

In a different domain, the cognitivist thesis that human cognition is computable is verifiable in the limit, for similar reasons. Each finite chunk of behavior is compatible with some computer program (cognitive theory), but each such theory is outrun eventually by uncomputable behavior. To verify the hypothesis of computability, enumerate all possible computable accounts and reject the computability hypothesis only when the first program compatible with human behavior is refuted. If behavior is computable, eventually the right program is first and the method converges to "yes". Otherwise, each program is eventually refuted and the method says "no" infinitely often.

3 PROCEDURAL REGRESSES

The procedural outlook just described is subject to its own empirical regress problem. Many empirical problems are solvable, even in the limit, only if certain empirical presuppositions are satisfied. For example, knowing that a curve is polynomial allows one to infer its degree in the limit (from increasingly precise data), but if the question is expanded to cover infinite series, the answer "infinite degree" is only refutable in the limit. If empirical presuppositions are necessary for success, how can one determine whether they are satisfied? By invoking another method with its own empirical presupposition? And what about that one? So it seems that one is left with a regress of methods checking the presuppositions of methods checking the presuppositions of methods.... The point of a method guaranteed to converge to the truth is fairly clear. But what is the point of a regress of methods, each of which succeeds only under some material presupposition that might be false?

The basic idea developed in this article is a methodological *no free lunch principle*: the value of a regress can be no greater than the best single-method performance that could be achieved by looking at the outputs of the methods in the regress rather than at the data themselves. If the performance of the best such procedure is much worse than what could be achieved by looking at the data directly, one may justifiably say that the regress is methodologically *vicious*. If the best method that looks only at the outputs of the methods in the regress succeeds in the best feasible sense, then the regress is *optimal*.

4 FINITE REGRESSES

Consider the empirical problem (H_0, K) of determining the truth of a given hypothesis H_0 over serious possibilities K. Fix a given sense of success (e.g., refutation with certainty, verifiability in the limit, etc.). Every method M_0 directed at assessing H_0 succeeds in the given sense over some set (possibly empty) of serious possibilities (input streams). The *empirical presupposition* H_1 of a given method M_0 for assessing H_0 is just the set of all serious possibilities (input streams) over which M succeeds (i.e., H_1 is just the empirical proposition "M_0 will succeed in the specified sense"). So let meta-method M_1 be charged with assessing the presupposition H_1 of method M_0. Meta-method M_1 reads from the *same* input stream as M_0, but instead of trying to determine the truth of the original hypothesis H_0, M_1 tries to determine the truth of H_1, the empirical presupposition of M_0. With respect to the question H_1, M_1 has its own empirical presupposition H_2, of which M_2 determines the truth value of H_1 under empirical presupposition H_2, and so forth.

For example, let H_0 denote Lyell's uniformitarian hypothesis, discussed earlier. After the Stonesfield find, Lyell declared victory for H_0, which might be interpreted as *halting* definitively with "yes".[6] Lyell would also have had a reason to scoff at the progressionists if fossil types expected by a certain epoch (e.g., "missing links") stubbornly refused to appear. On the other hand, perfect correspondence between the proposed schedule of progress and the actual fossil record would hardly be happy news for Lyell. So one might crudely reconstruct Lyell's method M_0 as something like the following: until the currently fashionable schedule is *instantiated* (i.e., the earliest geological appearance of each fossil type matches the schedule exactly), scoff at the "anomalies" in the progressionist paradigm and say "yes" to uniformitarianism without halting.[7] While the currently fashionable schedule is instantiated but not refuted, concede "no" without halting. Finally, when the schedule is refuted outright by a fossil that appears ahead of the current schedule (as the Stonesfield find did), announce victory (i.e., halt with "yes"). So M_0 retracts at most twice, starting with "yes", as the geological data pour in. But no possible strategy converges to the truth about uniformitarianism with just two retractions, since any number of successively accelerated schedules of first appearance for the various fossil types might appear perfectly instantiated for arbitrarily long periods of time before being shot down by a new find ahead of schedule. So M_0 finds the truth only under some empirical presupposition H_1. Assuming that the serious possibilities K at the time were just those compatible with uniformitarianism and progressionism and that the earliest time of appearance in the fossil record is eventually observed for each fossil type, the presupposition H_1 of M_0 is that progressionism *implies* A_1 (where A_1 is the auxiliary hypothesis that fossil types will first appear according to the currently fashionable schedule) and, hence, that uniformitarianism is true if A_1 is not (i.e., $H_1 = H_0 \vee A_1$). For given H_1, M_0 really does converge to the truth with just two retractions in the worst case and if H_1 is false, M_0 converges to the false conclusion that uniformitarianism is true.

In Kuhnian terms, the auxiliary hypothesis A_1 is an "articulation" of the progressionist paradigm and in the face of the uniformitarian competitor, the Stonesfield find constituted an anomaly for this particular articulation. As Kuhn takes pains to emphasize, the anomaly does not logically compel rejection of progressionism, since the schedule can be revised to accommodate the

6 Of course, I oversimplify. He declared victory for a tangle of reasons that would defy any elegant logical representation, but the Stonesfield find seems to have been a significant rhetorical blow to progressionism (Ruse [21]).

7 More realistically, output "?" until the "anomalous" failure to find the missing fossils percolates into a "crisis". That detail doesn't really change anything in the following analysis.

anomaly. But Lyell's method M_0 halts with "yes" when A_1 is refuted, so M_0 fails to find the truth when progressionism is true in virtue of some revised schedule.

When the progressionists responded by revising their schedule to a new schedule A_2, Lyell's method was called rhetorically into question, for if the new schedule were true, his method would halt with "yes" even though progressionism is true. Since H_1 is an *empirical* hypothesis, Lyell might have responded to the challenge with a meta-method M_1 that checks the truth of H_1. With A_2 on the table, it would be rhetorically pointless for Lyell to respond with a method that still presupposes H_1; he must at least entertain the revised schedule A_2 as a serious possibility, so that the presupposition of the meta-method M_1 is

$$H_2 = H_1 \vee A_2 = H_0 \vee A_1 \vee A_2.$$

Now over these extended possibilities, in which possibilities does M_0 succeed? Only in possibilities in which the revised schedule A_2 is false, since Lyell's premature halting with "yes" upon the refutation of A_1 would be rescued by the falsity of A_2. So M_1 should say "yes" until A_2 is instantiated, followed by "no" until A_2 is refuted, and should halt with "yes" as soon as A_2 is refuted. Notice that meta-method M_1 is pretty similar in spirit to Lyell's original method since it still ungenerously entertains only finitely many possible schedules of progress. And like the original method, the meta-method converges to the truth with two retractions starting with "yes", given that its empirical presupposition is true. The regress can be extended to any finite length, where meta-method M_i has presupposition $H_{i+1} = H_i \vee A_i$.

Say that a finite regress $(M_0, \ldots, M_n)$ *succeeds regressively* (relative to empirical problem (H_0, K)) in a given sense (e.g., verification with certainty) just in case there exist propositions $H_1, \ldots, H_n$ such that for each i no greater than n:

1. H_{i+1} is the presupposition of M_i with respect to H_i according to the specified sense of success and
2. K entails H_n.

So assuming that the relevant possibilities in the geological example are exhausted by H_n, the Lyellian regress may be said to succeed regressively concerning the uniformitarian question over serious possibilities $K = H_n$ in the sense of convergence with two retractions starting with "yes".

But sequential success is a far different matter from success with respect to the original question. How are the two related, if at all? The worry is that infinite regresses, like circles, do nothing at all but beg or forestall the original question under consideration. One way to answer this question is to construct a *regress collapsing function* $\Phi(a_1, \ldots, a_n) = a$ that takes a sequence of n possible answers to a single answer. The collapsing rule Φ may be thought of as converting the regress $(M_0, \ldots, M_n)$ into a single method M^* such that for each finite input history e,

$$M^{\star}(e) = \Phi(M_0(e), \ldots, M_n(e)).$$

Then if M^* succeeds in some ordinary, single-method sense in problem (H_0, K), one can say that the regress is no worse in value than a single method that succeeds in that sense. In other words, the *methodological value* of a regress is the best single-method performance that could be recovered from the successive outputs of the constituent methods in the regress without looking at the inputs provided to these methods. The regress is *vindicated* if the best single method performance that can be achieved by collapsing it is also the best possible single method performance. Otherwise, the regress is *vicious* (a term often employed with no clear sense). Viciousness now comes in well-defined degrees, depending on how far short of optimal performance the best-performing regress collapse falls.

For illustrations of vindication and viciousness, turn once again to the Lyellian regress (M_0, M_1) of length 2. Assuming that $K = H_2$, this regress succeeds *regressively* with two retractions starting with "yes", since that is what each method achieves given its presupposition. Now consider the best sort of method one could build from this regress without peeking at the inputs. Let e be an arbitrary, finite data history compatible with K. When $M_1(e)$ converges to "yes", whatever $M_0(e)$ converges to is true and when $M_1(e)$ converges to "no", whatever $M_0(e)$ converges to is false. So it is sensible to define the collapsing rule so that the first answer a_1 is repeated or reversed, depending on whether M_1 answers "no" or "yes":

$$\begin{aligned}\Phi(a_1, \text{``}yes\text{''}) &= a_1;\\ \Phi(a_1, \text{``}no\text{''}) &= \text{reverse}(a_1).\end{aligned}$$

Over possibilities in $K = H_2$, the following histories may occur:

1. neither schedule is instantiated;
2. schedule A_1 is instantiated but not schedule A_2;
3. schedule A_1 is refuted but schedule A_2 is not instantiated;
4. schedule A_1 is refuted and schedule A_2 is instantiated;
5. both schedules are refuted.

In such an input stream, method $M^*(e) = \Phi(M_0(e), M_1(e))$ retracts at most four times, starting with "yes". Notice that this is the *sum* of the worst-case retractions of the constituent methods in the regress (suggesting a general pattern) and that the initial answer is the same as that of the constituent methods.

Does M^* achieve the best possible single-method performance in the problem under consideration? Let M be an arbitrary method that converges to the truth about uniformitarianism over serious possibilities in $K = H_2$. Nature can withhold instantiation of both schedules until M is forced (on pain of failing to converge to the truth) to say "yes", since otherwise both schedules are false, so according to H_2, uniformitarianism is true. At that point, Nature is free to continue to present data instantiating but never refuting schedule A_1 until M concludes "no", since otherwise A_1 is true and implies that uniformitarianism is false. Nature is free to continue to present data refuting schedule A_1 without instantiating schedule A_2 (since A_2 is faster than A_1) until M concludes "yes". Nature is now free to continue to present data refuting schedule A_2 and instantiating schedule A_2 without refuting it until M concludes "no". Finally, Nature is free to continue to present data refuting A_2, forcing M to conclude "yes". So an arbitrary method that converges to the true answer in the problem also requires at least four retractions starting with "yes". Starting with "no" would require yet another retraction and starting with "?" would still require four (even not counting the change from "?" to "yes" (by arguments similar to the one just given). So the best possible single method performance in this problem is four retractions starting with "yes". Hence, Lyell's regress is *vindicated*, since it can be collapsed into a single method with the best possible performance.

Vindication is not trivial. For example, Lyell might have been a lunatic who reversed his answer every day, whereas his meta-method (physician) might have been perfectly rational and said "no" *a priori* concerning insane Lyell's success. This regress succeeds regressively in the strongest possible sense (each method succeeds over its respective presupposition with no retractions) but it is *entirely vicious* because both methods ignore the data entirely, precluding all attempts to collapse the regress into a method that even converges to the truth in the limit.

The preceding example illustrates that even extremely strong regressive success does not suffice for vindication. That is because the crazy method fails in an unnatural way. Real science loves to "frame" messy questions to appear crisper than they really are by specifying evidential triggers for when to reject a hypothesis or paradigm that is not really refuted (as in the case of Lyell's identification of progressionism with a particular schedule of progress). This opens the door to some failures to converge to the truth, since the trigger for halting may be premature. But in spite of this obvious risk of failure, reliance on determinate empirical "triggers" for rejection or acceptance has a silver lining: it ensures that failure of the method occurs in an *orderly way* that has implications concerning the truth of the original hypothesis. In the jargon coined above, reliance on evidential triggers links mere *regressive success* to genuine *methodological value*, as in the Lyell example. In that example, the trigger for dumping progressionism is refutation of schedule A_1, which fails only when progressionism is true (over serious possibilities $K = H_0 \vee A_1 \vee A_2$). Hence, the meta-method's determination that the method fails has a bearing on the original question and that information is exploited by the collapsing function. This is a new explanation of why induction *should* proceed by means of crisp triggers or defaults, for otherwise empirical regresses would be methodologically worthless, as in the case of the insane regress.

More generally, say that method M *converges* with at most n retractions starting with "yes" just in case M never starts with any other answer and never retracts more than n times in *any possible input stream* in K. Due to its reliance on concrete, empirical "triggers", Lyell's method converges with at most two retractions starting with "yes" over all possibilities in K, even though it does not converge to the truth in all of these possibilities.

For concreteness, the discussion so far has focused on a particular example, but the conclusions drawn are far more general, depending only on the logical relationships between the various success criteria. To lift the discussion to this more general, methodological level, let R be a relation between regresses and problems (e.g., convergence and regressive success in a given sense) and let Q be a relation between single methods and problems (e.g., success in some other sense).

> Relation R is *methodologically collapsible* to relation Q if and only if for each problem p and for each regress $(M_0(e), \ldots, M_n(e))$ satisfying R with respect to p, there exists a collapsing function Φ such that the single method $M^*(e) = \Phi(M_0(e), \ldots, M_n(e))$ satisfies Q with respect to p.

It is also interesting to turn tables and investigate whether single-method success can be *stretched* into some notion of regressive success and convergence. A *stretching* function is a mapping $\Psi(a) = (a_1, \ldots, a_n)$ from answers to sequences of answers. The *stretching* of method M by Ψ is the regress defined by:

$$(M_0(e), \ldots, M_n(e)) = (\Psi(M(e))_0, \ldots, (\Psi(M(e))_n).$$

Then relation Q is *methodologically stretchable* to relation R if and only if for each problem p and for each method M satisfying Q with respect to p, there exists a methodological stretching function Ψ such that the regress

$$(M_0(e), \ldots, M_n(e)) = (\Psi(M(e))_0, \ldots, (\Psi(M(e))_n)$$

satisfies R with respect to p.

Now one can define methodological equivalence between regressive and single-method performance as follows:

> R for regresses is *methodologically equivalent* to Q for single methods if and only if R is collapsible to Q and Q is stretchable to R.

For example, a regress of two methods that converge and succeed sequentially with one retraction starting with "yes" (i.e., a regress of two refuters) is methodologically equivalent to one method that succeeds with two retractions starting with "yes".[8] More generally, the pattern hinted at earlier amounts to this:

Proposition 4.1 *The following are methodologically equivalent.*[9]

- *Regressive success and convergence under a finite retraction bound n_i for each constituent method M_i.*
- *Single method success under the sum of the bounds n_i starting with "no" if an even number of the M_i start with "no" and starting with "yes" otherwise.*

That settles the matter for finite regresses of methods with bounded retractions. Moving on to weaker senses of convergence, it is easy to see that any finite regress of methods that succeed regressively and converge in the limit is equivalent to a single method that decides in the limit:

8 Here's the trick. Both methods start out with "yes". Let the constructed method M start with "yes" because M_1 will succeed and M_1 currently says that M_0 will succeed and M_0 now says "yes". If M_1 ever says "no", then let M reverse whatever M_0 says because M_1 is right in saying that M_0 is wrong (since M_1 has already used its one retraction and has therefore converged to the truth). At worst, both retract and M retracts once each time. So M retracts at most twice, starting with "yes". Methodological equivalence requires that one can also produce a regress of two refuters M_0, M_1, from an arbitrary method M that succeeds with two retractions starting with acceptance. Here's how to do it. Let M_0 say "yes" until M retracts once and say "no" thereafter. Let M_1 say "yes" until M retracts twice and say "no" thereafter. Let H_1 be the proposition that M_0 successfully refutes H_0 with certainty. That is true just in case M retracts at most once. Thus, M_1 really succeeds in refuting H_1 with certainty over all possibilities M succeeds over, as required.

9 The proofs of all the propositions may be found in [8].

just accept if an even number of the methods in the sequence reject and reject otherwise. Regresses of methods that verify in the limit or refute in the limit are not reducible to any of our notions of success and may be thought of as a natural way to build methodological success criteria applicable to more complex hypotheses. The situation simplifies when all of the presuppositions of methods in the regress are entailed by H_0 or by its complement. Then one may speak of an *H_0-entailed* or *co-H_0-entailed* regress, respectively.

Proposition 4.2 *The following are methodologically equivalent:*

- *An H_0-entailed regress (M_0, M_1) such that M_1 refutes [verifies] in the limit the presupposition H_1 of M_0 as a limiting refuter [verifier];*
- *A single method M that refutes [verifies] H_0 in the limit.*

Proposition 4.3 *The following are methodologically equivalent:*

- *A co-H_0-entailed regress (M_0, M_1) such that M_1 verifies [refutes] in the limit the presupposition H_1 of M_0 as a limiting refuter [verifier];*
- *A single method M that refutes [verifies] H_0 in the limit.*

5 INFINITE REGRESSES

Suppose it is required that *every* challenge to an empirical presupposition be checked empirically, so that there is a potentially *infinite* regress of methods testing the assumptions of methods... The point of such practice is far less obvious than that of finite regresses, since finite regresses are "anchored" or "founded" by genuine success of the terminal meta-method. infinite regresses have no final "court of appeals" in this sense to anchor them. Are they, therefore, necessarily vicious? This is no longer a matter of mere philosophical opinion. It is a logically precise question about methodological equivalence that will now be explored.

Recall that in the Lyellian regress, each method covers more possibilities than its predecessor, for a method that did not cover more possibilities would hardly be an effective rhetorical response to skeptical objections. Say that such a regress is *nested*, since the presuppositions of the successive meta-methods get ever weaker.[10] Then

Proposition 5.1 *The following are methodologically equivalent:*

- *An infinite, nested regress $(M_0, \ldots, M_n, \ldots)$ of sequential refuters;*
- *A single method M that decides H_0 with at most two retractions, starting*

[10] This does not imply that H_0 entails H_1, since H_0 is not a presupposition.

with acceptance, over the disjunction $(H_1 \vee \ldots \vee H_n \vee \ldots)$ *of all the presuppositions of the methods in the regress.*

In the special case in which the infinite refuting regress is H_0-entailed, it is equivalent to a single method that really refutes H_0 over the disjunction of presuppositions of constituent methods in the regress. More generally, if M_0 succeeds with n retractions, the refuting regress is equivalent to a single method that succeeds with one more retraction, starting with the same initial conjecture as M_0.

Several points should be emphasized. First, the collapsing construction used to prove the preceding results is not a single, infinitary collapsing function $\Phi(M_0(e), \ldots, M_n(e), \ldots)$ that looks at the outputs of all the methods in the regress at once. It is, rather, a sequence of finitary collapsing functions of increasing arity that are invoked at successive stages of inquiry

$$\Phi_0(M_0(())),$$
$$\Phi_1(M_0((e_0)), M_1((e_0))),$$
$$\Phi_2(M_0((e_0, e_1)), M_1((e_0, e_1)),$$
$$\vdots$$

so the collapsed output at a given stage of inquiry is constructed out of only finitely many of the outputs of the methods in the regress. Hence, the equivalences hold even if the infinite regress is built up through time in response to specific, skeptical challenges, instead of being given all at once. Second, no method in the regress has a presupposition as weak as the presupposition of the regress itself, so appealing to a regress is a way to weaken presuppositions of inquiry overall *after* a method with given presuppositions has been chosen. Third, although such regresses yield greater reliability, they are feasible only for hypotheses that are decidable with two retractions, which falls far short of dealing with Duhem's problem, which gives rise to problems that are only verifiable in the limit.

The last point is ironic for Popper's [18] "falsificationist" philosophy of science. Popper started with the common insight that universal laws are refutable but not verifiable. But his "falsificationist" philosophy was not that naïve. He was aware of Duhem's problem of blame-assignment and of the fact that an isolated hypothesis can be sustained come-what-may by twiddling other auxiliary hypotheses. He held that this "conventionalist stratagem" of preserving a pet hypothesis at the expense of changes elsewhere is a bad idea because it ensures convergence to the wrong answer if the hypothesis is false. Better, he thought, to *stipulate* crisp conditions under which the (non-refuted) hypothesis should be rejected in advance. Of course, the stipulation involves another hypothesis: that the rejection is not in error. But one can also set up falsification conditions for that hypothesis, etc. Carried to its logical conclusion, this recommendation amounts to an infinite refutation regress. I am unaware whether Popper somewhere addressed the question of nesting, but it would be quite natural for someone vaguely concerned with truth-finding to add this requirement. Now the whole point of Popper's philosophy was to find the truth in the face of Duhem's problem. But the preceding result shows that Popper falls far short. Questions involving even concrete auxiliary hypotheses like unformitarianism's schedules of progress are not even decidable in the limit, but an infinite Popperian regress of nested refuters exists only when the question is decidable with just two retractions.

The irony is worse than that. For Popper, the falsificationist, *could* have addressed Duhem's problem had he been a regressive *verificationist* rather than a regressive falsificationist. Say that a convergence concept *converges to rejection* if and only if (1) the concept entails refutation in the limit and (2) allows for rejections to be retracted. Verification with certainty converges to rejection. Indeed, among the success concepts under discussion that entail refutation in the limit, the only one that does *not* converge to rejection is refutation with certainty.

Proposition 5.2 *The following are methodologically equivalent:*

- *An infinite, directed regress* $(M_0, \ldots, M_n, \ldots)$ *of methods that converge and succeed in senses that converge to rejection.*
- *A regress* (M_0, M) *such that* M_0 *succeeds regressively in the same sense as before and* M *refutes the presupposition* H_1 *of* M_0 *in the limit over the disjunction* $(H_2 \vee \ldots \vee H_n \vee \ldots)$ *of all the other presuppositions in the regress.*

Recall that regresses of limiting methods are irreducible to simpler success criteria. If the regress is H_0-entailed or co-H_0-entailed, however, then one obtains the following, cleaner results.

Proposition 5.3 *The following are methodologically equivalent:*

- *An infinite,* H_0*-entailed, directed regress* $(M_0, \ldots, M_n, \ldots)$ *of methods that converge and succeed in senses that converge to rejection.*
- *A single method* M *such that* M *refutes* H_0 *in the limit over the disjunction* $(H_2 \vee \ldots \vee H_n \vee \ldots)$ *of all the other presuppositions in the regress.*

Proposition 5.4 *The following are methodologically equivalent:*

- *An infinite, co-H_0-entailed, directed regress* $(M_0, \ldots, M_n, \ldots)$ *of methods that converge and succeed in senses that converge to rejection.*
- *A single method M such that M verifies H_0 in the limit over the disjunction* $(H_2 \vee \ldots \vee H_n \vee \ldots)$ *of all the other presuppositions in the regress.*

To illustrate proposition 5.3, recall Lyell's uniformitarian hypothesis. Extending the finite regress discussed earlier without end, one obtains an infinite regress in which M_n says "yes" for H_n if the $(n + 1)$st schedule of progress is not instantiated, "no" if the schedule is instantiated but non-refuted, and "yes" otherwise, thereby presupposing $H_n = H_0 \vee A_1 \vee \ldots \vee A_{n+1}$. This is, evidently, an H_0-entailed nested regress of methods that converge and succeed regressively with two retractions starting with "yes" and, hence (by proposition 5.3), is equivalent to a single limiting refutation procedure M for H_0 that succeeds over the disjunction of the presuppositions, which is in turn equivalent to the disjunction of the two competing paradigms (i.e., uniformitarianism $\vee$ progressionism). Here is how to construct M in this particular case. Method M maintains a queue of the methods added to the regress so far. Each time a new method is added to the regress, it gets added to the end of the queue (the regress is only "potentially" infinite). If the method at the head of the queue says "yes", it is placed at the end of the queue (ahead of any new methods added at that stage). Each time the method at the head of the queue is shuffled to the back, M says "yes". Otherwise, M says "no". Suppose that some proposition H_i is true. Let H_n be the first such. Suppose that $n > 0$, so that H_0 is false. Then H_{n-1} is false. Since H_n is true, M_n converges correctly to "no". If $k < n$, then H_{k-1} is false and H_k is false, so M_k converges incorrectly to "yes". If $k > n$, then H_{k-1} is true and H_k is true (by nesting), so M_k converges correctly to "yes". Hence, M_n is the unique method in the sequence that converges to "no". So eventually M_n comes to the head of the queue after it has converged to "no" and M converges to "no" at that stage, as required, since H_0 is false. Now suppose that $n = 0$, so that H_0 is true. Then all of the presuppositions are also true, by nesting, and the hypothesis H_0 is true, so all of the methods converge correctly to "yes". Hence, M says "yes" infinitely often. So M refutes uniformitarianism in the limit, in accordance with proposition 5.3. Proposition 5.4 is illustrated by progressionism in the same example, if one exchanges "yes" with "no".

Observe how the collapsing construction in this example unwinds the *rhetorical* game of sequentially responding to challenges with methods that entertain more possibilities into a single, ongoing process of inquiry that finds the truth over all the possibilities covered by the constituent methods in the regress. This is a new and interesting model of how rhetorical and reliabilist conceptions of science can be reconciled and systematically compared.

6 CONCLUSION

Scientific method may be conceived as a justifying argument or as a procedure aimed at finding a correct answer. Both conceptions raise a natural question about the propriety of infinite empirical regresses, whether of "evidential justification" or of methods checking methods checking methods. Since it is hard to say what evidential justification is for, it is hard to bring the notion of infinite regresses of justification under firm theoretical control. The procedural concept of methodological equivalence, on the other hand, allows one to "solve" for the best single-method performance that a given kind of regress is equivalent to. Some motivated conditions on regresses result in nontrivial regresses that achieve sufficient power to address Duhem's problem.

ACKNOWLEDGMENTS

The author is indebted to C. Glymour for detailed comments on the penultimate draft of this paper.

REFERENCES

[1] Berger, W. and Labeyrie, L. (1987). *Proceedings of the NATO Advanced Research Workshop on Abrupt Climatic Change: Evidence and Implications*, Dordrecht: Reidel.
[2] Gold, E.M. (1965). "Limiting Recursion", *Journal of Symbolic Logic* 30, 28–48.
[3] Gold, E.M. (1967). "Language Identification in the Limit", *Information and Control* 10, 447–474.
[4] Hinman, P. (1978). *Recursion Theoretic Hierarchies*, New York: Springer-Verlag.
[5] Hitchcock, C. (ed.) (2004). *Contemporary Debates in the Philosophy of Science*, Oxford: Blackwell.
[6] Jain, S., Osherson, D., Royer, J.S. and Sharma, A. (1999). *Systems That Learn: An Introduction to Learning Theory*, 2nd ed., Cambridge (Mass.): MIT Press.
[7] Kelly, K. (1996). *The Logic of Reliable Inquiry*, Oxford: Oxford University Press.
[8] Kelly, K. (2000a). "Naturalism Logicized", in Nola, R. and Sankey, H. [16], 177–210.
[9] Kelly, K. (2000b). "The Logic of Success", *The British Journal for the Philosophy of Science* 51, 639–666.
[10] Kelly, K. (2004). "Uncomputability: The Problem of Induction Internalized", *Theoretical Computer Science* 317, 227–249.

[11] Kelly, K. and Glymour, C. (2004). "Why Probability Does Not Capture the Logic of Scientific Justification", in Hitchcock, C. [5], 94–114.
[12] Kuhn, T.S. (1970). *The Structure of Scientific Revolutions*, Chicago: University of Chicago Press.
[13] Laudan, L. (1980). "Why Was the Logic of Discovery Abandoned?", in Nickles, T. [15].
[14] Martin, E. and Osherson, D. (1998). *Elements of Scientific Inquiry*, Cambridge (Mass.): MIT Press.
[15] Nickles, T. (ed.) (1980). *Scientific Discovery, Logic, and Rationality*, Dordrectht: Reidel.
[16] Nola, R. and Sankey, H. (eds.) (2000). *After Popper, Kuhn and Feyerabend*, Dordrecht: Kluwer.
[17] Osherson, D., Stob, M. and Weinstein, S. (1986). *Systems That Learn: An Introduction to Learning Theory for Cognitive and Computer Scientists*, Cambridge (Mass.): MIT Press.
[18] Popper, K. (1968). *The Logic of Scientific Discovery*, NewYork: Harper and Row.
[19] Putnam, H. (1963). "'Degree of Confirmation' and Inductive Logic", in Schilpp, P.A. [22].
[20] Putnam, H. (1965). "Trial and Error Predicates and the Solution to a Problem of Mostowski", *Journal of Symbolic Logic* 30, 49–57.
[21] Ruse, M. (1979). *The Darwinian Revolution: Science Red in Tooth and Claw*, Chicago: University of Chicago Press.
[22] Schilpp, P.A. (ed.) (1963). *The Philosophy of Rudolf Carnap*, LaSalle (Ill.): Open Court.

9

TRADE-OFFS

CLARK GLYMOUR
Department of Philosophy, Carnegie Mellon University, Pittsburgh, PA 15213, U.S.A., cg09@andrew.cmu.edu

Abstract: Statistical inference turns on trade-offs among conflicting assumptions that provide stronger or weaker guarantees of convergence to the truth, and stronger or weaker measures of uncertainty of inference. In applied statistics—social statistics, epidemiology, economics—these assumptions are often hidden within computerized data analysis procedures, although they reflect broader epistemological issues. I claim that the content of the trade-offs, and of the epistemological issues they manifest, is clarified by placing them within the framework of formal learning theory.

My concern here is not with any formal application of learning theory, or with presenting any new learning theoretic results—which is a good thing, since I have none. Rather, I want to show the heuristic value of the distinctions the framework makes, with how they help us to think about the trade-offs facing scientists in practice, and how the framework of learning theory develops the half of epistemology neglected in philosophy. In the tradition of philosophy, I will do some defining but no proving. I will start with three made-up but realistic examples of difficult decisions illustrating these trade-offs, remark on how the decisions are implicit even when the trade-offs are not recognized, describe a general framework for learning theory, and suggest how the methodological ideas of learning theory can help to clarify, if not to resolve, the trade-offs, and I will claim that the cases illustrate more general trade-offs in epistemology.

1. Scientist A and scientist B, in possession of the same data, are each out to discover the effects of smoking on lung disease. Scientist A has

M. Friend, N.B. Goethe and V.S. Harizanov (eds.), Induction, Algorithmic Learning Theory, and Philosophy, 219–232.
© 2007 *Springer.*

a strategy of estimation from finite data samples that is guaranteed to converge (as the data increase without limit) to a correct estimate under assumptions for which neither A nor B has good warrant. Scientist B has a strategy of estimation from the same data samples which is guaranteed to converge either to a correct estimate or to no answer at all, under assumptions that are much weaker than those used by A, and that both regard as warranted. On the present data, B's procedure returns no answer. Which procedure should be used? What *attitude* ought A and B and anyone else have towards estimates of the effect by A's method?

2. Scientists C, and D, in possession of the same data, have respective strategies they wish to apply to discover the influence of poverty on crime. C has a procedure that converges to the truth and yields confidence intervals for the estimates, but under assumptions for which C has no good warrant. D's procedure provably converges to the truth under weaker assumptions that all of the scientists regard as warranted, but it provides no confidence intervals and no degrees of belief. Which procedure should be used? What *attitude* ought the scientists or anyone else have towards estimates of the effect?
3. Scientists E and F are interested in whether among nations political development, indicated by a number of measures, causes economic development, indicated by a number of other measures, or *vice versa.* They have different strategies for obtaining detailed models, methods that apply under the same assumptions. To compare their strategies, they apply them to a number of simulated data sets. E's strategy always produces a unique model, but is correct only 14% of the time; F's strategy produces an average of three alternative models, but contains the true one 90% of the time. Which strategy is to be preferred?

These are not idle questions. Usually without explicit formulation, they are answered every day in applied science, and lives and health and the fortunes of nations turn on the way they are answered. Science goes on amidst conflicting desires, and the hard, logical fact that the desires cannot be reconciled. One conflict is between the desire to articulate and proceed by methods that yield justified beliefs—or their degrees or probabilities—*now*, and another desire, to articulate and proceed by methods that guarantee true results *eventually*. The conflict would dissolve if there were methods guaranteed to yield true results *now*, on present data, or on any finite sample one could specify and obtain, but almost always there are not. A second conflict is between the desire for informative answers and the desire to avoid false answers. It, too, would dissolve if there were methods that are both informative and known to be correct, and sometimes there are, but in many problems there are not, and we are forced to choose. A third is a conflict between the desire to proceed by methods that converge on the truth eventually under the weakest assumptions possible, and the desire to have measures of the uncertainty of our conjectures along the way. The same conflicts are at the heart of philosophical disputes over knowledge: the skeptic believes nothing on the grounds that nothing is certain, or even probable; the realist makes assumptions (e.g., that there is an external world), on the grounds that without assumptions nothing can be learned or predicted; the dogmatist believes everything with certainty on the grounds, roughly, of why not?

We implicitly resolve many of these conflicts by the adoption of a method of inference, almost always nowadays by a computerized method whose properties are themselves a matter of estimation. Given data on whatever—to take a very simple example, let us suppose the subject is ophthalmology and the data are from the reflectance properties of laser light bouncing off layers of retinal tissue—I will suppose the scientist wants to discover a rule to separate cases with glaucoma from cases without. (Examples in which the goal is to identify causal relations rather than merely to classify will follow later.) To do so she uses a program, usually a "statistical" classifier, that has definite properties, and thereby her inquiry becomes limited by those properties. What sorts of properties?

The program and the computer apply a mathematical function. The input to the function is a list of cases, in our example each case is a list of variable values, one of which is a yes/no variable that specifies whether the subject described by the case has glaucoma. The output of the function is, normally, another function, and usually some numbers. This second function is from the values of some or all of the listed properties to the value of the glaucoma variable; it constitutes the prediction rule found by the program. The significance of the numbers output by the program varies with the particular program; typically, the function output is described by values of a set of parameters, and the numbers output indicate probability features of the estimates of the parameters, or probability facts about the result of applying the rule to the cases that are input. Typically, too, for some inputs the program may give no output at all, or give a rule but fail to give numbers representing probability features.

The program has known (to experts, anyway) and unknown (to anyone) *convergence* properties. Consider all mathematically conceivable infinite sequences of cases. There are of course many probability measures on various sets of such sequences. For some of these probability measures, it is known that the program, if applied forever, will with probability 1 converge to a rule that separates the glaucoma and non-glaucoma cases as well as can be done by any rule, or by any of some general class of rules. For most other probability measures, nothing of the kind is known.

The program has (or may have) *finite sample statistical properties*, the probabilities or confidence intervals attached to the parameter values of the output rule. These probabilities are almost always calculated on the assumption that the probability distribution on the various subsets of the set of infinite sequences of cases is among a family of distributions with the known convergence properties. The program also has what I will call *independence* properties: for example, for most such programs, the output does not depend on the order in which the cases are listed; it may depend on what the output of the program would be on subsets of the sample.[1] And of course the program has *procedural* properties. For example, unlike people, such programs are seldom incremental, that is, they do not learn what they can from the first (or first and second, or first and second and third, etc.) case, then forget the case but remember the conclusion, and go on to the next case. Again, all programs have computational complexity properties, and so on.

Almost always, the scientist has a vague rule that says that if the numbers are within some range of values, then publish. The justification, the rationale, for doing so must be based on the properties of the program. Some of these properties, such as the computational complexity of an algorithm and its independence properties, are absolute, but many of them, such as the convergence properties, are conditional—if the data of a class has such and such features, then the program output has various other features. Essential to the rationale, then, is the grounds for paying attention to some of these conditional properties, and for endorsing their antecedents. To do so requires considering the goals and sub-goals of inquiry, and their interaction with methods of truth seeking and their limitations. The statistical literature addresses special cases of these questions, specifically convergence properties of various estimation procedures, assuming various families of probability distributions, and the finite sample properties of such probabilities. For a more general viewpoint, we must turn to another subject, formal learning theory.

The motivation for formal learning theory is a general picture of inquiry, idealized to be sure, but no more so than is common in methodological tracts in many disciplines. A scientific community observes the world and treats some of its happenings as data relevant to an inquiry. Data comes in discrete lumps in a sequence over time. Data can be passively observed or produced by experimental interventions, which do not change the world but do change

[1] As with cross-classification procedures that test rules against various subsets of the data.

the data observed. The data can sometimes be erroneous, a complication accounted for in some learning theory frameworks, but which I will ignore. The sequences of data may be finite or infinite. Hypotheses purport to describe features of the world, and they imply restrictions on the possible data sequences. Inquiry proceeds by some method, which conjectures hypotheses, or sets of hypotheses, or probabilities for sets of hypotheses, based on the data seen so far, and then investigators may conduct experiments to produce novel data. There is not simply one set of possible streams of data, but many separate ones, for separate problems of inquiry. For any particular problem, the scientific community makes assumptions that limit the set of alternative hypotheses and their connections with the data, assumptions that may also impose another structure, for example probability measures on hypotheses, or probability distributions for finite samples of data given various hypotheses. Different problems may be closely related, differing only in the assumptions made—as by different members of the community. There are standards of success in inquiry, although different investigators may have different standards, but all of them have something to do with eventually conjecturing the right answer and sticking with that conjecture. The task of learning theory is to formalize this picture, and to investigate mathematically the relations between assumptions, methods, and criteria of success.

Here is a rather more formal version.

> A *problem* is given by the following structures: (1) a set T of possible trees (in the graph theoretic, not the botanical, sense), each vertex of which is an element from a common countable vocabulary. Each tree is a *world*. (A tree is a representation of possible data sequences of some kind in a world; a path through the tree is a possible data sequence of that kind in that world.) (2) With each vertex in a tree there is a parameter, p, coding the branches at that node. (3) A set H of *hypotheses*: each hypothesis is associated with a probability distribution on subsets of T. (Very often, the probability distribution is $0/1$ valued, that is, a hypothesis allows a particular set of trees, and disallows all trees not in that set.) The set of such trees corresponding to a subset J of H will be denoted $T(J)$. A *course* is a path through a tree.
>
> A *method*, M, for a problem specifies, probabilistically (and again, $0/1$ measures are allowed) for each such initial segment of each tree in the problem a value of the parameter p for the last vertex in the segment. (p is the parameter by which a method decides after each datum is received what experiment next to make, if any.) A *course of M* through a tree is any path through T determined by the root of the tree and the values M assigns to p at each vertex. The set of courses through a tree

allowed by the values of p assignable by M is the *M range* of the tree. Further, M specifies a random (in the sense that its output is given by a probability measure) partial function from all finite initial segments of courses in the M ranges of trees in T to a probability measure on sets of H, measurable in some specified measure. A *strategy* is a mapping from problems to methods.

Intuitively, the method is applied to a problem in some (unknown) world or other, and, in particular, there is some set of jointly true hypotheses in that world. Initial segments of one of the sequences for these true hypotheses are fed to the method, which in response produces a sequence of conjectures, which may be probability assignments to hypotheses, or passes, and the method determines a next experiment—a branch of the tree at that vertex. There is no restriction on logical connections among the hypotheses, although most of the mathematical results in the subject are for sets of mutually incompatible hypotheses. A hypothesis can be a full fledged theory of some kind, or merely a set of sequences. Logical or probabilistic structure can be imposed on the hypotheses. The data sequences (courses) can have any structure one pleases; they can be recursive or recursively enumerable, for example, or not. The sequences that are courses within a tree are unrestricted in order and in length, although commonly they are thought of as infinite. The mechanism of methods is unrestricted in the framework; it could, for example, consist of many separate sub-methods conjecturing and voting by some rule; it could use Bayesian updating on probability measures imposed on the hypotheses; it could have a random feature, for example the output in some circumstances could be determined by whether a particular atom has or has not decayed.

With this simple set-up we can consider various powers and limitations for methods and strategies, we can consider a variety of success criteria for inference, we can if we like impose measures on the sets of trees, and we can compare properties of methods and problems with regard to various success criteria. Consider first some possible powers and limitations of methods. The most interesting methodological power is to influence the initial segments presented as arguments to the method. Formally, this requires no change to the set-up just sketched; informally, we can think of the method as having an internal device that selects experiments based on the data available, and the selection changes the continuation of the sequence. The most radical version of this power—the sort that philosophers who talk of conceptual schemes and relativism perhaps have in mind—requires a generalization in which a method can take an initial sequence fed to it from one world to an initial sequence of another world, and by doing so can change which hypotheses are true. The limitations of methods are of two kinds: computational and methodological. Computational restrictions restrict a method to some computational complexity class, for example to Turing computable functions, or primitive recursive functions, or finite automata or to somewhere above Turing computability in the arithmetical hierarchy. From my perspective, the most interesting limitations of this kind are those that are reasonable idealizations of least upper bounds on our computational abilities, aided by machines, that is, something like finite automata, but certainly methods limited by Turing computability. There is nothing we can compute, and arguably nothing we can do, in science that exceeds Turing computable bounds, while more limited models (e.g., a specific automata model or lower complexity bounds) may be violated by algorithms or technological developments, or may just be arbitrary. Methodological limitations are the kind of methodological injunctions that philosophers of science advocate. *Consistency*, for example, says that the output of a method on an initial segment must always be consistent with H and the initial segment. *Conservatism* says that a hypothesis, h, once conjectured on an initial segment s, must continue to be conjectured on all extensions of s that are consistent with h. *Determinacy* excludes random methods. *Singularity* requires that $0/1$ probability measures be output. When the hypotheses are not mutually exclusive, but some are logical consequences of others, *closure* requires that the output of the method be deductively closed over H. *Informativeness* limitations restrict the range of determinate methods, for example to singletons. And so on.

Success criteria are various desiderata one might have in inquiry, properties of pairings of methods and problems. For example: *Finite decidability* is satisfied by a method M for a problem hypotheses if and only if for all hypotheses h of the problem, all sequences w for h, and all initial segments s of w, $M(s)$ is undefined or is h, and for some initial segment s of w, $M(s)$ is h, and for all extensions s' of the first such s in w, $M(s')$ is h. *Decidability in the limit* is satisfied, under similar conditions with the same quantifiers, if the output of M is h for all but a finite number of initial segments of w. Special cases arise where there are but two hypotheses, h and $\tilde{}h$, in which case one can distinguish the verifiability and falsifiability of h, both finitely and in the limit. Various of these and other definitions have been shown to correspond to complexity classes of hypotheses. When hypotheses specify probability measures on initial sequences, or when methods assign probabilities to hypotheses, various probabilistic convergence criteria can constitute success criteria, and these are the subject of conventional convergence theorems in probability theory, including Bayesian convergence. When the method can result in a change of worlds, a number of alternative success criteria are possible, for example, one can require that eventually only true hypotheses always be output, but that the worlds may change and never stabilize. Many other success criteria are possible. Indeed, it would be possible to form success criteria that merely required convergence to a probability for the true hypothesis greater than some value less than 1 and greater than $\frac{1}{2}$, or sufficiently greater than that for any false hypothesis.

A problem is said to be *solvable* by a success criterion subject to a limitation if there exists a method satisfying that limitation that also satisfies the success criterion for that problem. The fundamental facts of inductive life are mathematical relations between problems and their solvability or unsolvabilty by various success criteria. We would, most of us, like a method of inquiry to come with a guarantee that for a specified sample size, N, the method will return the truth from a sample of size N, or, failing that, to have a guarantee of an objective frequency of truth for each such N. For many problems that is simply impossible as a matter of logic. The alternatives are therefore to accept weaker success criteria—convergence in the limit for example, with a guarantee that the truth will be reached and kept sometime, but no guarantee as to when—or to change the problem, commonly by excluding alternative hypotheses without good grounds. These are some of the trade-offs of inquiry, but not all of them.

A success criterion amounts to a preference relation among strategies, but there are other desiderata besides success. One is informativeness of methods. Suppose we are considering methods that are determinate, p is constant, and the success criterion is convergence in the limit to a set of hypotheses containing the true one. Given two methods M and N for the same problem, M is the more informative if the output of M never has larger cardinality than that of N on the same data, and is sometimes smaller. Other conceptions are clearly possible, e.g., in terms of inclusion relations on the output sets from the same data, and the definition can be generalized to methods for different problems, most naturally if the hypothesis space for one problem is a proper subset of that of another. Another, quite different, but in practice quite influential, desideratum is a measure of uncertainty for hypotheses conjectured by determinate methods. The most familiar measures of this kind are statistical confidence intervals.

For any specified success criteria and other desiderata, and for specified limitations on methods, many of the great questions of epistemology become purely mathematical issues—many of which, for a framework as general as this, are unsolved, but many of which are solved for various specializations of the framework and variants of it. When is a problem underdetermined by a methodological restriction—that is, when are there incompatible hypotheses in the problem such that no method in accord with the limitations exists that can separate the hypotheses, satisfying the success criterion for one of them and not the other according to which is true and which not? When do alternative strategies solve (according to a specified success criterion) the same class of problems? When do alternative strategies solve the same class of problems, but each according to different criteria of success? Are various limitations on methods, whether computational or methodological, *restrictive* in the sense that there are problems that can be solved (according to whatever success criteria) but cannot be solved by any methods in accord with the limitations? There is a wealth of answers to these questions for particular cases, some of them shocking to philosophical consciences, but there are better reviews of the literature than I can provide. Back to the problems of my alphabetic scientists.

Consider the task of inferring whether one variable X is a cause of another, Y, in the sense that manipulations that vary the first will reliably produce co-variations in the second variable. We will suppose that other variables, $Z_1, \ldots, Z_n$, with unknown causal connections are also measured. X is correlated with some of the Z variables. We will suppose that the data are such that the values of these variables have a probability distribution for each case, each patient or subject, and the probability distributions for any two cases are the same and the probability of any value (or measurable set of values) of any variable for any one case is independent of the values of any variables for any other case or cases. Let us further suppose that the probability distributions are all Normal. Scientist A has a strategy: perform a simultaneous regression of Y on X and $Z, \ldots, Z_n$, determine if the regression coefficient[2] for X is statistically significant (i.e., if the hypothesis that the regression coefficient is zero is rejected by a statistical test at some conventional level, say .05), and if it is significant, estimate its value by a method known to converge with probability 1 to the true value, e.g., maximum likelihood. Scientist A gets results. Scientist B, on the other hand, reasons as follows: Scientist A has assumed a family of probability distributions on samples—initial sequences. Given that assumption, for many

[2] Informally, a regression of variable Y on variable X for a finite set of data points gives the function of the form $Y = aX$ that minimizes the sum of squared differences between Y and aX. The real number a is the regression coefficient. In multiple regression, Y is estimated by a linear function of several variables and the values of linear coefficients that minimize the sum of squared differences between Y and its predicted values, and the values of those linear coefficients are the partial regression coefficients. In logistic regression the variable to be predicted is binary and the prediction rule is a linear function of the prediction variables specifying a function of the ratio of the probabilities of the two values of the binary variable. For proofs of the claims made subsequently about regression methods, and the description of alternative methods with the properties claimed, see [1].

problems, the strategy of scientist A does not, with probability 1, converge to the correct answer about the causal relations—the success criterion–between X and Y in the limit as the sample size increases without bound. Indeed, for some problems the strategy of scientist A converges with probability 1 to the wrong conclusion (a failure criterion!). This can happen in several ways. For example, suppose that X influences Y only through Z_1, so that if an intervention held Z_1 constant while varying X (an experiment that could conceivably be done at any point, but is not part of A's method), Y would not covary with X. In the large sample limit, multiple regression would almost surely yield the false result that X is not a cause of Y. B has another objection. Suppose there is an unmeasured variable G that influences both X and Y, but X does not cause Y. Then multiple regression of Y on X and the other measured variables will yield (again, large sample limit, almost surely) a significant regression coefficient for X, and A will conclude, wrongly, that X causes Y. B goes on: even if he and A were quite sure, now or eventually with more data, that there is no variable such as G, suppose X and Z_1 are correlated, either because X causes Z_1 or because they have an unrecorded common cause, and either Y causes Z_1 or Z_1 and Y have an unrecorded common cause. In this case, again, multiple regression will result in the judgement that X causes Y, even if there is no such causal connection. B's reasoning is in fact correct.

A and B are essentially debating what kind of discovery problem they face. If it is assumed that there are no unrecorded variables that influence any of the recorded variables, and that none of X or the Z variables are influenced by Y, and X does not influence Y through its influence on any of the Z variables, then A's procedure solves the problem by A's and B's success criteria. Without those assumptions, it does not. B proposes another strategy, which solves problems with weaker assumptions and a weaker success criterion. B's procedure will not assume that there are no unrecorded common causes, and it will not assume that Y does not cause any of the other recorded variables. It does assume, in addition to A's distribution assumptions, only that the causal relations among the variables do not perfectly cancel their effects, so that X is uncorrelated with Y even though X causes Y. Under those assumptions, B claims, correctly, methods that instantiate the strategy will return a class of causal hypothesis, almost surely (again in the large sample limit) including the true hypothesis. Furthermore, B claims, again correctly, that no alternative strategy that solves the problem is more informative than hers for this class of problems—no alternative strategy will converge in every problem of this kind to a set of hypotheses, with the true hypothesis as a member, that is never larger, and in some cases smaller, than that produced by the strategy B recommends. Unfortunately,

B admits, in some cases the method will return a class that includes the hypothesis that one variable causes another, and the hypothesis that it does not. So they try it, with the result that *B*'s method cannot determine in this problem whether *X* causes *Y*. Should they go with the results of *A*'s method, or with *B*'s, that is, pass?

Wittingly or not, this is a conundrum faced every day by myriad social scientists and epidemiologists, a conundrum one can think about most clearly, I suggest, by placing statistical/causal inference in the context of learning theory. I have presented it as a problem about linear regression, but it applies with many other assumptions about the probability distributions, and many other methods of regression and data analysis. Almost universally, the choice among social scientists is to go with *A*'s strategy, sometimes because they do not know of its limitations, sometimes because they do not know of *B*'s alternative, sometimes because, well, they want *some* result. Since most social science data is ignored, that may do no harm. But social science conclusions are often used for political ends—as they should be if they are sound—and epidemiological conclusions affect lives and health. The mistaken conclusions about hormone replacement therapy, obtained essentially by methods akin to *A*'s, are a sad example. So are those of the famous, or infamous, book by Murray and Herrnstein (1994) arguing that intelligence tests measure innate abilities.

Scientists *C* and *D* have a different but related difficulty. In this case, scientist *C* uses multiple regression, and obtains a conclusion: *X* causes *Y*. Scientist *D* uses *B*'s strategy and obtains the same conclusion. Scientist *C* prefers her analysis, because her method produces confidence intervals: given her estimate of the parameters, she can specify an interval of possible values for the influence of *X* on *Y* such that 95% of the samples of the same size will give estimates of that influence within that interval. Further, her method of estimating confidence intervals will almost surely reduce the width of the confidence intervals to 0 as the sample size increases without bound. *B*'s strategy by contrast, produces no such confidence intervals, and provably no strategy exists that solves the problem under *B*'s assumptions, with the informativeness of *B*'s strategy, and that will produce such confidence intervals. Scientist *D* prefers *B*'s strategy because it gives the conclusion under much weaker assumptions.

This kind of difficulty also arises outside of causal inference, for example in classification problems, where methods that allow confidence intervals (logistic regression, for example) vie with methods that converge to the truth under weaker assumptions (neural nets, for example) but for which no confidence intervals are available. In causal inference, scientists almost always choose the methods that allow confidence intervals; in classification, that is not the favored choice. The preference for confidence intervals in causal inference is largely due, I think, to the emphasis of traditional statistics on *measuring* the uncertainty of inferences. But those measures are only obtained under restrictive assumptions that are dispensed with in strategies that do not yield such measures. Assumptions have no measure, but they contribute, or ought to contribute, to uncertainty.

Scientists *E* and *F* have the most direct problem: they are faced with a trade-off between reliability and informativeness, on the same assumptions.

I have no resolution of the first two conundrums, only preferences: I prefer methods that are asymptotically reliable and in many cases informative under weak assumptions to methods that are almost always informative under much stronger assumptions. I prefer methods that converge to the truth under weaker assumptions but give no measures of uncertainty to methods that converge to the truth only under much stronger assumptions and, under those assumptions, give measures of uncertainty. Caution before counting. In the third example, where there are no prior probabilities for the hypotheses, the sensible thing would seem to be to compare the methods by dividing the percentage of correct outputs for *F*'s methods by the number of alternative hypothesis contained in their outputs, which in the story would result in a clear preference for *F*'s strategy.

Learning theory shows that what philosophers too often describe as a single issue typically resolves into a plethora of distinct questions, with different success criteria, different problems, different limitations, and different results. That's progress. Learning theory, in part the creation of a philosopher, ought to be at home in the drawing room of philosophy of science rather than, as it is, lost in the philosophical wilderness. By and large, the reaction of philosophers to the subject, or at any rate of those who know of it, has been either that learning theoretic questions are too abstract, too idealized, to be of interest, (although one must suspect the real objection is that too much mathematical effort is required to answer them: the philosophers make no such objection to statistics), or that the *real* issue is how hypotheses are invented (the news that vast classes of hypotheses can be searched without being explicitly formulated seems not to have arrived in many philosophy departments), or else that only probabilistic methods are of interest. Interest is a matter of the interested, and 20th and 21st century philosophers have increasingly narrowed what they take to be the interests of philosophy, but remarkable theoretical and practical facts are thereby ignored, and heuristics that ought to aid philosophical assessment of the sciences are neglected. The Bayesian strategy, for example, is restrictive for Turing computable learners. Again, there are real scientific problems that have a Bayesian solution but that can be solved by other methods more easily without any recourse to probabilities and their calculation.

The practical questions of life and science are always what to believe *now* and what to do *now*. We can make up probabilities, assume a finite set of alternatives and the costs of each action under each alternative, and apply a decision theoretic rule, but we may be wrong or disappointed if our assumptions are false. We can form beliefs according a rule of inquiry that is certain to be correct under a set of assumptions, but we may be wrong (and, absent luck, will be) if the assumptions are false. We can settle for forming beliefs according to a rule of inquiry that is guaranteed to converge to the truth eventually, with no guarantee when, under a set of assumptions, which again may be false. We can conjecture by some rule without caring whether it finds the truth. We can use no rule, no procedure, of inquiry, but form beliefs and choose actions on whim. Traditionally, philosophy has been in the business of developing positions on these questions. Skeptics reject all assumptions, and hence doubt everything. Plato claims there is nothing to be learned: all that we know we are born knowing, and we need only to have our recollection stimulated. Kant, on one reading, claims we can only experience worlds in which specific assumptions are true. Realists such as Herman Helmholtz and, in his later work, Bertrand Russell, urge that we in fact do and must make the minimal assumptions necessary to learn about the world with a rather small sample. Limiting theorists, such as Richard Von Mises and Hans Reichenbach, claim we should conjecture by procedures that converge in the limit to the truth under the weakest possible assumptions. Methodologists, such as Karl Popper and innumerable philosophers of science, claim we should form beliefs or conjectures according to rules or constraints that have nothing to do with finding the truth, but may have to do with avoiding falsehood. Relativists, such as Thomas Kuhn, claim truth, and even meaning, change according to our conjectures, and conclude there is no sense to finding the truth or to getting closer to it. Scientists, too, take positions on answers to these questions, but more often less reflectively, because the positions are built into their data analysis procedures and standards of practice and publication.

The wonderful contribution of learning theory is to explore the trade-offs implicit in general procedures for believing and acting—trade-offs of truth finding, of timing of truth finding, of changes of beliefs or conjectures, of informativeness. Facts about trade-offs will disappoint those who are uninterested in hypothetical reasoning about inquiry, those who only want to know, now or real soon, what is and is not true. I have sometimes lectured to philosophical audiences about learning theoretic trade-offs about relativism, giving results due to Kevin Kelly, Cory Juhl and myself, that show that there are strategies that are guaranteed to converge to the truth in various senses in problems in which the truth depends in specified ways on the conjectures produced by those very methods. Always, I find great disappointment: the philosophers want to know what version of relativism is true, and why, not what the trade-offs may be. Epistemology has but two aspects, religion and mathematics. The philosophers want to know the true religion and want a justification for it—is there an external world, are there other minds, is true relative to belief, what is the best method of inquiry, what can we know? So do we all. But any reflective judgement on such matters should consider the ambiguities of the questions and the trade-offs implicit in endorsing one set of answers as against others. That is the mathematics, and it is learning theory.

REFERENCES

[1] Spirtes, P., Glymour, C. and Scheines, R. (2000). *Causation, Prediction and Search*, 2nd ed., Cambridge (Mass.): MIT Press.

10

TWO WAYS OF THINKING ABOUT INDUCTION

NORMA B. GOETHE
School of Philosophy, National University of Cordoba, 5000 Cordoba, Argentina, ngoethe@ffyh.unc.edu.ar

Abstract: Inductive inference has always been a major concern in philosophy. This paper considers two different ways of thinking about induction: the classical way and the new, learning-theoretic way. After a brief introduction, I discuss the classical style of thinking about induction, which conceives of inductive inference as an extension of deductive argumentation. Then, I focus on the new, alternative learning-theoretic approach which sees induction as a type of computational process that converges to the truth. I conclude the paper by considering some of the important philosophical consequences of this fundamental shift in the metaphor through which philosophers conceive of inductive reasoning.

1 INTRODUCTION

Formal learning theory, also known as "algorithmic learning theory" or "computational learning theory", is nothing more or less than a general framework for studying the difficulty of finding or converging to the empirical truth, in much the same way that the theory of algorithms and computability concerns the difficulty of computing the true answer to a formally posed problem. The idea is to determine, for a given empirical question involving the testing of a given theory or the choice among a possibly infinite range of alternative theories, the best sense in which an empirical or computational method could converge to the true answer and the most efficient possible methods that achieve that sort of success. The novelty compared to standard computability theory is that the correct theory is not uniquely determined by any finite sequence of input data, so that finding the

M. Friend, N.B. Goethe and V.S. Harizanov (eds.), Induction, Algorithmic Learning Theory, and Philosophy, 233–258.
© 2007 *Springer*.

truth demands "inductive leaps" beyond the data that nonetheless converge or stabilize to the correct answer in the limit of inquiry. As such, formal learning theory may be viewed as an extension of the practice and ethos of theoretical computer science to inductive inferences in the empirical domain. The concept of "learner" is, therefore, very abstract and might be more aptly expressed as "inductive method" or "empirical procedure". (The reference to "learning" may, therefore, be more confusing than helpful in fields outside of computer science. More apt terms include "computational epistemology" or "convergent methodology", but I shall follow convention by employing the original moniker.)

As such, formal learning theory would seem to be of deep and immediate relevance to such traditional philosophical topics as the underdetermination of theory by evidence, the justification and character of inductive method, the prospects for scientific success, the structure of scientific revolutions, the relevance of probabilistic reasoning to scientific justification, and so forth. And yet the reception of the subject within philosophy has been slower than one might expect.

One obvious impediment to the exchange of ideas is the highly formalized presentation characteristic of much of the learning-theoretic literature, which freely presupposes standard results from the theories of computable functions, computational complexity, probability theory, and topology. This is to some extent unavoidable as far as the proofs of the theory's results are concerned, but the basic motivation and import of the results is readily conveyed with no more formalism than is routine in philosophical discussions of confirmation theory or the philosophy of biology or physics. Furthermore, there are now readily available informal expositions of the basic results.[1]

Another explanation for philosophical inattention is the usual sort of terminological confusion occasioned by alternative disciplines confronting the same subject matter. For example, statisticians call methods that converge to the truth in the limit "consistent", whereas learning theorists reserve this term for methods that never output answers contradicting the input data. Learning theorists use the term "explanatory" to refer to methods that converge to a particular, correct program, as opposed to "behaviorally correct" inference, which eventually produces only correct programs without necessarily stabilizing to a particular one. Philosophers would contrast explanatory theories with instrumental or predictive theories, placing both explanatory and behaviorally correct inference on the side of instrumentalism. Other notational issues in the learning-theoretic literature include

[1] See, for instance, [26, 27, 31] and [37].

fanciful or non-standard use of such terms as "Popperian", "reliable", and "paradigm". But by and large, the notational divergences are so flagrant as to pose little in the way of subtle, interpretive difficulty for the interested reader and there are already presentations of the subject that employ more philosophically familiar terminology.[2]

A more important motive for philosophical caution is the insensitivity of some of the formal learning literature to the nuances of real scientific problems. In a typical, rather stylized setting, the input stream is generated by a Turing machine and the aim is to converge to the program of any such machine. Call this the *recursive (computable) function identification* setting. On the surface, one can imagine that the Turing machine is a theory and that the numerical inputs encode scientific data, but the fit with actual scientific problems is inexact and metaphorical. There is no sense in which the gears and wheels in a conjectured Turing machine are true of the world. There is no explication of scientific explanation in terms of such gears and wheels. There is no attempt to relate explanatory power to truth, so results in the function identification paradigm do not address the debate over scientific realism. There is no attempt to model in real time what happens when the background assumptions of the function identification problem come into question. Real scientific theories admit of infinitely many solutions, whereas a Turing machine computes a unique input stream. There is an extensive discussion on "theory-laden" data in the philosophical literature, but the recursive function identification paradigm assumes a clear-cut input protocol and a fixed relation of empirical adequacy for Turing programs with respect to the input stream of natural numbers. Perhaps the most telling concern among philosophers is that philosophers are interested in justifying particular theoretical preferences in the short run, whereas long-run convergence to the truth is compatible with any particular short-run preference.[3] But these interesting and legitimate concerns can and have been answered by means of some deft tweaks to the original function identification paradigm.[4] And

2 See, for instance, [47] and [29].

3 See [62] and [8].

4 For example, the learning problem can be modeled in a logical language, in which the aim is to find an axiomatization of the complete truth or, more generally, to find the true cell in a set-theoretically definable partition over the set of possible input environments. See [47] and [29]. Questions regarding background assumptions can be modeled as requests for methods that check the empirical presuppositions of methods that check the empirical presuppositions of methods.... (See Kelly's paper "How to Do Things with an Infinite Regress" in this volume.) Convergence to the truth from theory-laden data can be modeled in a paradigm in which truth and experience are functions of the state of the learner. See [33], and [36]. Short-run principles of rational belief revision can be criticized from a

in any event, such substantive concerns constitute a portal for fruitful engagement, rather than reasons for ignoring the subject altogether.

However, there remains a deeper and more questionable source of philosophical reluctance: formal learning theory constitutes a shift in the fundamental metaphor through which contemporary philosophers conceive of inductive reasoning. According to what I shall call the *classical* conception, inductive inferences are justified by *arguments* from evidential premises to theoretical conclusions. Since the conclusions of such arguments outrun the premises provided, they are invalid by the standards of deductive logic. Therefore, there must be an extended logic called *inductive logic* or *confirmation theory* that explicates the notion of empirical or inductive justification. Whereas deductive logic justifies its conclusions completely, inductive logic does so only to a degree reflecting the imperfection of the evidence. The justification conferred by inductive arguments is an end or duty in itself. Learning theory, on the other hand, conceives of inductive methods as procedures that produce theoretical conclusions in response to increasing evidence and understands such methods to be justified insofar as they converge to the truth efficiently, reliably, and in the best possible sense. Hence, methods are means rather than ends in themselves.[5] And whereas inductive logic is sought in intuitions primed by the consideration of well-chosen examples of good and bad science, learning-theoretic justification is no more subject to intuitive feelings of propriety than is the correctness of a computational procedure—all that matters is the mathematical structure of the method and of the problem it is applied to.

I shall argue that much of the philosophical reluctance toward to the learning-theoretic literature can be explained as the usual sort of resistance that arises when the dominant paradigm [38] of a subject is challenged. Indeed, the classical paradigm is so thoroughly entrenched in contemporary philosophy of science as to be in a position to define the very scope of philosophical concern. In consequence, formal learning theory is not so much recognized as a serious competitor as merely dismissed as irrelevant. In this paper, I trace the historical roots of this curious situation from Hume and

learning-theoretic viewpoint by showing that there are problems in which the principles of rationality cripple the truth-finding powers of computational or, sometimes, even arbitrary agents. See [47], [29], and [30]. Furthermore, recent learning-theoretic results explain in a novel way how a short-run preference for simple theories is necessary for minimizing retractions of opinion en route to the truth, something Bayesian philosophy can explain only by circular appeal to prior probabilities biased in favor of simplicity. See Kelly's paper "How Simplicity Helps You Find the Truth without Pointing at It" in this volume.

5 See [64] for a focused discussion of the difference between categorical vs. hypothetical imperatives.

Kant to the seminal work of Hilary Putnam, and close with some general morals about the relationship between formal learning theory and philosophy.

2 THE CLASSICAL WAY OF THINKING ABOUT INDUCTION: THE "PROPOSITIONAL APPROACH"

The classical way of thinking about induction conceives of it as analogous to, but weaker than, deductive reasoning (Carnap [2], p. 205). Both sorts of reasoning are understood to be (logical) relations between propositions. Both allow one to form arguments from premises to conclusions. But in opposition to deductive reasoning, inductive reasoning is *ampliative* in the sense that the conclusion goes beyond the content of the premises, whereas the conclusion of a correct deductive argument is only either as strong as, or weaker than, the premises.[6]

According to the classical scheme, deductive reasoning seems uninteresting from the point of view of acquiring *new* knowledge, but it has the advantage of being "truth-preserving", and in this sense it is "logically reliable": if the premises are true the conclusion is guaranteed to be true; while in the case of inductive reasoning, the premises only support or corroborate the conclusion making it more or less probable, at a given time, but leaving some (empirical) uncertainty as to the truth of the conclusion at some future time. Accordingly, what came to be known as "the problem of induction", or "Hume's problem", often read as a form of empirical skepticism concerning the logical validity of our empirical generalizations, is nothing but the fact that inductive inferences are not logically reliable, i.e., truth-preserving. As Peirce [51] and Carnap [2] point out, there are many kinds of inductive inferences.[7] However, the "universal inference", i.e., the inference from particular instances (given data) to a hypothesis of universal form (empirical extrapolation), has often been regarded by philosophers as the most important kind of inductive inference, so that the term "induction" was often restricted to this kind of inference. Philosophers also speak of the "underdetermination" of an empirical generalization by the evidence gathered at a given time. So, in this sense, one might also say that the conclusion of an inductive argument remains underdetermined by

[6] For an exposition of this view of induction, see [62]. See also [42].

[7] Among some of the most important kinds of inductive inference, Carnap ([2], pp. 207–208) considers the universal inference, the direct inference, the predictive inference, the inference by analogy, and the inverse inference.

the premises, which hold only at a particular time for all we know. This is what is meant by saying that in an inductive argument, the premises only offer support or warrant for the conclusion relative to the data gathered at the moment, but fall short of guaranteeing its truth. Thus, in the classical view of inductive inference, the price to pay for learning, i.e., acquiring new knowledge, is empirical *uncertainty* concerning future predictions.

A more radical reading of Hume is that one can't learn anything at all. That is because one is not entitled to make inductive inferences, as some of Hume's reflections in the *Enquiry* seem to indicate:

> Thus the observation of human blindness and weakness is the result of all philosophy, meets us at every turn, in spite of our endeavors to elude or avoid it. (Sec. IV, Part I, p. 26)

Most modern philosophical thought about induction and scientific enquiry is still gripped by Hume's "problem of induction". However, were these conclusions drawn by Hume concerning the supreme goal of modern inquiry, the advancement of learning? It is worth noting that in Hume's case, the so-called "problem of induction" itself appeared as the outcome of the strictures imposed by the deductive paradigm underlying the classical view of scientific demonstration. Against this background, Hume's arguments may be viewed as a critical challenge to the classical scheme rather than as an argument for skepticism. Moreover, such a challenge is the culmination of an important critical movement that acquired force in the early 17th century.[8] In order to make this point, let us turn briefly to the history of the classical view of science from the perspective of some of its critics.

2.1 The Advancement of Learning as the Goal of (Modern) Inquiry and the Strictures Imposed by the Classical View of Science

On the Aristotelian view of science, scientific knowledge is justified by its internal, deductive structure. It is part of this "classical" view that scientific knowledge ought to be demonstrative. Indeed, the paradigmatic examples of "genuine" demonstrations were to be found in the formal sciences, where the proofs were required to be ostensive (i.e., direct). This requirement was important. Only direct proofs follow the classical "*a priori* order of reasoning" from axioms (general principles/causes) to conclusions

[8] In the formal sciences such critical movement started much earlier. See, for instance, [45].

(particular instances/effects). Such proofs (called "causal proofs") possess a certain epistemic virtue in the sense that they are truly explanatory. Four different notions of cause were in play and this terminology was also applied to mathematical demonstrations.[9] To discuss the virtues of certain types of mathematical proofs in terms of the "causality" of proofs was still common usage among 17th century mathematicians such as Wallis and Descartes.[10]

In order to satisfy the said requirements, the classical view of science needs to make two basic assumptions. Firstly, there is the assumption of a deductive, explanatory order of truths (axioms/general principles) that can be taken as "ultimate reasons" underlying "causal proofs", i.e., from causes to effects. Secondly, there is the assumption of an "idealized cognitive agent" who is able to comprehend the crucial deductive relations.

The relationship between inductive and deductive reasoning, often entangled with discussions of analysis and synthesis, relates to complex issues in the history of science, though the latter terms have a knotty history of their own.[11] Let us focus here on the discussion as it appears in the early 17th century, when such methodological issues were of relevance to the development of analytic geometry, and the emergence of the new science. The topic of the advancement of learning and discovery was indeed one of the most important methodological issues being debated at the time. Descartes, for instance, argued against the Aristotelian view of science and logic. His goal was to establish methodological norms for the direction of a new type of "ideal inquirer" in the search after truth. Such rules were conceived of as methodological procedures for the resolution of problems by means of analysis. Descartes criticized both the axiomatic method of the ancient geometers for not providing insight into the actual procedure of inquiry used by them, and traditional logic for teaching rules that were unable to offer any useful guide in the search for truth. On Descartes' view, the norms of traditional methodology were but an impediment for those interested in learning. In particular, like Wallis and other 17th century mathematicians, he complained that the axiomatic (also called "synthetic") form of presentation

[9] According to Aristotle, the paradigm of scientific demonstration was to be found in the formal sciences. See [46].

[10] For a discussion of mathematical practice and its critics in the 17th century, see Mancosu [45].

[11] Timmermans [66] discusses the way analysis and synthesis were related to the notions of *a posteriori* and *a priori* reasoning in medieval methodology until Zabarella's teaching at the School of Padua. See also Garber [11].

of classical textbooks of geometry did not show the "true way of discovery", the method of analysis, which had been used for generating new results.[12]

What was thus being called into question at the time is the epistemic primacy of deductive structure in mathematical practice and scientific exposition. Descartes, in particular, refused to use the axiomatic form of textbook presentation in his *Géométrie* (1637), claiming that his analytical "proofs" did not require "synthetic" validation.[13] According to Descartes, such proofs are truly explanatory given that they show "the true way of discovery" which is what is required for understanding mathematical results. Much to the confusion of scholars, he decided to turn around the classical terminology and call his analytical proofs "*a priori*" and "causal proofs" while in the classical terminology "analysis" was associated with induction, that is to say, the *a posteriori,* "nondemonstrative" order of reasoning.[14]

To sum up, the different lines of attack launched by the 17th century critics were united by the perception that the classical deductive scheme was useless for those interested in learning and acquiring *new* knowledge.[15] This critical movement reached its culmination in the next century when, against the background of Newton's contributions to the new science of nature, Hume renewed the attack. His main target was twofold. On the one hand, he calls into question the idea of a causal-explanatory order of true principles—"waiting to be discovered" by an ideally omniscient agent. On the other hand, he radically challenges the assumption of an "unbounded cognitive agent" able to "reliably" relate to such objective principles establishing "necessary connections" which would be required for learning and understanding science.

Note that Hume does not explicitly speak of "the problem of induction".[16] When first presenting his challenge to the classical viewpoint in his early work, the *Treatise*, he attacks the classical conception of general principles as "ultimate reasons" (causes) which may be "discovered" without consulting any form of "experience".

To begin with, Hume [20] makes it clear that from the perspective of cognitively bounded agents like us "all causes are of the same kind" (Book I, Part III, Sec. XIV, p. 171). In particular, there is no foundation

[12] For a discussion of the originality of Descartes' view of analysis as method of discovery, see [66].

[13] For a discussion of this issue, see Gaukroger [12].

[14] See ([66], pp. 434–438).

[15] For the different lines of attack against traditional logic in the 17th century, see [44].

[16] Note that the expressions "induction" and "inductive inference" do not appear in the Analytical Index of Hume's *Enquiry* prepared by L.A. Selby-Bigge, editor of the 2nd edition [21].

whatsoever for us to distinguish between "efficient causes and causes *sine qua non*". Secondly, Hume argues against the possibility of establishing reliable relations between "causes and effects" by mere scrutiny of the isolated samples found in observation (here and now), i.e., without grounding our judgments on what has been learned from experience so far. This is of the greatest importance concerning our knowledge of nature, for we would never be able to predict future events without consulting past observation, which teaches us about the regularities and connections of events. For cognitively bounded agents like us, such connections between events are always given in time and are never "observed" or grasped at a glance. Thirdly, Hume insists time and again that there is no strictly logical (i.e., *demonstrative*) argument "why we should extend that experience beyond those particular instances, which have fallen under our observation" in the past (Book I, Part III, Sec. VI, p. 91). This is a case of "predictive inference", that is, the inference from given data to another set of data not overlapping with the former.[17]

Hume's main objection against tradition is that "causal relations" (in general) were conceived of as a type of *logical* relation of deducibility.[18] At this early stage of his career, however, Hume seems to pose his radical challenge to the traditional inquirer *tout court*, for there is no reason to assume that "perceptually bounded" agents like us should have "unbounded" access to the formal domain of inquiry. (I shall say more on this issue in Sec. 3.2.)[19]

[17] In Carnap's view ([2], p. 208), this is the most fundamental kind of inductive inference. It is more important than the universal inference, not only from the point of view of our practical decisions, but also from that of scientific reasoning. Carnap points out that a "singular predictive inference", the special case where the predicted data consist of only one individual, stands in close relation to the estimation of relative frequency.

[18] Hume [20] objected to the traditional way of conceiving reasoning. This requires the recourse to a third idea or "middle term." (See Book I, Part III, Sec. VII, pp. 96–97, n. 1.) He calls into question "a very remarkable error,... universally received by all logicians. (...) This error consists in the vulgar division of the acts of the understanding into *conception, judgment* and *reasoning*, and the definitions we give of them". In his view, reasoning does not need to be mediated by a third idea: "we infer a cause immediately from its effect; and this inference is not only a true species of reasoning, but the strongest of all others, and more convincing than when we interpose another idea to connect the two extremes".

[19] In his early work, Hume suggests the radical idea that establishing "necessary connections" in both empirical and formal inquiry ought to be seen as nothing but a form of experience, which is the "bounded" determination of cognitive agents like us. See Hume ([20], Book I, Part II, Sec. IV and Part III, Sec. XIV). See the editor's introduction to Hume ([21], pp. xiii–xvii).

Later in his career, Hume [21] summarizes this challenge to the (traditional) inquirer, selectively reinforcing his early arguments as specifically applied to the study of nature, when he writes in the *Enquiry*:

> Were any object presented to us, and were we required to pronounce concerning the effect, which will result from it, *without consulting past observation; after what manner, I beseech you, must the mind proceed in this operation? It must invent or imagine some event, which it ascribes to the object as its effect; and it is plain that this invention must be entirely arbitrary.* The mind can never possibly find the effect in the supposed cause, by the most accurate scrutiny and examination. (Sec. IV, Part I, p. 25, emphasis added)

Given that the "effect" is totally different from the "cause", Hume insists again, we cannot expect to "discover" a future effect by analyzing the notion of the supposed cause without relying upon what we learned from past observation. Moreover, there can be no *a priori* demonstrative argument to prove that the effect follows from a mere analysis of (the notion of) the cause so as to "reliably deliver" the correct prediction. Let us focus on what Hume is saying here. On the one hand, the traditional inquirer into nature who insists upon defending his scheme will have to concede that *his* "science of nature" is on an equal footing with works of fiction. However, scientific innovation and our predictions in general (in science and everyday life) are not arbitrary inventions as works of fiction might be; neither are they strictly derived from some "abstract general principles" following the rules of deductive reasoning. If we take such general principles to be modern "laws of nature", Hume will ask his question again: how do we come up with them ("without consulting past observation")? By "past observation" he means to include reasoning by analogy, experience and observation. Moreover, and perhaps more importantly, there is no further question to be asked as to the "ultimate" causes of our generalizations (i.e., laws of nature), as Hume writes in the *Enquiry*:

> (T)he utmost effort of human reason is to reduce the principles, productive of natural phenomena, to a greater simplicity, and to resolve the many particular effects into a few general causes, by means of reasoning by analogy, experience, and observation. *But as to the causes of these general causes, we should in vain attempt their discovery*; nor shall we satisfy ourselves (...) Elasticity, gravity, cohesion of parts, communication of motion by impulse; these are probably the ultimate causes and principles which we shall ever discover in nature. (Sec. IV, Part I, emphasis added)

On the other hand, once we grant Hume's point that there is no learning about the world "without consulting past observation"—a trivial point for us today, but far from trivial for the modern inquirer trying to shake off tradition—we are still faced with another challenge. For, whenever we draw conclusions based upon available data (past observations), if we seek something like a deductive argument from the given data to a prediction, deductive reasoning will fail us in the discovery of any reliable link between "past observations and future events". (This part of Hume's argument is what present-day philosophers call the "underdetermination" of theories by given evidence. However, note that the argument is as old as Sextus Empiricus (II, Ch. XV); it is therefore misleading to call it "Hume's problem".)

How is it possible, from the observations we have made, reliably to draw any conclusions beyond those past instances of which we have had experience? How should we reason based on past observations? Hume simply asks, "Why one prediction rather than another?" A reasonable prediction is one that accords with past regularities, which have established an expectation in us. The prediction we thus choose accords with our deeply entrenched beliefs and expectations. So, what is wrong with Hume's answer?

Our understanding of Hume's critical challenge is deeply influenced by Kant's reaction to Hume's answer. In the opening sections of the *Critique of Pure Reason* Kant shows a characteristic ambivalence between science and the philosopher's reflection on science when he suggests that the scientific enterprise itself was put at risk by Hume's arguments.[20] The issue of learning, as conceived by Hume, is on this view deeply problematic. Thus Kant took it upon himself to resolve a problem, which he thought Hume left unresolved. His solution is to distinguish between issues related to origination, learning and discovery, on the one hand, and the philosophically sanctioned justification for scientific knowledge on the other.

Kant introduces a sharp distinction between two sets of questions: on the one hand there is the empirical question concerning the origination of our knowledge claims (*quid facti*) and on the other hand there is the normative question concerning the justification of knowledge (*quid juris*)[21]. From this moment, the problem of justifying induction and, in particular, the question of our license for making predictions in natural science, becomes sharply dissociated from the problem of describing the way we draw inductive inferences. Accordingly, Hume's answer offers a description of the way we engage in our inductive practices; however, to trace origination does not amount to providing insight into the grounds for validity. As N. Goodman [16] once put it, Kant's distinction makes it appear that Hume was simply missing the point.

[20] In his *Critique*, when facing Hume's arguments Kant also speaks as if the true business of philosophy were at risk. See note 34 below.

[21] See Kant ([24], A84–85, B116–117).

On Kant's view, the relevant *philosophical* question to ask about both formal and empirical inquiry is not how to acquire or generate knowledge—issues concerning the advancement of learning, discovery and the growth of knowledge are merely historico-psychological questions—but how to ground scientific knowledge by offering reasons ("begründen"). Kant diagnosed that Hume, like the great mathematician-philosophers of the 17th century, failed to distinguish these issues, beginning with Descartes' famous methodological pronouncements.[22] In particular, Kant diagnosed that the roots of psychologism in (general) logic are to be found in the confusion between empirical and normative questions concerning the rules of deductive reasoning.[23]

This "Kantian" perspective became canonical for much of our contemporary reading of modern epistemology.[24] Moreover, as I shall argue in the concluding section of the paper, the sharp distinction that Kant introduced is still with us today as the classical/propositional approach to inductive inference which manifests itself variously in philosophy, statistics, and economics as confirmation theory, belief revision theory, Bayesianism, and rational choice. The persistence of Kant's distinction, I shall argue, is one of the main obstacles that stand in the way of the philosophers adopting the learning theory approach to induction which on this view provides a means for acquiring new knowledge, not a "justification", or rational warrant, for that knowledge that is an end in itself.

2.2 A New Approach to "the Old Problem of Induction and Probability"

In 1950, R. Carnap published his work on "inductive logic", a logical theory of induction. Although his work is no longer seriously entertained, it is fair to say that all philosophical work on induction since 1950 is,

[22] In his lecture notes, Kant ([25], p. 3) emphasizes that logic is a science that "cannot be an organon of truth or an art of discovery, but only a canon of truth". The logical principle of truth is consistency, which "can only serve for criticism or correction" not for discovery.

[23] Note that the distinction between "the empirical/subjective" and "the logical/objective", which is often traced to Frege's arguments against psychologism in logic, underlies Kant's characterization of the laws of logic as discussed in his logic lectures. See Kant ([25], pp. 4–6). Carnap [2] has his own up-to-date version of the distinction between "logical" and "psychological" issues. The distinction between the objective or logical and the subjective or psychological also plays a role in Popper [54]. For a contemporary version of this distinction, see Worrall ([67], pp. 28–29).

[24] Like Carnap, the young Russell thought that Kant did not altogether escape from the ambiguity between "purely logical" and "psychological" questions. See, for instance, [13].

to some extent, either a response to or a development of Carnap's basic perspective. Carnap's work on induction falls squarely within the classical paradigm or "propositional approach". Carnap's goal is the construction of "a (formal) system of inductive logic that can take its rightful place beside the modern, exact systems of deductive logic". The idea is that inductive arguments are approximations of deductive arguments, so that inductive logic is an extension of deductive logic. In a deductively valid argument, every world in which the premises are true also makes the conclusion true. In a good inductive argument that is no longer the case, but the set of worlds in which the premises are true and the conclusion false is sufficiently "small" that one can say that the premises confirm or partially entail the conclusion to the degree given by the conditional probability of the conclusion given the premises. The project is not how best to engineer methods for generating new knowledge but to consult intuition to explicate the concept of confirmation, which is the sole source of inductive justification—a Kantian end in itself. Since the theorems and principles of the resulting inductive logic hinge upon an explication or conceptual analysis, they are no less *analytic* than the principles of deductive logic.

Carnap's system was not merely the apotheosis of the classical conception of induction; it provoked the invention of a bold, new perspective that ultimately turns the classical paradigm on its head: the learning-theoretic way.

3 THE NEW, ALTERNATIVE WAY OF THINKING ABOUT INDUCTION: AGAINST THE BACKDROP OF COMPUTATION

The formal learning-theoretic conception of induction is based not upon the metaphor of partial support or justification, but upon the concept of a procedure for effectively computing the true answer to a question. As such, it may be viewed as a mathematically rigorous reversion to the pre-Kantian conception of logic as a truth-finding organon. In a pleasantly appropriate historical nexus, the idea was first proposed by Hilary Putnam [55] in direct response to Carnap's classically-motivated inductive logic in Schilpp's Festschrift celebrating Carnap's life work.

In his critique of Carnap's inductive logic, which includes reflections on the role of simplicity in inductive inference, Putnam argues that "one can *show* that no definition of degree of confirmation can be adequate or can attain what any reasonably good inductive judge might attain without using such a concept" (Putnam [55], p. 270). These lines might suggest that

Putnam's criticism contains an implicit defense of "informal rationality". But Putnam is explicit about his aim in the paper. He intends to show that the actual inductive procedure of scientists has features that cannot be represented by a "quantitative concept of degree of confirmation", of the sort proposed by Carnap (i.e., an *a priori* probability distribution of a particular type, called a "measure function" by Carnap [2]).

Putnam concedes that one should never abandon a logical project solely because of philosophical arguments, and the philosopher, he claims, "misconstrues his job when he advises the logician (or any scientist) to stop what he is doing" (Putnam [55], p. 287). However, he insists, it is not part of wisdom to continue with a logical venture, when relevant considerations can be advanced against it. Since Carnap's project is a logical venture, so must be the considerations advanced against it! Putnam intends to provide strict proof that there are basic features of the inductive judgments the scientist routinely makes that cannot be "modeled" by any measure function whatsoever. The logical considerations he advances against Carnap's theory of logical induction rely heavily upon considerations familiar to the mathematical theory of computability. In order to make his point, Putnam states (a) a condition of adequacy for induction; and shows that (b) no inductive method in Carnap's sense can satisfy it; however, (c) some methods "which can be precisely stated" can satisfy it (Putnam [55], p. 270).

Only a few months later, Putnam published a second article entitled "Probability and Confirmation" [56]. In light of his previous results, he discusses again some of the shortcomings of "the most recent work on the mathematical study of induction", that is, Carnap's system of inductive logic. This time he refers to Carnap's project as a design for a computing or "learning machine" in the following terms:

> We may think of a system of inductive logic as a design for a 'learning machine': that is to say, a design for a computing machine that can extrapolate certain kinds of empirical regularities from the data with which it is supplied. Then the criticism of the so-far-constructed '*c*-functions' is that they correspond to 'learning machines' of very low power. (Putnam [56], p. 297)

Already Carnap [2] had made use of the notion of a "learning machine", though rather marginally. It is only with Putnam's work that the idea of a "learning system" in the sense of a conceptual computing machine began to take shape. In the concluding remarks of his second paper, Putnam insists upon the limitations of Carnap's conception of inductive logics as learning devices or "machines that learn". Here, he is particularly insightful when he emphasizes the importance of developing more sophisticated mathematical models for human learning:

> In this paper I have stressed the idea that the task of inductive logic is the construction of a 'universal learning machine'. Present-day inductive logics are learning devices of very low power (...) *In the future, the development of a powerful mathematical theory of inductive inference, of 'machines that learn', and of better mathematical models for human learning may all grow out of this enterprise.* (Putnam [56], pp. 303–304, emphasis added)

Carnap's reply to Putnam in the Schilpp Volume does not acknowledge that a "radically new perspective" on the mathematical treatment of induction is being born, although he concedes that Putnam's mathematical arguments are ingenious.

While some philosophers take Putnam's [55] conclusions to contain negative philosophical morals about "mechanically" following computable prediction methods in science, formal learning theorists regard Putnam's article as a pioneer piece for having made a lasting contribution to this new field of research. What was his point?[25] We find three main themes in Putnam's arguments against Carnap that are relevant in this context. Firstly, Putnam exploits an important analogy between the concepts of computability and induction, which had thus far been ignored by most philosophers. Secondly, Putnam provides a pragmatic means-ends critique of Carnap's logical theory of inductive inference. In particular, he asks how well Carnap's method would perform at finding regularities in the data. More specifically, Putnam requires as a goal that if an inquirer sees an unbounded sequence of instances of a universal generalization, he should eventually adopt the generalization; and the more generalizations on which the inquirer succeeds, the better. Thirdly, his approach invokes the pragmatist idea of "convergence" on the truth in the limit of inquiry, i.e., arriving at the truth without arriving at certainty.

To formal learning-theorists, Putnam's results show how the study of *effective* rules of inductive inference can be fruitfully approached from a recursion-theoretic perspective without recourse to probability or degrees of confirmation as Carnap ([2,3]) proposed to do. Independently of Putnam, the cognitive scientist E.M. Gold ([14,15]) developed the formal learning-theoretic approach to induction to analyze the learnibility by children of various classes of grammars.

[25] Detailed discussions of the precise import of Putnam's argument may be found in [8] and [36].

3.1 A Way of "Acquiring Knowledge" by Analyzing Data and the Bounded Perspective of Human Learning

Let us focus briefly upon the three principal themes just mentioned. First, the relationship between induction and computation draws upon the important idea that both induction and deduction constitute ways of gaining knowledge. In sharp contrast, recall that the classical view conceives of inductive inference as embedded in the activity of advancing arguments, i.e., as something "like a logical argument" from given data (as premises) to empirical generalizations (as conclusions), albeit an argument that is not truth-preserving.

In opposition to the classical view of induction, formal learning theory understands inductive inference as "a way of gaining knowledge by analyzing data", that is to say, by a step-by-step procedure which is conceived of as algorithmic or computable. In particular, a unified formal treatment of induction and computability offers an alternative response to what present-day philosophers call the "underdetermination" of empirical generalizations by given data. This alternative response is guided by means-ends considerations concerning the putative goal of inquiry: truth-finding. That is Putnam's second idea. How should one reason from evidence (given data) to theoretical hypothesis in order to attain the aims of inquiry? What are the most useful norms of scientific inquiry for "cognitively bounded agents" like us? Are there any norms which may hinder us in successfully attaining the goals of inquiry? These are some of the questions formal learning theory examines.

In particular, the new conception of learning emphasizes that both computability and induction make assumptions concerning the existence of certain "epistemic limitations" in cognitive agents following an algorithmic procedure. From the point of view of our perceptual capacities, we are cognitively bounded agents, i.e., our inductive abilities will always be limited because the evidence ("data stream") we are presented with can only be observed one glance at a time. If the entire evidential future could be seen so-to-speak "all at once", empirical prediction would never pose a problem and induction as a form of "gaining knowledge by analyzing data" would terminate with *certainty*. There would be no successive moments of observation and thus no problem of "foreseeing the future". Similarly, a computing machine's computational abilities are also limited by its inability to write upon or scan an infinite memory store at once. As Kelly and Schulte [35] point out, Turing's 1936 results concerning computable functions follow precisely from such assumptions about the perceptual limitations of human agents following an algorithm. In 1936 Turing himself makes this analogy between machine learning and human observation explicit, when he writes:

> (T)he computation is carried out on one-dimensional paper, i.e., on a tape divided into squares. I shall also suppose that *the number of symbols which may be printed is finite.* ...The difference from our point of view between the single and compound symbols is that the compound symbols, if they are too lengthy, cannot be observed at one glance.
>
> The behavior of the computer at any moment is *determined by the symbols which he is observing*, and his "state of mind" at that moment. We may suppose that there is a bound B to *the number of symbols or squares which the computer can observe at one moment.* If he wishes to observe more, he must use *successive observations.* (1936, emphasis added)[26]

The computer is fed more and more data, and the computer's potentially infinite, ever extendable memory can be seen as a second, internalized data presentation, only some finite segment of which can be "observed" at one glance, at a given time.

Finally, following the third theme exemplified in Putnam's work, both formal computability and induction can be conceived of as operating with "criteria of success" or "convergence to the truth" (the "right answer"). One of the leading ideas of the new conception of induction is that inductive inference, just as deductive inference, is based upon a methodology that "reliably" converges to the truth.[27] According to the new way of thinking about induction, some scientific methods of inquiry are guaranteed to eventually converge to the truth, but unlike deductive inference, inductive procedures need never terminate in finite time with the right answer; in other words, there is no guarantee that after some finite time the inquirer will be certain that she is in possession of the truth. As Schulte writes in this volume, "An inquirer can be in possession of the truth without being certain that she is". The idea of truth as "success in the limit of inquiry" ("convergence to the right answer") echoes Peirce's suggestion that "all the followers of science" aim at finding the truth "in the limit of inquiry", simply because:

> It is unphilosophical to suppose that, with regard to any given question (which has any clear meaning), investigation would not bring forth a solution of it, *if it were carried far enough.* (Peirce [50], p. 58, emphasis added)

[26] I owe this quotation to Kelly and Schulte ([35], p. 160).

[27] What philosophers call here "logically reliable inductive inference" corresponds to the computer scientist's notion of "BC-learning". (See the second section in the introduction to this volume; see also Schulte's paper "Logically Reliable Inductive Inference" in this volume). The computer scientist's notion of "reliable" learning is quite different—another example of cross-disciplinary notational pitfalls.

On the other hand, Putnam was Reichenbach's student, and Reichenbach's "pragmatic vindication of induction" most likely also played a role in motivating Putnam's work on the methodology of inquiry.[28] (For a discussion of the general concepts and definitions of reliable convergence to the truth, which takes into account Putnam's conception of "empirical success" for inductive methods, and for some of the standard objections, see Schulte's paper "Logically Reliable Inductive Inference" in this volume.)

3.2 The "Empirical" and the "Formal Domain" of Inquiry

The new way of thinking about induction is clearly not committed to the distinction between the empirical and the formal domains of inquiry underlying the classical paradigm. Accordingly it is free of the Kantian distinction between "empirical questions" and "questions of justification" where the first concerns how we acquire new knowledge, and the other concerns the philosophically approved grounds for or justification of that knowledge.

The classical understanding of inductive inference has long had difficulty coming to grips with formal uncertainty. Even in its most recent, Bayesian incarnation, the classical paradigm attributes formal omniscience to its ideally rational agents but empirically ignorant agents, in spite of the fact that uncomputable problems (will the computation never halt?) look for all the world like problems occasioning Hume's problem of induction (will bread continue to nourish?). Indeed, Hume himself offers support for this idea when, late in his career, he seems to adopt the classical view of demonstrative science in the *Enquiry*: "Of the first kind are the sciences of Geometry, Algebra, and Arithmetic; in short, every affirmation which is either intuitively or demonstratively certain". This is so, Hume now claims, because the formal sciences make no reference to facts:

> Though there never were a circle or triangle in nature, the truths demonstrated by Euclid would for ever retain their certainty and evidence. (Sec. IV, Part I)

At this stage of his career, Hume also affirms a sharp dichotomy between two kinds of objects of inquiry, "relations of ideas" and "matters of fact" which fits squarely into the classical perspective:

[28] In the context of his interpretation of probability, Reichenbach takes into account Peirce's idea that inquiry might converge on the truth without providing a clear sign that this is the case. See [27].

> All reasonings may be divided into two kinds, namely *demonstrative reasoning*, or that concerning relations of ideas, and *moral reasoning*, or that concerning matters of fact and existence. (Sec. IV, Part II, emphasis added)

(In seventeenth century terminology, "moral reasoning" refers to probabilistic reasoning, and "moral certainty" designates the kind of certainty that may be attained by "cognitively bounded agents" like us when engaged in "moral reasoning".)

However, the view that the "cognitively bounded perspective" belongs only to the empirical domain, while in the formal domain of inquiry we operate with the notion of an "ideal agent" does not seem to apply to the younger author of the *Treatise*. Interestingly enough, it does not even apply to the great mathematician Leibniz, who is associated with the culmination of rationalist epistemology.[29]

Note that Hume introduces the sharp dichotomy between "relations of ideas" and "matters of fact" late in his career, so to speak, "as a last minute solution" at the same time that he drops his early analysis of the notions of (the *infinite* divisibility of) space and time, in spite of the central role the notion of time ("successive observations") plays for inductive reasoning. (And it is precisely with the notion of "infinity" that the bounded perspective becomes an issue for human learning.)[30]

It should be noted, however, that when in the *Treatise* Hume inquires about our foundation for establishing reliable ("necessary") connections in reasoning, his critical considerations do seem to apply to both the formal and the empirical domains of inquiry:

> Upon the whole, necessity is something, that exists in the mind, not in objects; (...). Either we have no idea of necessity, or *necessity is nothing but that determination of thought.* (Book I, Part III, Sec. XIV, emphasis added)

In this early work Hume's arguments suggest the radical idea that establishing "reliable relations" in both empirical and formal inquiry ought to be seen as nothing but a form of "experience", a determination of thought deeply marked by our bounded perspective (outer and inner experience, respectively). This radical idea deeply influenced Kant's intellectual

[29] For a discussion of the "bounded perspective" of the human agent and the idea of "limits" in Leibniz see [61]; for Leibniz's reflections on this idea see [41].

[30] See Hume ([20], Book I, Sec. IV) for a discussion of the infinite divisibility of space and time. For a discussion of Hume's early views on mathematical notions, see also Cassirer ([4], Ch. V.I). See also Larvor [39].

development, for it motivated him to question his pre-critical views of an "ideal agent" with unbounded cognitive capacities. But Hume needed to be set right. It is part of our received view that Kant proposed a synthesis between "empiricist" epistemology and "formal inquiry". All forms of human learning start with "experience": Kant agrees with Hume that this is how learning originates. But the *grounds* of objectivity and thus *rational* justification ought to be "independent of experience". Such a requirement is at the basis of Kant's own negative characterization of the notion of "a priori". Kant thus breaks with the traditional terminology, while holding fast to the classical ideal of *a priori* certainty.[31]

The learning-theoretic viewpoint breaks with both Hume and Kant by recognizing that uncomputability is a sort of internalized instantiation of Hume's problem that arises because an algorithmic reasoner's finitely bounded attention does not extend to the future of the uncompleted computation it directs [32]. The purely formal question whether a computation will never halt is, therefore, quite analogous in this respect to the empirical question whether bread will always nourish. In the formal learning-theoretic literature, this analogy is deftly handled in a uniform way using the concepts and techniques of the modern theory of computable functions. Some learning problems are unsolvable due purely to empirical underdetermination, some are unsolvable for purely formal reasons and some are unsolvable due to an interaction between both sorts of considerations. Relations of ideas are simply matters of internal fact.

4 WHAT STANDS IN THE WAY OF ADOPTING THE LEARNING-THEORETIC WAY OF THINKING ABOUT INDUCTION IN PHILOSOPHY?

In order to understand what stands in the way of adopting the new learning-theoretic way of thinking about induction in philosophy, I shall begin by summing up some of the assumptions underlying the classical conception.

First, inductive reasoning is fundamental for gaining knowledge and, thus, for the "advancement of learning".

[31] We recall here that the classical scheme is supposed to guarantee the "*a priori* order of reasoning", a movement from principles/axioms to consequences, i.e., from causes to effects.

Second, as with deductive reasoning, inductive reasoning is conceived of as being embedded in the activity of advancing arguments.

Third, deductive reasoning (strictly speaking, *a priori* demonstrative reasoning), in turn, is conceived of as embedded in the activity of scientific explanation. On the classical view of science, scientific knowledge ought to be *demonstrative*. Classical geometry is the paradigmatic example of *a priori* demonstrative science. The mode of textbook presentation of classical geometry, often called "the synthetic mode" of presentation, is axiomatic.

Finally, against the backdrop of the *new* science of nature (in the 18th century), Hume's radical challenge concerning the ultimate foundation of our inductive practices calls into question the classical, "deductive paradigm".

Kant's attempt to take into account Hume's challenge without giving up the classical paradigm consists in the introduction of a sharp distinction between the empirical issue of "learning" about the world (the questions of "discovery" in science, and the "origin of ideas" in the case of individual cognitive development) and, the normative issue of "grounding" knowledge by offering reasons for truth.

We can place Carnap's attempt to construct a system of inductive logic, "a new approach to the old problem of induction" as fitting into the Kantian concern to reconcile the empiricist view of learning with the formal requirement of certainty. Carnap's project is motivated by the goal to do for inductive reasoning what Frege had accomplished for deductive reasoning by the design of a system of deductive logic. Frege's goal to ground mathematical knowledge by offering purely *logical* proofs also motivated Carnap's previous projects aimed at grounding empirical knowledge. Frege's mathematical treatment of deductive inference is *axiomatic*. It is with his pioneering work in mathematical logic that the classical paradigm underlying scientific inquiry witnessed a powerful revival. Moreover, the idea of justification, that is, of *grounding* knowledge by proof becomes central. Frege's project was influential: it was a leading motivation for Carnap's conception of a (formal) system of inductive logic, as well as of Carl Hempel's theory of confirmation.[32]

It is with this line of research that the Kantian distinction between the ways of learning (origination/discovery) and the ways of grounding knowledge witnessed a powerful revival. Popper [54], for instance, claims

[32] Between 1910 and 1914 Carnap was a student at the University of Jena where he attended Frege's lectures on logic and mathematics, which he recorded in his Jena Notes [1]. See also Reck & Awodey [59], Introductory Essay. Carnap's Jena Notes include his lecture notes on Frege's course "Logic in Mathematics" taught in 1914.

that the ways theories are generated "neither call for logical analysis" nor are susceptible of it and also quotes Kant's distinction between the *quid facti* and the *quid juris*.[33]

In sharp opposition to the classical picture, the "new way of thinking" about induction, on the other hand, is guided by three principal ideas exemplified in Putnam's criticism of Carnap.

First, the learning-theoretic way of understanding induction draws upon the mathematics of computational theory. Second, this new way of thinking conceives of inquiry as ongoing processes governed by means-ends considerations. Putnam, for instance, gave a pragmatic or means-ends critique of Carnap's logical theory of inductive inference. Third, it gives up the idea of being certain about something in the short run, and in this sense, it illustrates the pragmatist idea of arriving at the truth without arriving at certainty. As to this third theme, among Putnam's philosophical forerunners are the American pragmatists Ch. S. Peirce [50] and W. James [23], who explicitly advocated this idea.

We finally return to the question with which we began: what is it that stands in the way of philosophers adopting the learning theory way of understanding induction in philosophy? Much of contemporary philosophy of science was born out of a dialectical sequence of responses to the classical views of Carnap. Formal learning theory, the new way of thinking about induction, does not fit into the classical scheme. Indeed, the move from the classical conception to the learning-theoretic one shares many of the features of a Kuhnian [38] paradigm shift [34]. First, significant conceptual retooling is required, for the classical stress upon analysis of the concept of confirmation gives way to objective mathematical analyses of convergence and efficiency in specific empirical problems. Since some philosophers view conceptual analysis as the very *raison d'etre* of their discipline, this shift in

[33] Popper ([54], pp. 31–32) explicitly refers to Kant, when presenting his view on the role of logic in scientific inquiry. (See also notes 21–23 above.) Contemporary philosophers of science retain a similar distinction. The *quid juris* of scientific belief is called the context of justification, whereas the *quid facti* is called the context of discovery. Note that the terminology was coined in a somewhat different sense by H. Reichenbach (1938). According to Reichenbach's own account, the context of discovery invites "rational reconstruction", an exercise in which the successive cognitive stages of discovery are chained together or "intercalated" with justified inferences. Hence, there could be a logic of discovery in Reichenbach's sense. Indeed, it is ironic that Reichenbach should be viewed as an advocate of the doctrine that there is no logic of discovery, as his entire (normative) theory of induction was based upon a rule for discovering chances (the so-called "straight rule of induction"). He thought all legitimate scientific reasoning could be reduced to that rule. See ([26, 27]). See note 28 above.

the requisite skill set is more than a matter of inconvenience or inertia: it is a basis for judgments of philosophical irrelevance.[34] Second, the very data to be explained are colored by the respective paradigms. In the classical picture, one sees theories in need of empirical justification. Learning theorists, on the other hand, see strategies or methods whose truth-finding effectiveness are at issue. Third, there is a characteristic difference in emphasis or foreground vs. background. If justification is something like inductive logic or confirmation, then rules for generating hypotheses—logics of "discovery"—must be superfluous to justification, itself, which is a relation between theory and evidence.[35] But if justification is just a method's truth-finding efficacy, then justification applies to discovery methods and testing methods equally (just as computational analysis applies equally to testing the relation $x + y = z$ and to computing $x + y$).[36] Finally, and perhaps most importantly in this case, is the fact that the classical paradigm has come to define the legitimate domain of philosophical discourse about inductive inference: e.g., "One must ask what is specifically philosophical about studying the genesis of theories?" (Laudan [40], p. 182)[37] So thoroughly has the classical paradigm established itself as the arbiter of philosophical relevance that formal learning theory is not even recognized as a potential competitor—it is viewed as something else, like empirical psychology, even though it has no more kinship with the peculiarities of specifically human thought than does the efficiency analysis of a long division algorithm in a computer science class.[38]

[34] For a related point, see Larvor's paper "Between History and Logic" in this volume. In his work, Larvor focuses on the way philosophers are currently drawn away from formal models by history of science and science studies. He also (in private correspondence) suggests that they are afraid of losing their field of research. His idea is as follows: a historian can never lose his topic of research. Renaissance diplomatic history may go out of fashion, but it is still there to study. But a philosopher could be left without any expertise if his chosen problem is superseded in the course of a problem-shift. Does this idea seem too farfetched? Note that this seems to come very close to Kant's deep concerns about the future of philosophy when faced with Hume's destructive arguments. As we read in the Introduction to the *Critique of Pure Reason*, such concerns were a leading motivation for this work.

[35] See [40].

[36] See Kelly ([26, 28]).

[37] See Kelly ([29], Ch. 9) for a discussion of discovery from the perspective of formal learning theory.

[38] Interestingly, Popper [54] objected, along similar lines, that the Logical Positivists had no category for inductive rules or strategies that are neither psychological generalizations nor analytically valid logical truths.

It is true that initial confrontations between traditional philosophy and formal learning theory occasion substantial challenges of notation, of technical background, and of modeling real scientific questions with the new machinery. But these difficulties are surmountable and are dealt with to some extent by the papers in this volume. The deeper difficulty is the grip that the traditional way of thinking about inductive reasoning has maintained on philosophical thought, from Aristotle to Hume through Kant and the logical positivists to present-day confirmation theory; understanding learning theory requires breaking free from this philosophical picture.

ACKNOWLEDGEMENTS

The author is indebted to Kevin T. Kelly, Michèle Friend, Oliver Schulte, and Brendan Larvor for detailed comments and proofreading of the paper.

REFERENCES

[1] Carnap, R. (1910–1914). *Frege's Lectures on Logic: Carnap's Jena Notes*, in Reck and Awodey [59].

[2] Carnap, R. (1950). *Logical Foundations of Probability*, Chicago: University of Chicago Press, 2nd ed., 1962.

[3] Carnap, R. (1952). *The Continuum of Inductive Methods*, Chicago: University of Chicago Press.

[4] Cassirer, E. (1907/22). *Das Erkenntnisproblem in der Philosophie und Wissenschaft der neueren Zeit*, Vol. II, reprint of 3rd ed., Darmstadt:Wissenschaftliche Buchgesellschaft, 1991.

[5] Clark, P. and Hale, B. (eds.) (1994). *Reading Putnam*, London: Blackwell.

[6] Cottingham, J. (ed.) (1994). *Reason, Will and Sensation: Studies in Cartesian Metaphysics*, Oxford: Oxford University Press.

[7] Dalla Chiara, M.L., et al. (eds.) (1997). *Logic and Scientific Methods*, Dordrecht: Kluwer.

[8] Earman, J. (1992). *Bayes or Bust*? Cambridge (Mass.): MIT Press.

[9] Fichant, M. (1991). *G.W. Leibniz, De l'Horizon de la doctrine humaine* (1693). Paris: Vrin.

[10] Garber D.(2001), *Descartes Embodied*. Cambridge: Cambridge University Press.

[11] Garber, D. and Cohen L. (1982). "A Point of Order. Análisis, Síntesis, and Descartes' Principles", *Archiv für Geschichte der Philosophie* 64, 136–147; reprinted in [10].

[12] Gaukroger, S. (1994). "The Sources of Descartes' Procedure of Deductive Demonstration in Metaphysics and Natural Philosophy", in Cottingham [6], 47–60.

[13] Goethe, N.B. (2007). "How Did Bertrand Russell Make Leibniz into a 'Fellow Spirit'?", in Phemister and Brown [53], 195–205.

[14] Gold, E.M. (1965). "Limiting Recursion", *Journal of Symbolic Logic* 30, 28–48.

[15] Gold, E.M. (1967). "Language Identification in the Limit", *Information and Control* 10, 447–474.

[16] Goodman, N. (1954). *Fact, Fiction, and Forecast*, 1st ed., University of London: Athlone Press; Cambridge (Mass.): Harvard University Press, 4th ed., 1983.

[17] Grayling, A.C., Pyle, A. and Goulder, N. (2006). *Continuum Encyclopedia of British Philosophy* 3, 2106–2109.
[18] Grosholz, E. and Breger, H. (eds.) (2000). *The Growth of Mathematical Knowledge*, Dordrecht: Kluwer.
[19] Hitchcock, C. (ed.) (2004). *Contemporary Debates in the Philosophy of Science*, Oxford: Blackwell.
[20] Hume, D. (1739). *A Treatise of Human Nature*, Vol. 1, Selby-Bigge, L.A. (ed.), (reprinted from the original edition), Oxford: Clarendon Press, 1958.
[21] Hume, D. (1748). *Enquiries Concerning Human Understanding and Concerning the Principles of Morals*, Selby-Bigge, L.A. (ed.), (2nd ed. of 1902, reprinted from 1777 ed.), Oxford: Clarendon Press, 1957.
[22] Jain, S., Osherson, D., Royer, J.S. and Sharma, A. (1999). *Systems That Learn: An Introduction to Learning Theory*, 2nd ed., Cambridge (Mass.): MIT Press.
[23] James, W. (1948). "The Will to Believe", in Thayer [65].
[24] Kant, I. (A, 1781/B, 1787). *Kritik der reinen Vernunft*, Göttingen: Akademie Ausgabe, AA: IV (1781)/AA: III (1787).
[25] Kant, I. (1800). *Logik* (Jäsche), AA: IX, English translation in: *Introduction to Logic and his Essay on the Mistaken Subtlety of the Four Figures*, Abbott, T.K. (ed. & trans.), London, 1885.
[26] Kelly, K. (1987). "The Logic of Discovery", *Philosophy of Science* 54, 435–452.
[27] Kelly, K. (1991). "Reichenbach, Induction, and Discovery", *Erkenntnis* 35, 123–149.
[28] Kelly, K. (1993). "Learning Theory and Descriptive Set Theory", *Logic and Computation* 3, 27–45.
[29] Kelly, K. (1996). *The Logic of Reliable Inquiry*, Oxford: Oxford University Press.
[30] Kelly, K. (1998). "Iterated Belief Revision, Reliability, and Inductive Amnesia", *Erkenntnis* 50, 11–58.
[31] Kelly, K. (2000). "The Logic of Success", *The British Journal for the Philosophy of Science*, Special Millennium Issue 51, 639–666.
[32] Kelly, K. (2004). "Uncomputability: The Problem of Induction Internalized", *Theoretical Computer Science* 317, 227–249.
[33] Kelly, K. and Glymour, C. (1992). "Inductive Inference from Theory Laden Data", *Journal of Philosophical Logic* 21, 391–444.
[34] Kelly, K. and Glymour, C. (2004). "Why Probability Does Not Capture the Logic of Scientific Justification", in Hitchcock [19], 94–114.
[35] Kelly, K. and Schulte, O. (1997). "Church's Thesis and Hume's Problem", in Dalla Chiara, et al. [7], 159–177.
[36] Kelly, K., Juhl, C. and Glymour, C. (1994). "Reliability, Realism, and Relativism", in Clark [5], 98–161.
[37] Kelly, K., Schulte, O. and Juhl, C. (1997). "Learning Theory and the Philosophy of Science", *Philosophy of Science* 64, 245–267.
[38] Kuhn, T.S. (1970). *The Structure of Scientific Revolutions*, Chicago: University of Chicago Press.
[39] Larvor, B. (2006). "Philosophy of Mathematics", in Grayling et al. [17].
[40] Laudan, L. (1980). "Why Was the Logic of Discovery Abandoned?", in Nickles [48], 173–183.
[41] Leibniz, G.W. (1693). *De l'Horizon de la Doctrine Humaine*, in Fichant [9].
[42] Lipton, P. (1991). *Inference to the Best Explanation*, Cambridge: Routledge (expanded 2nd ed., 2004).
[43] Löwe, B., Peckhaus, V. and Räsch, T. (eds.) (2006). *Foundations of the Formal Sciences*

IV, The History of the Concept of the Formal Sciences, College Publications, London 2006 [Studies in Logic 3].

[44] Maat, J. (2006). "The Status of Logic in the Seventeenth Century", in Löwe et al. [43], 157–167.

[45] Mancosu, P. (1996). *Philosophy of Mathematics and Mathematical Practice in the Seventeenth Century*, Oxford: Oxford University Press.

[46] Mancosu, P. (2000). "On Mathematical Explanation", in Grosholz and Breger [18], 103–119.

[47] Martin, E. and Osherson, D. (1998). *Elements of Scientific Inquiry*, Cambridge (Mass.): MIT Press.

[48] Nickles, T. (ed.) (1980). *Scientific Discovery, Logic, and Rationality*, Dordrectht: Reidel.

[49] Osherson, D., Stob, M. and Weinstein, S. (1986). *Systems That Learn: An Introduction to Learning Theory for Cognitive and Computer Scientists*, Cambridge (Mass.): MIT Press.

[50] Peirce, C.S. (1878). "How to Make Our Ideas Clear", *Popular Science Monthly* 12 (January); reprinted in Peirce [52], 32–60.

[51] Peirce, C.S. (1878). "Deduction, Induction and Hypothesis", *Popular Science Monthly* 13 (August); reprinted in Peirce [52], 131–153.

[52] Peirce, C.S. (1998). *Chance, Love, and Logic, Philosophical Essays*, Lincoln: University of Nebraska Press.

[53] Phemister, P. and Brown, S. (eds.) (2007). *Leibniz and the English-Speaking World*, The New Synthese Historical Library, vol. 62, Dordrecht: Springer.

[54] Popper, K. (1958). *The Logic of Scientific Discovery*, New York: Harper.

[55] Putnam, H. (1963). "'Degree of Confirmation' and Inductive Logic", in Schilpp [63], 761–784; reprinted in Putnam [58], 270–292.

[56] Putnam, H. (1963). "Probability and Confirmation", in *The Voice of America Forum Lectures, Philosophy of Science Series* 10, Washington, D.C.; reprinted in Putnam [58], 293–304.

[57] Putnam, H. (1965). "Trial and Error Predicates and the Solution to a Problem of Mostowski", *Journal of Symbolic Logic* 30, I, 49–57.

[58] Putnam, H. (1975). *Mathematics, Matter, and Method*, Cambridge: Cambridge University Press.

[59] Reck, E. and Awodey, S. (eds.) (2004). *Frege's Lectures on Logic: Carnap's Jena Notes* 1910–1914, translated and edited, with an introductory essay by the editors, Chicago-La Salle, (Ill.): Open Court.

[60] Reichenbach, H. (1938). *Experience and Prediction: An Analysis of the Foundations and the Structure of Knowledge*, Chicago: University of Chicago Press.

[61] Rescher, N. (2004). "Leibniz's Quantitative Epistemology", *Studia Leibnitiana*, Band XXXVI/2, 210–231.

[62] Salmon, W.C. (1966). *The Foundations of Scientific Inference*, Pittsburgh: University of Pittsburgh Press.

[63] Schilpp, P.A. (ed.) (1963). *The Philosophy of Rudolf Carnap*, La Salle (Ill.): Open Court.

[64] Schulte, O. (2000). "Review of Martin and Osherson's 'Elements of Scientific Inquiry'", *The British Journal for the Philosophy of Science* 51, 347–352.

[65] Thayer, H.S. (ed.) (1982). *Pragmatism*, Indianapolis: Hackett.

[66] Timmermans, B. (1999). "The Originality of Descartes's Conception of Analysis as Discovery", *Journal of the History of Ideas* 60, 433–447.

[67] Worrall, J. (2000). "Lakatos and After", Centre for Philosophy of Natural and Social Science, Discussion Paper Series, London: London School of Economics.

11

BETWEEN HISTORY AND LOGIC

BRENDAN LARVOR
School of Humanities, University of Hertfordshire, de Havilland Campus, Hatfield AL10 9AB, Hertfordshire, U.K., b.p.larvor@herts.ac.uk

Abstract: In this paper, I argue that there cannot be a general and comprehensive account of scientific enquiry, for two reasons. The first is that our understanding of scientific enquiry depends on disciplines with incompatible standards of rigour and modes of explanation. I focus on logic and history. The second reason is that the rigour and success of empirical science depends on the particular features of a) domains of enquiry and b) research programmes. Consequently, our understanding of enquiry is more like a wisdom-tradition than a science, to which formal learning theory is a valuable addition.

Some people think that normative philosophy of science is finished. That is, the project of identifying correct methods for empirical enquiry is simply unfeasible. On the other hand, formal learning theory is a rigorous investigation of the formal constraints on empirical enquiry (both self-conscious enquiry by scientists and spontaneous learning by children). In this paper I shall explore the relationship between formal learning theory and the *fin de méthode* view.

I shall argue that reports of the death of normative philosophy of science are exaggerated. However, what we cannot hope for is a science of science, that is, a wholly general and unified account of enquiry as such. This is partly because our understanding of enquiry must draw on a variety of disciplines that have incompatible standards of rigour and therefore cannot be unified into a single discipline (here I shall discuss logic and history, though we could easily include psychology and sociology). Partly, it is because enquiries are too diverse to constitute the domain of a science.

First, I shall review the arguments that are supposed to have done for normative philosophy of science. Then I shall briefly examine some familiar

M. Friend, N.B. Goethe and V.S. Harizanov (eds.), Induction, Algorithmic Learning Theory, and Philosophy, 259–278.
© 2007 *Springer.*

arguments (made by Wesley Salmon, Larry Laudan and Clark Glymour) about Bayesianism. These show that Bayesianism cannot be the whole story about scientific enquiry. Finally, I shall examine formal learning theory in the light of our prior experience with philosophy of science in general and Bayesianism in particular. I shall argue that much of our accumulated wisdom about enquiry makes essential reference to features that vary in structure from case to case, and therefore cannot be abstracted into a general account. The same points tell against any general logic of enquiry. Consequently, normative philosophy of science has to be a tradition of methodological wisdom rather than a science of enquiry. Success, for formal learning theory, means finding a place in that tradition.

This plan requires quite a long run-up before I get to say anything directly about formal learning theory. However, in inter-disciplinary discussions, a review of the journey to here taken by one of the disciplines can prevent a lot of talk at cross purposes. Since much of what follows will have a rather deflationary tone, let me make clear from the outset that in my view formal learning theory is a Good Thing. Others in this volume explain why it is a Good Thing. My aim is to explore its limits and its relations with some of the other Good Things in the philosophy of science larder. However, this voluntary division of labour does not mean that there are no disagreements in this volume. Clark Glymour ends his contribution with the slogan "Epistemology has but two aspects: religion and mathematics". He argues this claim by reporting that philosophers, when they hear about formal learning theory, usually want to set it aside and return to 'religious' (that is, familiar philosophical) questions.[1] I do not doubt that his report is true—because it sounds like the reaction that philosophers tend to give to any alternative to their familiar topics. They may not want to hear about mathematical epistemology, but they are scarcely more ready to hear about the history of science and mathematics, except of the most anecdotal 1066 *And All That* variety. Epistemology has *many* aspects—of which mathematics is one.

1 THE END OF THE MYTH OF METHOD

Very few philosophers of science now believe that there is a single logical shape called 'the scientific method' that distinguishes science from other kinds of enquiry and accounts for its successes and failures. The collapse

[1] "Is there an external world, are there other minds, is truth relative to belief, what is the best method of enquiry, what can we know?" (see Glymour's paper "Trade-Offs" in this volume).

of the myth of method is popularly associated with Kuhn and Feyerabend but it is less widely appreciated that what was, in their hands, a scandalous heresy has become an orthodoxy. The basic claim was made in a very sober form at the 1974 meeting of the Philosophy of Science Association. At that time the point was still widely contested, so the author had to proceed carefully, "It may sound strange, if not heretical to suggest, as I wish to do, that there is no such thing as the rationality of science. At best we can talk about rationalities of science." ([28], p. 191) Two decades later the cautious tone had given way (in another writer) to something close to sarcasm:

> ... there was a view of science that commanded widespread popular and academic assent... I shall call it "Legend"... Champions of Legend acknowledge that there have been mistakes and false steps here and there, but they saw an overall trend toward the accumulation of truth, or, at the very least, of better and better approximations to truth. Moreover, they offered an explanation of both the occasional mistakes and for the dominant progressive trend: scientists have achieved so much through the use of SCIENTIFIC METHOD. ([13], p. 3)

Kitcher's sarcasm is directed at the Myth of Method ("Legend"), not at science itself. His book is subtitled "Science without Legend, Objectivity without Illusions", and aims to show how we can understand progress in science without having to subscribe to this "Legend".[2] Examples of similarly dismissive attitudes to "Legend" could be multiplied without difficulty, not least because the story of the fall of the myth of method has become standard material for writers of undergraduate textbooks.[3]

How did this shift happen, and why did it happen just then, in the second half of the twentieth century, twenty-three centuries after *Posterior Analytics* and three and a half centuries after Bacon and Descartes? The crucial development seems to be the coming to maturity of the historiography of science. This is a twentieth-century development.[4] There were historians

[2] The motto of Kitcher's book is Shakespeare's Sonnet 130 "My mistress' eyes are nothing like the sun".

[3] For example John Losee's *Historical Introduction to the Philosophy of Science* [21] ends in a debate over the very possibility of normative philosophy of science. Larry Laudan's 1987 article [19] reverses the rhetoric of Skolimowski's 1974 piece [28]; in arguing that philosophers should continue to seek *the* scientific method, Laudan consciously sets himself against the grain of current opinion. It should be noted that "philosophy of science" here means philosophy of science *written in English*. The French, for example, have never really forgotten the lessons taught by Duhem and Bachelard.

[4] See [15] for an indication of the rate of growth of research in the history of science.

of science in earlier centuries, but they were isolated and did not constitute a discipline. Perhaps as a consequence, their work tended to be synoptic and lacked consistent attention to detail. Moreover, such histories were rarely philosophically disengaged. Think, for example, of Whewell's *History of the Inductive Sciences* [30]. It is certainly philosophically-motivated (a decade earlier Whewell wrote *The Philosophy of the Inductive Sciences, Founded Upon Their History* [29]). Since his history covered the entire history of empirical science in two volumes it could not hope to pay close attention to the fine detail of every episode. Only in the last century have we seen the establishment of degree programmes, peer-reviewed journals and the rest of the institutional apparatus that allows pioneering and programmatic work to give way to normal research. Only under these circumstances is it likely that scholars will devote themselves to the intense study of narrow domains. And when present-day historians write synoptic histories, they can root their accounts in the monographs and articles of specialists on *this* science in *that* period. No such specialist literature existed in Whewell's day.

How did the development of the history of science as a mature discipline tell against the Legend of Method? This is a large question since neither the history nor the philosophy of science is a monolith, so the story of their encounter cannot be simple. As the case of Whewell shows, they were tangled up together before the history of science emerged as an independent discipline (and indeed before normative methodology became a specialism). Among the philosophers, there was little understanding that history is a separate discipline with its own characteristic standards of rigour. The most naïve philosophers hoped that a simple appeal to the facts of history would bear out their models of the logic of science. Others were more sophisticated—Lakatos, for example, drew on his earlier life as a Marxist to argue that every historical narrative presupposes an ideology (or in the case of the history of science, a methodology), so the competition between methodologies becomes a competition between their respective associated historical narratives.[5] On the whole, though, most English-speaking philosophers of science had little to say about the philosophy of history. Those who engaged with the philosophy of history at all usually assumed that historical explanations are like explanations in natural science or, if they are not, they ought to be.[6] Others seemed to believe

[5] "History of Science and its Rational Reconstructions" in Lakatos ([17], vol. 1).

[6] See for e.g., Hempel [11]. Of course, Hempel did not *thoughtlessly* assume that historical explanations must appeal to "covering laws". Rather, he formed a view about explanation in general that drew its most compelling examples from natural science, and then insisted that historical explanations must be of this same sort. In doing so (I think) he articulated

that historians do no more than set facts in chronological order, ready for philosophers to use.[7] This naivety about the nature of historiography may explain the optimistic hope that the history of science could serve as an objective test-bed for philosophical theory. If an account of method fails to capture the practice of the greatest scientists in history (went the thought), then it cannot be correct. After all, the great scientists must have been doing it right—the philosophers' job is to work out the logic of what they were doing. Thus, history was to supply philosophy with an objectivity that it is usually supposed to lack. History would teach philosophy by examples. For most English-speaking philosophers of science, this 'historical turn' became problematic in the shape of Thomas Kuhn.[8]

Kuhn claimed that there is something in the very activity of history-writing that is incompatible with the traditional quest of normative philosophy of science for general methodological precepts. His account of historical method is rather sketchy, and the majority of his critics in philosophy of science had little interest in the philosophy of history. It is hardly surprising, therefore, that Kuhn's attempt to argue the incompatibility of rigorous history and normative methodology generated more heat than light. Nevertheless, from Kuhn and his sources I extract the following.

First, historians treat events, including scientific developments, in their own terms rather than as preparations for the present. Further, they understand as good humanists[9] that a word or deed means what it means in virtue of what is said and done around and about a given occasion of speech or action. This entails that we cannot extract antique theories from their contexts for the purpose of comparison with later theories, because when so extracted they cease to be themselves. We cannot, for example, compare what Aristotle had to say about bodies under gravity with what Newton thought because Aristotelian bodies, strictly speaking, do not encounter gravity. It does not follow that meaningful comparisons cannot be made, but they require such a labour of translation and explication that we do not get from them regimented data suitable for testing philosophical theses. All we get is a plausible account of the journey from there to here. So we cannot, for example, do a statistical

and defended the tacit assumption of most English-speaking philosophers of science at the time.

7 Example [19].

8 I argue for my reading of Kuhn in [18].

9 That is to say, academics with the mental habits characteristic of the traditional "humane" disciplines of history, literature and philosophy. Good humanists in *this* sense need not be humanitarians, nor need they be atheists.

survey of past science to see what logical approaches have worked best.[10] To do so, we would have to decontextualise episodes in the history of science to such a degree that they would become unrecognisable to the historians who know them best. Such surveys of 'the historical record' are not on because the historical record is not and cannot be a body of regular, standardised data like Wisden.

Second, making sense of the science of the past often consists in setting it in its proper time and place. In other words, historians of science typically deny *internalism* (the view that the growth and trajectory of science can be explained without reference to anything except the encounter between argument and evidence). For example, in the most recent edition of *Isis*[11] the leading article is "Wonderful Secrets of Nature: Natural Knowledge and Religious Piety in Reformation Germany" by Kathleen Crowther-Heyck. Internalist philosophers of science would insist that the effects of religious piety on German science must have been short-lived, inessential and probably regrettable. Historians would consider it unrigorous to approach the question with such prejudices in hand (perhaps religious piety played an essential, long-lasting and laudable role). To take another leading history of science journal, *History of Science*, the March 2003 edition contained the following major articles: "'Purifying' Science: E.C. Slater and Postwar Biochemistry in the Netherlands" (Ton van Helvoort); "Herbert Spencer and the Disunity of the Social Organism" (James Elwick); and "'Men of Science': Language, Identity and Professionalization in the Mid-Victorian Scientific Community" (Ruth Barton). The rejection of internalism is less obvious from these titles (though the scare quotes around "purifying" and "men of science" are significant[12]). Nevertheless, it is clear that van Helvoort, Elwick and Barton are not simply recording successive encounters between hypotheses and data. These examples were chosen arbitrarily (simply by taking the most recent editions at the time of writing) and could easily be multiplied.

Third, historians are wont to historicise everything, including the logical categories that philosophers need in order to do normative methodology.

[10] I have in mind Laudan's suggestion in his [19]. He proposes treating methodological precepts as hypothetical imperatives, so that the differing cognitive goals of present-day and historical scientists are taken into account.

[11] June 2003. George Sarton founded *Isis*, an international journal dedicated to the history of science, in Belgium in 1912. After World War I, Sarton and the journal moved to the United States. Today, *Isis* is edited by Margaret Rossiter at Cornell University, and published and distributed quarterly by the University of Chicago Press.

[12] "Men of science" is the nineteenth-century transitional phrase between "natural philosopher" and "scientist".

Here is Kuhn on distinctions such as that between the contexts of discovery and justification[13]:

> For many years I took them to be about the nature of knowledge, and I still suppose that, appropriately recast, they have something important to tell us. Yet my attempts to apply [these distinctions] even *grosso modo*, to the actual situations in which knowledge is gained, accepted, and assimilated have made them seem extraordinarily problematic. Rather than being elementary logical or methodological distinctions, which would then be prior to the analysis of scientific knowledge, they now seem to be integral parts of a traditional set of substantive answers to the very questions upon which they have been deployed.[14]

Any historian worthy of the name will want to know how the tradition that Kuhn mentions ("a *traditional* set of substantive answers") arose and developed, in response to what and against what opposition. In this historicist light, the apparently innocent distinctions employed by philosophers suddenly seem question-begging. Kuhn's importance, in my view, is to have articulated (however obscurely) the stance of professional historians towards the objects of their studies.

Historicism may be natural to historians but it is alien to most philosophers of science, and, more importantly, inimical to their philosophical projects. Some philosophers embraced historicism: history (they insist) teaches that everything is relative to its moment, to its place in the great flux. The search for timeless abstractions called "correct methods of enquiry" is therefore pointless or perhaps even impossible. However, this historicist, relativist, quietist line fails to acknowledge that questions about scientific procedure and propriety are not mere philosophical cobwebs. Sometimes, scientists (and science funding bodies) have decisions to make that involve normative methodological judgments. We cannot be content to say "Do the done thing for your time and place", because current practice may be mistaken or undetermined. It is precisely in cases of underdetermination or doubt that normative issues become pressing. Then there are those occasions when scientific controversy achieves a wider public significance, such as the struggles over creationism, food safety and global warming. In such

[13] In the 'context of discovery' hypotheses are invented; in the "context of justification" they are evaluated. In the twentieth century it was widely held (by Popper and most of the Vienna Circle, for example) that there is no logic of hypothesis-invention but there is a logic of hypothesis-evaluation.

[14] [14], p. 9

disputes, each side inevitably accuses the other of being unscientific and these accusations are not always empty or philosophically naïve. What is more, philosophers have accumulated a rich store of methodological ideas plus some impressive formal theories (such as formal logic, probability and the formal learning theory discussed in this volume). It is trite to dismiss all this with a single historicist gesture. The Legend of the One True Method may have been discredited, but the original question remains: *how should we enquire*?

The history of science, then, is not the friendly source of methodological morality tales for which normative methodologists once hoped. On the contrary, historians deride such tales as "whiggism". If methodologists are to address the enduring questions about the nature of enquiry, they must first disengage themselves from the history of science, at least as it is written up by professional historians. This is relatively easy to do. Notice first that the historicist dictum—that historical phenomena owe their identities to their places in the great flux, and must therefore not be abstracted for fear of distortion—is not a conclusion of historical enquiry. It is, rather, a methodological precept. Since philosophers do not share the aims of historians, they need not share all of the historians' premises and methods. Historians usually want to know why *this* scientist at *this* time and place wrote, spoke and acted as he did. Philosophers have no professional interest in such specifics—their interest is in the argument as such (if such a thing can be identified). In any case, the history of science may not be as relevant as philosophers once thought. Larry Laudan argues that methodological precepts should be read as hypothetical (rather than categorical) imperatives: 'If you have cognitive goals $\{A, B, C\}$, then use method $\{a, b, c, d...\}$'. Most of the great scientists of history had different cognitive goals to scientists of the present day (Newton and Boyle, for example, saw the construction of a natural theology as a central task of science).[15] Consequently, the successes and failures of their methods are irrelevant to us. This allows Laudan to square our intuition that the great scientists of the past were instrumentally rational (that is, they effectively matched means to ends) with Feyerabend's observation that they rarely acted in conformity with methodological precepts that seem compelling to us. In any case, there is no *a priori* reason to suppose that the best methods of enquiry lie in the recorded past at all. Perhaps they have not yet been discovered.

Disengagement from the history of science returns methodologists to the problem that reference to the historical record was supposed to solve. How are methodological questions to be resolved? Philosophical argument

[15] These examples, and the argument here reported, are taken from [19].

rarely resolves anything, and philosophical intuitions are hardly objective. The history of science was supposed to provide an objective test-bed for philosophical theories. If the history of science is not suitable for this purpose, then where is normative methodology to find its objectivity? Studies of present-day scientific practice are unsuitable for many of the same reasons that prevent history from grounding philosophical arguments. One attractive option is to abandon the search for empirical-historical tests of methodological theory altogether, in favour of an *a priori* account. Philosophers cannot, nowadays, pretend to offer substantive *a priori* knowledge. But mathematicians can. We come at last (by a lengthy but, I hope, dialectically satisfying route) to the present popularity of formal accounts of enquiry.

2 THE FORMAL TURN: BAYESIANISM

Insofar as normative methodology is turning to mathematical models of enquiry, it is returning to where it was before its dalliance with history. English-speaking philosophers of science used to look to formal logic and mathematical accounts of induction in their attempts to specify the "logic of science".[16] Then, from Kuhn and others, we learned to suspect such models. These logics seemed so abstract as to have lost sight of actual scientific practice altogether, and depended for their intelligibility on a set of arbitrary and question-begging distinctions. Now, having learned that it is hard to sleep with historians without catching their historicist fleas, many philosophers of science seem ready to turn back to mathematics. Dialectical journeys often lead back to a more sophisticated and acutely self-aware version of the starting-point. The trick, of course, is not to forget what was learned *enroute*.

The most popular mathematical model of enquiry at the time of writing seems to be Bayesianism.[17] It is worth taking a glance at Bayesianism, since its difficulties are typical of those suffered by formal accounts of enquiry.

[16] And in the case of the Vienna Circle, German-speaking philosophers too.

[17] Bayes' theorem allows us to evaluate the conditional probability $P(T/E.B)$ that a given hypothesis T is true provided both our background beliefs B and some new piece of evidence E are true:

$$P(T/E.B) = \frac{P(T/B) \times P(E/B.T)}{P(E/B)}$$

To use (even this simple version of) the theorem, we must know: (i) the conditional probability of the hypothesis being true given only the background beliefs; (ii) the conditional probability of the evidence E given that both the background beliefs B and the hypothesis T are true; and (iii) the conditional probability of the evidence E given that the background beliefs B are true.

Moreover, the sort of argument that I wish to make with respect to formal learning theory has already been made in the case of Bayesianism by Wesley Salmon, Larry Laudan and Clark Glymour, among others.[18]

In order to use Bayes' theorem one must have identified a small (or at any rate, a finite) number of "serious" or "plausible" hypotheses. One must already know what counts as a plausible explanation, what counts as relevant evidence and what counts as background belief. These judgments of relevance and plausibility are only possible if one already knows (or at least, believes that one knows) a great deal about the domain under investigation. In case anyone should suppose that relevance and plausibility can be decided by pre-scientific common sense, recall that only a few centuries ago, serious, intelligent people thought that the number of planets is related to the number of holes in the human head. In other words, the infinity of possible worldviews and underlying metaphysical schemes must somehow be cut down to a manageable handful *before* the Bayesian story can start. Therefore, Bayesianism cannot supply a complete account of scientific enquiry.[19] Salmon suggests that the Bayesian algorithm be supplemented with a Kuhnian account of how relevance and plausibility are fixed. Indeed, Salmon hoped for a synthesis between the historical and logical sides of philosophy of science. In the end, though, the Kuhnian supplement comes to dominate the Bayesian element in his account: "The [Bayesian] algorithms are trivial; what is important is the scientific judgment involved in assessing the probabilities that are fed into the equations."[20]

Laudan's argument[21] focuses on just one methodological thought: a good theory ought to address all the phenomena in its field. Other things being equal, it is a shortcoming in a theory T to say nothing about some phenomenon p in its domain of enquiry. In Bayesian terms, T says nothing about p if the conditional probability of p given T equals the prior probability of p. In this case, the likelihood of T in the face of p is exactly the same as its probability prior to p. That is to say, a true Bayesian's confidence in T will be unaffected by T's failure to address p, even though p is among the phenomena that a good theory in this domain ought to address. A Bayesian might reply that T could still be true—but we do not want scientific theories that are merely true, we want ones with deep explanatory power. Of course, T may offer such significant explanatory benefits on some

18 [7, 20, 26]

19 Bayesians know this. Most Bayesians regard Bayesianism as no more than a set of rational norms on belief formation and modification.

20 [25], p. 287

21 [20], p. 173

other front that it effectively re-defines the domain so that p falls outside it. Then T's failure to address p ceases to be an issue. This, though, is not a judgment that Bayesianism can formalise. What loss of scope is a price worth paying for a gain in explanatory power? Like Salmon's judgments of relevance and plausibility, this part of scientific thinking falls outside of the Bayesian account. It remains to be seen whether any formal account can model such judgments.[22]

Clark Glymour offers a generalised version of Laudan's argument. Glymour maintains that:

> There are a variety of methodological notions that an account of confirmation ought to explicate and methodological truisms involving these notions that a confirmation theory ought to explain: for example, variety of evidence and why we desire it, *ad hoc* hypotheses and why we eschew them, what separates a hypothesis integral to a theory from one 'tacked on' to the theory, simplicity and why it is so often admired, why 'de-Occamized' theories are so often disdained, what determines when a piece of evidence is relevant to a hypothesis...[23]

Glymour goes on to explain that, in his view, Bayesianism can explicate some but not all of the items in this list, and that "There are elementary but perfectly common features of the relation between theory and evidence that the Bayesian scheme cannot capture at all without serious—and perhaps not very plausible—revision."[24] He goes on to explain that he considers Bayesianism pertinent for statistical reasoning and that it captures some principles of ordinary reasoning. I report his view here but not his arguments for it, since these would put off consideration of formal learning theory even further.

My point is that contemplation of actual scientific practice, for all it led philosophers to an unwelcome brush with historicism, did leave us with a fund of methodological wisdom. Not the One True Method, but a collection of methodological "notions and truisms", together with a respect for the sensitivity to the particular situation that is required to use them well. Bayesianism cannot be a comprehensive account of scientific reason unless it can model and account for all these notions and truisms. This point tells only against those who imagine that Bayesianism is the whole

[22] In this volume, Clark Glymour argues that formal learning theory can articulate some of these trade-offs, even if it cannot calculate them.

[23] Glymour in [25], p. 293.

[24] Glymour in [25], p. 294.

story—no-one denies that it has something important to say. Most Bayesians regard Bayesianism as a set of constraints on the relative strengths with which we hold our beliefs, but insist that how those beliefs are formed and evaluated within these constraints is another question. Bayesianism cannot by itself explain scientific successes (though it may explain those failures in which its formal constraints are violated) nor can it offer advice about the choice of concepts, tools, background metaphysics, problems to address and experiments to attempt.

3 FORMAL LEARNING THEORY

The simple claim that I have just made with respect to Bayesianism seems to come for free in the case of formal learning theory. For, learning theory asks under what conditions a problem is solvable "in the limit", that is, does the sequence of conjectures produced by a given scientist-function stabilise in the long run? This question matters to us even though "in the long run we're all dead", because "if you can't know the truth in the long run, you can't know it in the short run either."[25] So learning theory offers us constraints of the form: learners with *these* computational bounds will never solve problems of *that* logical type. It might seem that we could stop here with the conclusion that learning theory provides theorems about what is possible in the limit that leave a lot of latitude for the exercise of informal expertise (like Salmon attempt to yoke Bayes and Kuhn together). However, this will not do, first because learning theory does not entirely respect the short-run/long-run distinction, in the sense that it considers questions of efficiency (which scientist-functions stabilise quickest?). Second, learning theory is able to formalise *some* of our common stock of methodological notions and maxims (such as consistency, conservativeness and decisiveness, for example) in order to assess their effects in the limit. An obvious question arises: can it formalise all of them? If not, what part of our accumulated methodological wisdom escapes formalisation, and why?

To make the question more pointed, consider this programmatic passage from Martin & Osherson *Elements of Scientific Inquiry*:

> Whatever the motives for studying inquiry, we must begin by appreciating the complexity of the subject matter. As a form of human behavior, science involves a wide range of activities, both in and out of laboratories. Surely such a phenomenon cannot be understood without substantial idealization. By limiting attention to

[25] [8], p. 282

> just a few, salient aspects of science we may hope to understand their interaction within the larger scheme, and eventually illuminate further variables that can be added to our model at a later stage.[26]

We are given no reason to suppose that enquiry (considered as a complex) has a logically modular structure that would allow us to isolate and model some of its 'elements' without reference to the whole activity—notice that this atomism is incompatible with the holism proper to historical studies. Nor is it obvious that anything recognisable as human enquiry remains after the 'substantial idealization' required for formalisation. Indeed, it may be that a mathematically precise model of enquiry falsifies the phenomenon. Some case-studies seem to show that ambiguity is necessary for progress.[27]

To return to our question, let us see if there are any methodological notions and precepts that cannot be included in a sequence of progressively more complex formal models as envisaged by Martin & Osherson.

The first thing that leaps out of the literature on formal learning theory is the notion of a "paradigm". This term is clearly owed to yet distinct from the concept introduced by Kuhn.[28] One reason to adopt this Kuhnian terminology is this: learning theory has from its inception tacitly accepted the Kuhnian view that enquiry can only get going when there is a consensus on all the hard metaphysical and methodological questions. In Gold's [9], published only five years after *The Structure of Scientific Revolutions*, the class of possible languages is specified in advance. In later expositions, enquiry is modelled as a game played between Nature and a scientist. The players are given a set of possible realities. Nature chooses one of these possibilities as the actuality, and the scientist has to identify this from the data-stream, or "environment" supplied by Nature. Technically, a "paradigm" in formal learning theory has five elements: the set of potential realities; a problem; for each possible world a set of data-streams or "environments"; some scientist-functions; and a criterion of success.[29] All this has to be fixed

26 [22], p. 1. Notice that this title suggests more than a study of formal constraints on inquiry.

27 I have in mind [16], in which the elasticity of concepts is essential to the growth of mathematical knowledge. Also, see Grosholz [10], who argues that mathematical knowledge sometimes grows by hybridisation, and that these hybrids "often admit an instability or inconsistency that is however held in place or made tractable by the rational relatedness provided by the abstract structure that holds the domains together" (p. 88). Her case study suggests that instability or inconsistency may be more than regrettable flaws; they may be essential for progress.

28 I am tempted to label this new sense of "paradigm" thus: paradigm$_{23}$, in honour of the twenty-two senses of the term that Margaret Masterman discerned in [14].

29 [22], pp. 2–3

in advance, as in Kuhnian normal science (and Bayesianism: recall the seven stars/seven apertures). There is an important difference, in that in Kuhnian normal science the paradigm is socially agreed rather than explicitly stated. Moreover, the social agreement is not a set of rules but rather a collection of paradigmatic examples; and the question of the similarity of new cases to these examples is forever open-ended. Nevertheless, the Kuhnian view that scientists have to fix their methods and metaphysics *before* they can do any science is unusual among formal accounts of enquiry.

However, learning theory says nothing about the central Kuhnian question: how do we get from a failing paradigm to a more promising one? Of course, philosophy of science does not have an agreed, established answer to this question. It does, however, have some methodological notions and truisms. One we have already had from Laudan is that a theory (or paradigm or research programme or tradition of enquiry—the exact unit of analysis is not important) is in trouble if it cannot explain phenomena that clearly lie within its domain.[30] Such problematic phenomena need not contradict the theory; they need only embarrass it by showing that its scope is insufficiently broad. This, as we noted in the case of Bayesianism, raises the whole problem of the unity and limits of a domain of enquiry. How do we establish the domain-boundaries that allow us to say that a theory fails to address some of its proper objects? For to do so, we must have some way of identifying the domain that is independent of the theory in hand.

This question is especially knotty because domain-boundaries can shift under the influence of a dramatically successful new theory. For example, Galileo at once narrowed an Aristotelian domain, as he separated the science of motion from the general problem of change; but at the same time he unified the physics of the Earth and the Heavens. Nevertheless, insofar as a domain of enquiry is a natural unit, it is so in virtue of commonalities and connections in its subject matter. There is a science of physics in virtue of commonalities among physical processes. There is a science of primatology in virtue of commonalities among primates. It seems to me unlikely that we will ever have a general logical account of the unity of domains because the commonalities that unify a domain are inevitably specific to that domain. The commonalities that unite primates into a domain of enquiry seem to be of a quite different sort from the commonalities among physical processes that render a science of physics possible. If I am right about this, then *no* formal theory of enquiry can ever fully articulate those methodological notions and truisms that refer to the borders between domains of enquiry. Those borders

[30] Laudan calls these "non-refuting anomalies" [20], p. 166; Lakatos calls them "heuristic falsifiers" [16], p. 82.

are established and maintained by means and criteria that are local to those domains, and will therefore resist formal modelling. Indeed, for the same reason, no general account (formal or otherwise) can go far beyond the truism that, other things being equal, a theory should address all those phenomena in its domain. The case of Galileo shows why: sometimes, progress requires that domain boundaries be redrawn.

Next, there is the obvious passivity of the "scientists" (that is, scientist-functions) in learning theory. They do not need to carry out experiments, because the data is supplied to them by the "environment". In formal learning theory, the "environment" is nothing but a stream of data. This is a more significant idealisation than may at first appear. An experiment is a complex of physical processes (bombarding *this* with a stream of *those*, mixing some of *that* with a tincture of the *other*, *etc.*). It takes considerable scene-setting to turn such manipulations of matter into arguments. Doing or seeing something does not, by itself, generate propositions. It is at this point that familiar considerations about the theory-laden-ness of data enter the story. The conduct and interpretation of experiments normally requires expert judgment, tacit knowledge and a trained eye (and if the experiment is to be robust it must be repeated in a variety of different laboratories, requiring a whole community of trained eyes).[31] All our accumulated sensitivity to the logical and phenomenological subtlety and fallibility of experimentation is abstracted away in formal learning theory. Instead, we have an optimistic sort of empiricism in which the 'environment' supplies indubitable facts, ready to use, to a passive observer. This is probably as it should be, as it is not obvious how the experimenter's skill and tacit understanding could be formally modelled. Here again, if I am right, we have an essential aspect of enquiry that resists formal modelling.[32] It is not just that the results of experiments are empirical and therefore less than wholly certain. It is that navigating these uncertainties requires hands-on know-how that is nevertheless part of the logic of empirical enquiry.

The 'scientists' of formal learning theory can usually get away with passively contemplating whatever data they are fed because it is normally stipulated that the 'environment' drops all the facts into their laps eventually. In the case of language-learning, Gold's 1967 paper assumed that every

[31] See, e.g., [6].

[32] This connects with an issue in artificial intelligence and philosophy of mind. Some philosophers (under various influences including Heidegger, Merleau-Ponty, Kuhn and the later Wittgenstein) deny that know-how can be comprehensively analysed into propositional knowledge-that. If they are correct then experimenters' skills cannot be entirely coded in a logic-system. See [4].

string in the target language is presented to the learner at least once.[33] This is a reasonable assumption if the point is to establish impossibility results, because a language that is unlearnable with this generous assumption is certainly unlearnable without it. However, deciding what data to collect is a crucial part of scientific judgment, not least because it is often the case that data cannot be collected passively. Much of our information about the world cannot be gathered without the use of specialist machines, and will not be gathered unless some scientist believes that it can be gathered and is worth gathering. No-one committed to humour-medicine would bother to go looking for germs. No-one committed to perfect, unchanging Heavens would bother to set up a screen to display sun-spots. Moreover the available technology sets limits to the range of possible experiments and observations. In learning theory the available data is specified when the "environment" is fixed, but (reasonably) the "environment" is not changed by the "scientist's" learning. The "environment" sends its data in a pre-determined stream regardless of the "scientist's" state of knowledge. In practice, new knowledge raises new questions, which in turn elicit data that would not otherwise have been forthcoming.

Now, learning theorists know that their "scientists" are unnaturally passive. Therefore they developed the concept of an "oracle" that collects data from the "environment" and feeds it to the "scientist". The oracle decides what data to collect next as a function of the information received so far.[34] This, though, merely raises questions of efficiency (that is, which oracles help scientists to reach their success-points quickest?). So far as I know, learning theory does not model the fact that a scientist may never see certain data because his false beliefs prevent him from looking for them. The oracle stands between the "scientist" and the "environment", and determines the order in which the data arrives. The oracle is not modified by what the "scientist" learns (though in advanced versions of learning theory the

[33] [9], p. 448. This presentation can take two forms: "text" and "informant". A "text" for a language L is "a sequence of strings $x_1, x_2, \ldots$ from L such that every string of L occurs at least once in the text" (*ibid.* p. 450). An 'informant' for L "can tell the learner whether any string is an element of L, and does so at each time t for some string yt" (*op. cit.*). In the case of learning from an informant rather than from a text, it is assumed that every finite string of the alphabet common to the possible languages will come up at least once. No other strings are presented, and each string is accompanied by information as to whether the string is correct or incorrect. He briefly considers 'request informants' that answer queries about strings chosen by the learner, but quickly proves that these are equivalent to the 'arbitrary informant' that rules on all the possible strings in an arbitrary order (*ibid.* p. 467).

[34] [22], pp. 87–90

"scientist" may be modified by what *it* learns). Real scientists change their search-strategies as well as their hypotheses in the light of new information. In learning theory, the real scientist has been split into a "scientist" that forms beliefs and produces hypotheses, and an oracle that decides what question to ask next. The point here is that changes to the "scientist" ought to induce changes in the oracle. This brings us back to the first point: there are no paradigm-shifts (or problem-shifts, or changes of background metaphysics) in learning theory.

A third methodological notion that learning theory may struggle to articulate is that of the *ad hoc* hypothesis. In fact there are two principal senses in which a hypothesis may be *ad hoc*. One is straightforward: a hypothesis is *ad hoc* if it has been cooked up solely to save a theory from one particular counterexample, but offers no explanation or insight into anything beyond that one case ("All swans are white—obviously, this pink one has been dyed."). The other sense of *ad hoc* requires reference to a research programme: a hypothesis is *ad hoc* in this sense if it is developed using resources from outside the proper repertoire of the research programme. A hypothesis could be *ad hoc* in the second sense but not in the first. That is, a hypothesis could be an insight and a step in the right direction, and yet be *ad hoc* with respect to a research programme that makes use of it ("The flow of humours through a wound may be blocked by tiny animals, so we should boil our instruments and bandages to kill them." Here, germ theory is *ad hoc* with respect to humour-medicine, but is in itself an insight). It may be possible to formalise the first sense of *ad hoc*. However, general accounts of methodology, including formal learning theory, will struggle to articulate the second notion of *ad hoc*-ness because it requires the notion of a sustained and unified enquiry (or paradigm, or research programme, *etc.*) as a heuristic whole. Here, the point is the same as I argued above for domains of enquiry. The principle of unity of a research programme is specific to that programme. If the unity of a research programme cannot be formally modelled, then it must also be impossible to model the case in which a given hypothesis violates that unity (excluding trivial models in which the unity of the programme is an unanalysed black box).

The pattern of argument should be obvious now if it was not before: I have deliberately picked out methodological notions and truisms that require some reference to the specifics of the domain of enquiry or the unity of the enquiry itself. It is, ultimately, the unity of these complex wholes (domains of enquiry and research programmes) that resists formalisation. There is another notion that depends on them, which for the sake of a label we might call the "Whewell bonus". This is the case in which a theory predicts or explains a phenomenon that it was not originally intended to address;

the case in which even the theory's champions are pleasantly surprised by its success. This notion is notoriously difficult to formalise because as it stands it involves the intentions of the theory's originator. These intentions (being psychological items) ought to be irrelevant if the Whewell bonus is an objective virtue of theories. The challenge is to capture the thought that a theory has reached beyond its original scope without making any psychological reference. Here again, I would argue for particularism: there is no general logical account of the original scope of theories. If in any given case it is possible objectively to determine the original scope of a theory, it is only by paying attention to the details in hand. Apart from anything else, it may be that the proper scope of a theory is in part a function of the history of the discipline in question prior to the formulation of the theory. The same point comes up if we consider unification by consilience, in which a pair of distinct theories in hitherto separate domains of enquiry come to reinforce each other (for example, if studies of population DNA and the evolution of language groups produce the same hypotheses about human kinship and migration).

The Whewell bonus raises another class of methodological notions and truisms, namely, those that exploit the insight that theories are as much tested against each other as they are against nature. We accept our current scientific orthodoxies because they are our least lousy theories so far. Therefore, we cannot fully formalise a set of criteria for accepting or rejecting theories without articulating some principle(s) of comparison between theories. In particular, a Whewell bonus is only a real bonus if the phenomenon has not already been adequately explained by some other theory. I shall not insist on this point, however, as it may be possible to modify formal 'scientists' so that they watch each other's results as well as their own and modify their enquiries accordingly. Indeed, there are already some results about teams of "scientists", though these do not answer the present point.[35]

4 CONCLUSION

So far I have argued that there are methodological notions and truisms that formal learning theory has not captured, and probably will not ever capture. This is essentially the same claim that Glymour, Laudan and Salmon made about Bayesianism. Since this is a negative claim, let me now allay some natural fears. The point is not to "attack" formal learning theory, but rather to learn a lesson from the history of our own discipline. When philosophers acquire a new tool, there is a danger that the rising generation of researchers may be so impressed with it that much of what was known before is forgotten.

[35] [24]

The more impressive the tool, the more intense the danger. Something of the sort seems to have happened to the philosophy of science and mathematics with the development of formal logic from Frege onwards, and it took a rather wrenching "historical turn" for philosophers to rediscover the sensitivity to history and practice that previous generations took for granted. Let us not make the same mistake again. Let us learn what we can from formal learning theory without losing sight of what we have learned from historical, sociological and phenomenological studies of science. These studies have shown us that much of the rigour of science is bound up with the subject-matters and specialist techniques of individual sciences. Dendrochronology, for example, depends on truths about logic, trees and the weather, plus the judgment of experts and the reliability of their computers. To isolate the logical aspect and forget the rest makes a mystery of the effectiveness of the technique, and this is to invite mystification and scepticism. It is precisely because the rigour of each special science depends on its characteristic objects and techniques that enquiry is not the sort of thing of which there can be a science.

Philosophy ought not to ignore developments in logical theory. But the alternative to mathematical accounts of enquiry is not mere "religion", and philosophy ought not to ignore historical and phenomenological studies of scientific practice either. Philosophy cannot (as Salmon hoped) simply combine these approaches because their standards of rigour are incompatible. Abstraction is a methodological sin among historians, while historicism is anathema to logicians. Philosophers, caught in the middle, are at once chided by mathematical logicians for lacking the clarity of the exact sciences, and by historians for concocting abstractions while neglecting concrete reality. The point of this article is to explain to exponents of formal learning theory that, if philosophers do not embrace it with both hands, it may be because one hand is already carrying a rather unwieldy load of historical and phenomenological insight.

REFERENCES

[1] Breger, H. and Grosholz, E. (eds.) (2000). *The Growth of Mathematical Knowledge*, Dordrecht: Kluwer.

[2] Clark, A. and Millican, P. (eds.) (1996). *Machines and Thought: The Legacy of Alan Turing*, Oxford: Clarendon Press.

[3] Cohen, R.S. et al. (eds.) (1976). *Boston Studies in the Philosophy of Science* XXXII, Dordrecht: Reidel.

[4] Dreyfus, H.L. (1992). *What Computers Still Cannot Do: A Critique of Artificial Reason*, Cambridge (Mass.): MIT Press.

[5] Feyerabend, P. (1993). *Against Method*, 3rd ed., London: Verso.

[6] Galison, P.L. (1987). *How Experiments End*, Chicago: University of Chicago Press.

[7] Glymour, C. (1981). "Why I Am Not a Bayesian", in *Theory and Evidence*, Chicago: University of Chicago Press, 63–93; reprinted in Papineau, D. [25].

[8] Glymour, C. (1996). "The Hierarchies of Knowledge and the Mathematics of Discovery", in Clark, A. and Millican, P. [2].

[9] Gold, E.M. (1967). "Language Identification in the Limit", *Information and Control* 10–5, 447–474.

[10] Grosholz, E. (2000). "The Partial Unification of Domains", in Breger, H. and Grosholz, E. [1], 81–91.

[11] Hempel, C.G. (1963). "Reasons and Covering Laws in Historical Explanation", in Hook, S. [12], 143–163.

[12] Hook, S. (ed.) (1963). *Philosophy and History*, New York: New York University Press.

[13] Kitcher, P. (1993). *The Advancement of Science*, Oxford: Oxford University Press.

[14] Kuhn, T.S. (1970). *The Structure of Scientific Revolutions*, 2nd ed., Chicago: University of Chicago Press.

[15] Kuhn, T.S. (1986). "The Histories of Science: Diverse Worlds for Diverse Audiences", *Academe* 72–4, 29–33.

[16] Lakatos, I. (1976). *Proofs and Refutations*, Worrall, J. and Zahar, E. (eds.), Cambridge: Cambridge University Press.

[17] Lakatos, I. (1978). *Philosophical Papers* 1–2, Worrall, J. and Currie, G. (eds.), Cambridge: Cambridge University Press.

[18] Larvor, B. (2003). "Why did Kuhn's Structure of Scientific Revolutions Cause a Fuss?", *Studies in History and Philosophy of Science* 34–2, 369–390.

[19] Laudan, L. (1987). "Progress or Rationality? The Prospects for Normative Naturalism", *American Philosophical Quarterly* 24, 19–31; reprinted in Papineau, D. [25].

[20] Laudan, L. (2000). "Is Epistemology Adequate to the Task of Rational Theory Evaluation?", in Nola, R. and Sankey, H. [23], 165–176.

[21] Losee, J. (1993). *Historical Introduction to the Philosophy of Science*, 3rd ed., Oxford: Oxford University Press.

[22] Martin, E. and Osherson, D. (1998). *Elements of Scientific Inquiry*, Cambridge (Mass.): MIT Press.

[23] Nola, R. and Sankey, H. (eds.) (2000). *After Popper, Kuhn and Feyerabend*, Dordrecht: Kluwer.

[24] Osherson, D., Stob, M. and Weinstein, S. (1986). *Systems That Learn: An Introduction to Learning Theory for Cognitive and Computer Scientists*, Cambridge (Mass.): MIT Press.

[25] Papineau, D. (ed.) (1996). *The Philosophy of Science*, Oxford: Oxford University Press.

[26] Salmon, W.C. (1990). "Rationality and Objectivity in Science or Tom Kuhn Meets Tom Bayes", in Savage, C.W. [27], 175–204; reprinted in Papineau, D. [25].

[27] Savage, C.W. (1990). *Scientific Theories, Minnesota Studies in the Philosophy of Science* 14, Minneapolis: University of Minnesota Press.

[28] Skolimowski, H. (1974). "Evolutionary Rationality", in Cohen R.S. et al. [3], 191–213.

[29] Whewell, W. (1847). *The Philosophy of the Inductive Sciences, Founded Upon Their History*, 2nd ed., vol. 1 and 2, London: J.W. Parker.

[30] Whewell, W. (1857/1873). *History of the Inductive Sciences, from the Earliest to the Present Time*, 3rd ed., vol. 1 and 2, New York: D. Appleton.

Index

LOGIC, EPISTEMOLOGY, AND THE UNITY OF SCIENCE

Editors:
Shahid Rahman, *University of Lille III, France*
John Symons, *University of Texas at El Paso, U.S.A.*

1. S. Rahman, J. Symons, D.M. Gabbay and J.P. van Bendegem (eds.): *Logic, Epistemology, and the Unity of Science*, Vol. 1. 2004 ISBN 1-4020-2807-5
2. D. Vanderveken (ed.): *Logic, Thought and Action*, Vol. 2. 2005 ISBN 1-4020-2616-1
3. J. van Benthem, G. Heinzmann, M. Rebuschi and H. Visser (eds.): *The Age of Alternative Logics*. Assessing Philosophy of Logic and Mathematics Today, Vol. 3. 2006 ISBN 1-4020-5011-9
4. J. Faye, P. Needham, U. Scheffler and M. Urchs (eds.): *Nature's Principles*, Vol. 4. 2005 ISBN 1-4020-3257-9
5. B. van Kerkhove and J.P. van Bendegem (eds.): *Perspectives on Mathematical Practices*. Bringing Together Philosophy of Mathematics, Sociology of Mathematics, and Mathematics Education, Vol. 5. 2006 ISBN 1-4020-5033-X
6. A. Fagot-Largeault, S. Rahman and J.M. Torres (eds.): *The Influence of Genetics on Contemporary Thinking*, Vol. 6. 2007 ISBN 978-1-4020-5663-5
7. Catarina Dutilh Novaes: *Formalizing Medieval Logical Theories*, Vol. 7. 2007 ISBN 978-1-4020-5852-3
8. S. Rahman, T. Tulenheimo and E. Genot (eds.): *Unity, Truth, and the Liar.* The Modern Relevance of Medieval Solutions to the Liar Paradox, Vol. 8. Forthcoming 2008.
9. M. Friend, N.B. Goethe and V.S. Harizanov (eds.): *Induction, Algorithmic Learning Theory, and Philosophy*, Vol. 9. 2007 ISBN 978-1-4020-6126-4

springer.com

MIX
Papier aus verantwortungsvollen Quellen
Paper from responsible sources
FSC® C105338

If you have any concerns about our products,
you can contact us on
ProductSafety@springernature.com

In case Publisher is established outside the EU,
the EU authorized representative is:
Springer Nature Customer Service Center GmbH
Europaplatz 3, 69115 Heidelberg, Germany

Printed by Libri Plureos GmbH
in Hamburg, Germany